Inorganic Ion Exchangers in Chemical Analysis

Editors

Mohsin Qureshi, Ph.D.
(Deceased)
Emeritus Scientist
Council of Scientific and Industrial Research
Retired Dean
Faculty of Engineering and Technology
Aligarh Muslim University
Aligarh, India

K. G. Varshney, Ph.D.
Department of Applied Chemistry
Faculty of Engineering and Technology
Aligarh Muslim University
Aligarh, India

CRC Press
Taylor & Francis Group
Boca Raton London New York

CRC Press is an imprint of the
Taylor & Francis Group, an **informa** business

CRC Press
Taylor & Francis Group
6000 Broken Sound Parkway NW, Suite 300
Boca Raton, FL 33487-2742

First issued in paperback 2019

ISBN 13: 978-0-367-45067-0 (pbk)
ISBN 13: 978-0-8493-5526-4 (hbk)

Visit the Taylor & Francis Web site at
http://www.taylorandfrancis.com

and the CRC Press Web site at
http://www.crcpress.com

Library of Congress Cataloging-in-Publication Data

Inorganic ion exchangers in chemical analysis / editors, Mohsin
Qureshi, K. G. Varshney.
p. cm.
Includes bibliographical references and index.
ISBN 0-8493-5526-5
1. Inorganic ion exchange materials. 2. Ion exchange.
I. Qureshi, Mohsin. II.Varshney, K. G.
QD562.I63I56 1990
543'.0893—dc20

90-39194
CIP

Library of Congress Card Number 90-39194

DEDICATION

To all friends and colleagues in the international community of chemists who have made 3 decades of inorganic ion exchangers such a pleasure.

K. G. Varshney, Ph.D.

PREFACE

Since the publication of C. B. Amphlett's monograph *Inorganic Ion Exchangers* in 1964, these materials have gained renewed interest and a well defined position in analytical chemistry. As a result, in 1982 A. Clearfield edited a CRC Press publication entitled *Inorganic Ion Exchange Materials* which gave an excellent description of these various types of materials as elucidated by authorities in the field. The main emphasis of these publications has, however, been on physical chemistry, particularly of the crystalline materials. The amorphous or poorly crystalline inorganic ion exchangers were not properly dealt with and their analytical aspects were not fully explored.

In view of this, the present project was undertaken in order to give a comprehensive overview of the diverse studies done so far on inorganic ion exchangers and their analytical applications. An attempt has been made to highlight the various aspects of these studies with the help of a renowned group of scientists working in the field of chemical analysis. It is hoped that the information presented in this publication will benefit analytical chemists seeking some simple and novel means of separations with the help of the easy-to-prepare chromatographic materials.

K. G. Varshney, Ph.D.

THE EDITORS

Professor Mohsin Qureshi, Ph.D., was an Emeritus Scientist of the Council of Scientific and Industrial Research of India. He retired as the Head of the Chemistry Section and Dean of the Faculty of Engineering and Technology, Aligarh Muslim University on March 31, 1984. He was appointed as a Lecturer of Chemistry in the Chemistry Department of Aligarh Muslim University in 1944, joined the Louisiana State University, Baton Rouge as a graduate student in 1955, and received his Ph.D. degree in 1958 from that University under Professor P. W. West. On his return to India in 1959, Dr. Qureshi started a new course in Analytical Chemistry leading to the M.S. degree in Aligarh Muslim University. In 1963 he was appointed Reader in Analytical Chemistry and in 1968 Professor in the Chemistry Section, Z. H. College of Engineering and Technology, Aligarh Muslim University, Aligarh. Professor Qureshi died on July 17, 1988 in a road accident.

Until his death, Professor Qureshi had been an active worker and was engaged in three research projects sponsored by the Council of Scientific and Industrial Research (CSIR), the University Grants Commission (UGC), and the Department of Science and Technology (DST). His research interests included chromatography, ion exchange, synthetic inorganic ion exchangers, detection and determination of organic functional groups, solid state chemistry, and environmental chemistry.

Professor Qureshi published more than 130 research papers in leading international journals and guided about 30 students for the Ph.D. degree. He was a Fellow of the Chemical Society of England, chaired numerous sessions in national and international conferences, and was invited to deliver talks by various Indian and foreign universities and research organizations.

Krishna Gopal Varshney, Ph.D., at present, a Reader in the Department of Applied Chemistry, Faculty of Engineering and Technology, Aligarh Muslim University, did his M.Sc. in 1963 and Ph.D. in 1968 specializing in Analytical Chemistry from Aligarh Muslim University, Aligarh (India) under the supervision of Professor Mohsin Qureshi. He was engaged in the various assignments of teaching and research before his placement as Lecturer in 1971 in his present department, where he started guiding research independently. In 1983 he went to Rome as a Visiting Professor to do postdoctoral research with Professor Aldo LaGinestra at the University of Rome on thermal and catalytic behavior of inorganic ion exchangers.

Thus far, Dr. Varshney has guided 13 students for their Ph.D. degree and 18 students for their M.Phil. degree, and has about 100 research publications to his credit all in journals of international repute. He has attended more than 20 conferences/seminars/workshops both of national and international character and has lectured/presided in some of them. His fields of main interest have been chromatography, ion exchange, electrophoresis, spectrophotometry, synthetic inorganic ion exchangers and environmental chemistry.

Dr. Varshney has been a member of the Advisory Board of CRC's *Handbook of Chromatography* (Inorganics) with Drs. Gunter Zweig and Joseph Sherma as its Chief Editors and Dr. Mohsin Qureshi as the Section Editor. He has also been a co-author in a chapter entitled "Thin Layer Chromatography of Inorganics and Organometallics" in the *Handbook of Thin Layer Chromatography*, edited by Drs. J. Sherma and B. Fried, a Marcel Dekker publication.

Dr. Varshney is a life member of the various learned scientific societies of India, such as the Indian Science Congress, Indian Chemical Society, Indian Council of Chemists, and the Indian Society of Analytical Scientists. He also particpates in panels of experts for various leading research journals of India.

The current research interest of Dr. K. G. Varshney lies in the area of environmental studies. He has been conducting research projects funded by the Council of Scientific and Industrial Research (CSIR), the University Grants Commission (UGC), and the U. P. Council of Science and Technology (UPCST) to explore the various applications of inorganic ion exchangers, such as the industrial effluent treatments, catalysis, and the analysis of alloys, minerals and pharmaceutical products.

CONTRIBUTORS

Cornelis J. Coetzee, D.Sc.
Professor
Department of Chemistry
University of the Western Cape
Bellville, South Africa

Alan Dyer, Ph.D., D.Sc.
Reader
Department of Chemistry
University of Salford
Salford, England

Mukhtar A. Khan, Ph.D.
Reader
Department of Chemistry
Jamia Millia Islamia
New Delhi, India

Milan Marhol, Ph.D.
Nuclear Research Institute
Rez u Prahy, Czechoslovakia

Vladimír Pekárek, C.Sc.
Department of Crystallization and Calorimetry
Institute of Inorganic Chemistry
Czechoslovak Academy of Sciences
Prague, Czechoslovakia

Mohsin Qureshi, Ph.D.
(Deceased)
Emeritus Scientist
Council of Scientific and Industrial Research
Retired Dean
Faculty of Engineering and Technology
Aligarh Muslim University
Aligarh, India

J. van R. Smit, D.Phil.
Hahn & Hahn
Hatfield, Pretoria
Republic of South Africa

Laszlo Szirtes, Ph.D.
Institute of Isotopes
Hungarian Academy of Sciences
Budapest, Hungary

Krishna Gopal Varshney, Ph.D.
Department of Applied Chemistry
Faculty of Engineering and Technology
Aligarh Muslim University
Aligarh, India

ACKNOWLEDGMENTS

I am grateful to my friend Dr. J. P. Rawat, Reader, Department of Chemistry, Aligarh Muslim University, Aligarh, for his valuable suggestions and critical evaluation of the manuscript despite his very busy schedules. My thanks are also due to the publishers who granted permission to reproduce information from the sources noted in the text, and to Mr. H. C. Saxena for his excellent typing skills. In conclusion, my sincere thanks to my family members for their great sacrifices and moral support during the preparation of this manuscript.

K. G. Varshney, Ph.D.

TABLE OF CONTENTS

Chapter 1

HISTORICAL BACKGROUND OF INORGANIC ION EXCHANGERS, THEIR CLASSIFICATION, AND PRESENT STATUS

Vladimír Pekárek and Milan Marhol

TABLE OF CONTENTS

I. HISTORICAL BACKGROUND

From a contemporary point of view, the development of ion exchangers and sorbents can be divided into several periods:

1. Up to 1850, i.e., the period of the first experimental observations and information. The principle of ion exchange had not yet been discovered.
2. The period from 1850 to 1905, characterized by discovery of the principle of ion exchange and the first experiments in the technical utilization of ion exchangers.
3. The period from 1905 to 1935, characterized by the use of inorganic ion-exchange sorbents and modified natural organic materials.
4. The period of artificial organic ion exchangers (1935 to 1940), characterized by rapid development of these materials. Inorganic ion-exchange sorbents were almost completely eliminated from all applications.
5. The period from the mid-1940s to the present, characterized both by the continued rapid development of artificial organic ion exchangers and by a renaissance in inorganic ion-exchange sorbents and their practical application.

The use of solid absorbing substances to improve water quality has been recorded since ancient times. Although most applications involved the removal of solid impurities through filtration, ion exchange was also inadvertantly employed in the exchange of dissolved salts between the water and the solid material.

The intentional use of ion exchange, without knowledge of its theoretical nature, based purely on empirical experience, occurred a number of centuries later; in 1623, Francis Bacon described a method for removing salt from sea water. Sometime later, Hales recommended that sea water be desalinated by filtration through stoneware.

It was not until the end of the 18th century, when the basis was formed for studying adsorption, that sorbents began to be employed in technical applications (in 1790, Lowitz purified sugar beet juice by passing it through charcoal). As experimental information increased, the first qualitative conclusions were drawn by de Saussure at the beginning of the 19th century.

The first half of the 19th century was characterized by the appearance of the first information leading to the discovery of the ion-exchange principle, based primarily on the work of chemists studying soil chemistry. Gazzeri (1819) found that soil and, especially, clay retain dissolved fertilizer particles. In 1826, Sprengel stated that humus frees certain acids from soil. Fuchs (1833) pointed out that the action of lime frees potassium and sodium from some clays.

By the middle of the 19th century, sufficient experimental information had been collected on the as yet unformulated "ion-exchange" principle.

In the middle of 1845, Thompson and Spence carried out studies of the behavior of ammonia salts in soils and came to a surprising conclusion which they published in their work, *The Adsorption Ability of Soils,* at the beginning of the 1850s. Their experiment was based on the following principle: If an ammonium sulfate solution is passed through a column packed with the test soil, the filtrate is found to contain calcium sulfate in place of the original ammonium sulfate. Thus, the soil adsorbs only the ammonium ions (and not the whole ammonium sulfate molecule) and simultaneously releases an equivalent amount of calcium ions. Unfortunately, neither of these workers was able to recognize and correctly interpret their experimental data and did not make the basic discovery to which correct interpretation would have led.

During 1850 to 1855, the agrochemist Way published a number of papers dealing with the behavior of soils in the presence of various cations. The results of his research can be summarized in the following points:

1. Soil exchanges various cations (K^+, Na^+, and Mg^{2+}) to varying degrees, in equivalent ratios; the concentration of anions in solution does not change. The exchange occurs primarily on the clay in the soil.
2. As the concentration of the salt in solution increases, the exchange rate also increases, although not proportionately. A maximum exchange rate is attained.
3. The exchange occurs relatively rapidly.
4. Cation exchange occurs on the clay particles in the soil.
5. Substances can be prepared on which cation exchange occurs, as on the clay particles. These substances, artificial aluminosilicates, can be prepared by precipitating solutions of sodium aluminate and silicate.

In the second half of the 19th century, agrochemists published a great many papers dealing with ion exchange in soils. The work of Eichhorn (1858) was very important and demonstrated that exchange processes in soils are reversible, i.e., an equilibrium is attained and is dependent on the concentrations of the exchanged ions. In 1859, Boedecker proposed an empirical equation describing the establishment of equilibrium on inorganic ion-exchange sorbents.

The discovery and development of the theory of ion exchange was reflected in practical applications. Harm (1896) and Rümpler (1903) proposed the use of natural and artificial aluminosilicates to purify beet syrup. Cvet, who developed the methods and theoretical basis of adsorption chromatographic analysis (1906), used both oxides and hydrated oxides of metals, various types of soils, silicates, and other inorganic substances (now considered as typical inorganic ion-exchange sorbents) as adsorbents.

Gans developed the basis for the synthesis and technical application of inorganic cation exchangers at the beginning of the 20th century. He termed the cation exchangers based on aluminosilicate "permutites", and felt that these ion exchangers would find broad application. However, the usefulness of the permutites prepared at that time, based on aluminosilicates, was limited by their low chemical and mechanical stability. They were useful only for exchange in neutral media and their exchange capacity was low.

Nonetheless, in 1917, Folin and Bell developed an analytical method based on this material for the determination of ammonia in urine (using zeolites).

During the period between the 1930s and 1940s, inorganic ion-exchange sorbents were replaced in almost all fields by the new organic high-molecular-weight ion exchangers.

The renaissance of inorganic ion-exchange sorbents began in the 1950s (although the beginning can be traced to the 1940s). A great impetus for their renewed use came from the field of nuclear research. At that time, there was a need for new ion-exchange materials and processes useful at high temperatures in the presence of high ionization radiation doses, and in highly acidic or oxidizing media. Common organic ion-exchange resins were found to be inadequate under these conditions, as they were greatly degraded and lost their ion-exchange properties.

One of the possible ways of solving this problem involved replacing the organic skeleton of the ion exchanger by an inorganic skeleton. The earlier research on the sorption properties of some hydrated oxides (silica gels and alumina) was neglected and intense research was carried out in the 1950s on the ion-exchange properties of oxides, hydrated oxides, and the insoluble salts of polyvalent elements. Pioneering work was carried out in this field by the research team at the Oak Ridge National Laboratory led by Kraus, and by the English team led by Amphlett. Potentially suitable ion-exchange sorbents that were studied included not only the oxides and insoluble salts of polyvalent elements, but also the salts of heteropolyacids, hexacyanoferrates, aluminosilicates, and zeolites.

Further extensive research and study of inorganic ion-exchange sorbents were carried out in the 1960s and 1970s. Research led from the original amorphous type of ion-exchange sorbents to the study of crystalline ion-exchange materials. A great deal of emphasis was placed on research on the structure of sorbents and on their physicochemical properties.

In the last decade, intense research has continued on the synthesis of crystalline ion-exchange materials, elucidation of their structure, and correlation with their physical, chemical, and ion-exchange properties. Great contributions were made in this area by Clearfield and co-workers. A number of new materials were prepared that had excellent properties. Interest in the industrial use of inorganic ion-exchange sorbents has also increased (especially in the nuclear industry and in environmental protection).

From a historical point of view, it is interesting to consider the growth in the use of inorganic sorbents (partly in analytical chemistry) as reflected by a review of works published from 1957 to 1973. Such a review was carried out by Soviet authors (1976) using the reference journal *Referativnyi Zhurnal Khimiya*. They analyzed two editions of this journal annually for the given period. A total of 34 selected editions contained 4800 references dealing with sorption on inorganic sorbents (oxides, hydroxides, salts, zeolites, natural sorbents, and active carbon). The percent of publications dealing with sorption on inorganic sorbents generally was practically constant throughout the year; from 1957 to 1965, the number in the global literature increased exponentially — doubled every 2.5 years. The number of references decreased after 1965; the increase in the total number was practically linear (about 9000 works annually). It can be estimated that the journal published a total of about 130,000 references dealing with this subject from 1957 to 1973.

Most of these works dealt with technological and laboratory applications of sorbents (45% in 1957 and 77% from 1961 to 1973). A smaller number of publications studied the sorption mechanism (37% in 1957 and 20% from 1961 to 1973) and sorbent synthesis (18% in 1957 and 3% from 1961 to 1973).

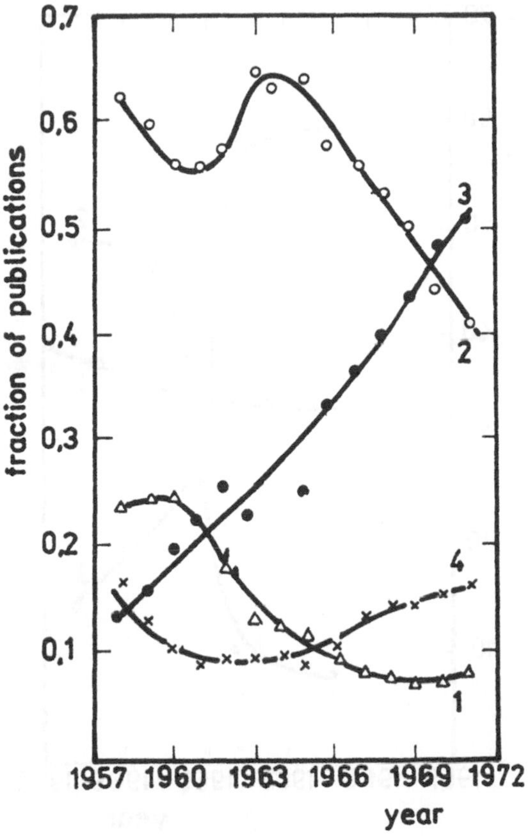

FIGURE 1. Fraction of publications for the time period 1957 to 1972 dealing with concentration (curve 1), purification and separation (curve 2), and sorption phenomena (curve 3), all related to the total amount of published papers in this field. Curve 4 deals with the analytical application of inorganic ion-exchange sorbents related to the total number of papers on sorption phenomena.

Sorbents have been used in analytical applications, especially for the concentration and partial separation of analyzed substances prior to the actual analysis and sorption phenomena accompanying coprecipitation. Liquid, gas, and thin-layer chromatography have been used for the separation of the studied substances.

Figure 1 depicts the fraction of the publications dealing with concentration (curve 1), separation (curve 2), and analysis of substances utilizing sorption (curve 3) for the time period 1957 to 1972. Curve 4 gives the fraction of works utilizing inorganic sorbents in analytical practice alone.

The curves in Figure 2 give the fraction of the published works utilizing inorganic sorbents in gas chromatography (curve 1), thin-layer chromatography (curve 2), and precipitation and coprecipitation methods (curve 3) for the time period 1957 to 1972.

In these years, research on the use of inorganic sorbents was centered primarily on the preparation of suitable ion-exchange sorbents with high selectivity and resistance to external conditions and the use of these sorbents for the chromatographic separation of mixtures that are difficult to separate.

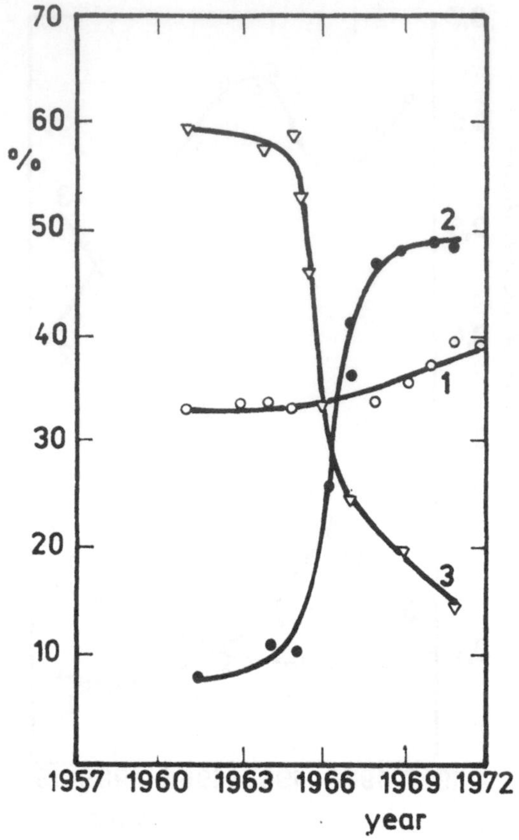

FIGURE 2. Percentage of publication for the time period 1957 to 1972 for applications in gas chromatography (curve 1), thin-layer chromatography (curve 2), and precipitation as well as coprecipitation (curve 3) related to the total number of papers dealing with the sorption phenomena.

II. THE VALIDITY OF THE CONCEPT "INORGANIC ION EXCHANGERS"

As mentioned in Section I, inorganic compounds were used in ancient times for ion exchange. The use of zeolites in detergents and in water-pollution control far outweighs in tonnage the amount of organic ion exchangers used annually. These inorganic sorbents have mostly been termed ion-exchanging materials. The concepts of cation and anion exchangers are, however, intimately connected with the development of organic ion exchangers and the theoretical basis for the ion-exchange process that followed from the properties of these macromolecular compounds.

Ion exchange is a process in which reversible stoichiometric interchange of ions of the same sign takes place between an electrolyte solution or molten salt and a solid phase. It should be emphasized that this definition of the concept of ion exchange assumes not only equivalence of the exchange and reversibility, but also that the ion exchanger be reproducibly prepared with given physicochemical characteristics, that chemical changes do not occur during the ion-exchange process that would substantially affect the sorption properties, and that the whole exchange process be characterized by an equilibrium state with a defined ion-exchange equilibrium constant.

The term ''inorganic ion exchangers'' is used in the title of the monograph by Amphlett, which describes the rapid development of these materials and their application in the 1950s and 1960s. Ion-exchange materials are divided into various categories with very different properties. While the ion-exchange process for some inorganic materials fulfills the definition conditions for the classical concept of ion exchange, although often for only a limited range of experimental conditions, there are a great many materials where one or more of the definition conditions for exchange are not fulfilled. Application of ion-exchange concepts to sorption processes is often doubtful in light of the sorbent properties. The term ion exchangers is inaccurate for this group of substances and the more general term ion-exchange sorbents would be preferable, as it describes more closely the studied group of materials.

Thus, the difference between organic and inorganic ion-exchanging materials should be clarified. There are a number of approaches that can be used and, hence, a number of the major differences will be considered here to determine if inorganic sorbents should be included among classical ion exchangers or whether it would be preferable to classify them as a specific group with specific ion-exchange properties.

III. COMPARISON OF THE PROPERTIES OF ORGANIC ION EXCHANGERS AND INORGANIC ION-EXCHANGE SORBENTS

A. CHEMICAL DEFINABILITY OF A SORBENT

It is first necessary to decide how inorganic sorbents are to be defined as chemical substances. Inorganic ion-exchange sorbents are usually well-defined chemical species, compared to synthetic organic ion exchangers that are not true compounds. Synthetic organic resins were used most widely because of the ease with which reproducible products could be prepared and because of their excellent mechanical and chemical stability and reproducibility. Thus, the bulk of recent ion-exchange literature has dealt with synthetic organic resins. Their availability, stability, and reproducibility probably are largely responsible for the popularity of ion exchange as a separation method.

Although inorganic sorbents are defined substances, their composition and degree of crystallinity are so important for the sorption properties of the product that the reproducibility of the preparation procedure is still a subject of considerable study. Depending on the method of preparation, a number of substances can exist in a number of forms, varying continuously from undefined amorphous substances to a defined structure with given crystallographic parameters, with the consequent variation in the ion-exchange properties of these sorbents. In general, the preparation of ion-exchange materials is very dependent on the conditions. In addition, the prepared materials do not always exist in an equilibrium state and aging is an important factor to be considered. The chemical and sorption properties of synthesized substances can change greatly with time. For example, amorphous $TiO_2 \cdot nH_2O$ is obtained by hydrolysis of an HCl solution of $TiCl_4$ with aqueous ammonia at room temperature. Aging of the amorphous precipitate under water at room temperature for 14 weeks produces no change detectable by X-ray diffraction methods. However, further aging of this precipitate for 30 weeks at room temperature yielded an X-ray diffraction pattern for anatase with different ion-exchanging properties. It is even more difficult to define the composition and stability of highly polymeric species, as equilibrium is attained extremely slowly. For example, it takes 15 years for polynuclear Fe^{3+} species to equilibrate in neutral or alkaline solutions. Thus, reproducibility is a critical factor that is hard to ensure for these substances.

The presence of inclusions in the sorbent solid phase is a further complicating factor in the synthesis of inorganic ion-exchange sorbents; these inclusions are generally difficult to remove. Precipitates of amphoteric oxides tend to contain appreciable amounts of the co-ion, which are difficult to remove by washing with water. Thus, some confusion can arise in relation to the ion-exchange properties because of the presence of these undesirable cationic or anionic impurities.

B. CHEMICAL, THERMAL, AND RADIATION STABILITY

Chemical stability is an additional criterion that emphasizes the basic difference between organic ion exchangers and inorganic ion-exchange sorbents. Organic ion exchangers are usually stable in a wide pH range, while inorganic materials differ widely in this respect. The insoluble salts of polyvalent metals are highly stable even in very concentrated acids. Crystalline antimonic acid is extremely stable to most reagents and is very difficult to dissolve, even by heating with concentrated hydrochloric acid or 1 *M* potassium hydroxide.

On the other hand, aluminosilicates are generally less resistant to acids than to bases. Some sorbents are dealuminated at pH 7 and a number of sorbents cannot be used at pHs of 1 to 2. Dealumination is accompanied by a decrease in the exchange capacity of the aluminosilicate. On the other hand, most zeolites are stable up to a pH of about 11. Decomposition occurs at high pH values. Compared to organic sorbents, which are incomparably more stable to acids, aluminosilicate sorbents are very unstable. Synthetic mordenite (Zeolon 900) is an exception, as it is chemically and thermally one of the most stable zeolites and can be used for sorption in the pH range of 1 to 12. For this reason, inorganic zeolites, which were used for all the early work on ion-exchanging materials, have become less popular over the years.

On the other hand, the resistance of many oxides and hydrous oxides to strong oxidizing reagents and organic solvents constitutes an advantage over organic resins. However, the presence of a complexing reagent can completely decompose natural or synthesized inorganic sorbents when the complexing agent attacks the basic structural unit of the sorbent (dealumination, sorbents containing metals that readily form complexes, e.g., Zr, Ti, U, Sn, and Fe). The definability of inorganic ion-exchange materials is complicated by the fact that the hydrolysis of some sorbents (e.g., zirconium phosphate gels) depends both on the method of preparation, i.e., the sorbent properties, and also on the ion loading. The thermal and radiation stabilities of inorganic ion-exchanging materials are great advantages.

The polymer matrices of organic resins break down at elevated temperatures, with a concommitant decrease in their exchange capacity. In contrast, inorganic sorbents are very stable at elevated temperatures; this is also true of zeolites, especially in the Na^+ or K^+ forms. For example, mordenite is stable up to 800°C and clinoptilolite up to 750°C.

Organic ion exchangers are very sensitive to exposure to high radiation doses, which cause significant changes in their capacity and selectivity, presumably through hydrolysis of their functional groups, chain scission, and changes in the cross-linking. Ionization radiation doses of the order of 10^5 to 10^6 Gy (1 Gy = 100 rad) significantly alter the properties of synthetic organic ion exchangers. The evolution of gases such as CO, O_2, CH_4, SO_2, and nitrogen oxides during irradiation practically prevents separation processes in the column. An absorbed dose of 10^7 Gy or more makes organic ion-exchange resins totally useless. Synthetic inorganic sorbents are, in general, more resistant to radiation degradation, compared to organic materials. However, there are limits beyond which these materials also undergo radiolytic degradation. Inorganic ion-exchange sorbents are stable up to accumulated doses of 10^7 Gy when irradiated in their virgin form. Zeolites used in cleanup operations in nuclear power stations have been found to function at doses of 2×10^9 Gy over a 10-year period. Moisture-free inorganic ion-exchange materials loaded with high-specific-activity radionuclides have been found to be stable for very long periods of time. This property can be used to advantage in the stabilization of waste forms for safe transportation and long-term storage.

C. MECHANICAL STABILITY

Organic ion exchangers and especially copolymers can be prepared with defined grain size, are resistant to abrasion, and are thus useful for use in packed columns. In contrast, inorganic ion-exchange sorbents exhibit poor hydrodynamic properties and are difficult to prepare in the form of particles with an acceptable particle-size distribution. These drawbacks can be eliminated to a certain degree by using the sol-gel method or supports or embedding materials for preparation of dimensionally more homogeneous grains; nonetheless, the abrasion resistance remains very low and the mechanical strength of the particles is unsatisfactory. For these reasons, the use of a number of inorganic sorbents for column separations is dubious. Zeolites are usually prepared as crystals with a size of 1 to 10 μm. Large crystals can also be prepared, but have less favorable sorption parameters. Thus, particles with suitable dimensions are prepared by using a porous ceramic binding agent such as kaolin, attapulgite, or halloysite. Thermal treatment then yields sorbent particles with spherical, tablet, rod-like, or some other suitable shape.

D. INTERNAL ORDER AND HOMOGENEITY IN INORGANIC SORBENTS

The structural arrangement of the sorbent is important for its sorption properties. Organic resins essentially have an elastic structure even for high degrees of cross-linking, and the hydration number of the exchanging ions may be nearly as large as that for the ions in the aqueous solution. Microcrystalline zirconium phosphates with essentially a two-dimensional layered structure exhibit similar behavior. In this structure, changes in the C-crystallographic dimension permit hydration of the exchanging ions by nearly the same number of water molecules as the ions in the aqueous solution. In both cases, the entropy of the exchangeable ion can be independent of the crystal radius. However, inorganic ion-exchange sorbents generally have relatively rigid structures which undergo only slight swelling or shrinkage on immersion in an aqueous solution, resulting in ion-sieve or steric effects for various ions. For example, antimonic acid has a highly rigid structure and does not undergo any dimensional change during ion exchange. When highly hydrated cations are sorbed on the weakly hydrated sites of the inner phase without loss of water, the number of degree of freedom may decrease in the system. However, when the ingoing cations are less hydrated ions, e.g., Cs^+ and Rb^+, the number of degrees of freedom in the system may increase.

The internal structural ordering of the ion-exchanging material is closely connected with a further property, the internal homogeneity. The internal solid phases of synthetic organic ion exchangers are relatively homogeneous with defined arrangement of the ion-exchanging sites, usually with exactly defined ion-exchange properties. Inorganic ion-exchange sorbents are usually unhomogeneous and contain ion-exchange sites with various exchange properties. Thus, the ion-exchange mechanism can vary widely. The ion-exchange properties often cannot be defined quantitatively, as they can vary over a rather wide range between two limiting values. They can be roughly compared to those of multifunctional organic ion exchangers.

E. ION-EXCHANGE PROCESSES

Obviously, all the above examples comparing organic and inorganic sorbents are most useful in analyzing the mechanism of ion exchange on various inorganic materials. Each type of inorganic material has specific characteristics for ion exchange, and thus the most important features of ion exchange on inorganic material will first be considered and compared with that of classical ion exchange on organic resins.

1. Aluminosilicates

Aluminosilicates constitute the largest group of inorganic ion-exchange sorbents. These substances consists of a lattice formed by $Si(OH)_4$ tetrahedra and $Al(OH)_6$ octahedra, forming a layered (clay, mica) or three-dimensional (zeolite, feldspar) structure. Isomorphic exchange can involve replacement of the silicon ions in the sorbent phase by ions with a lower charge, e.g., Al^{3+}, and trivalent aluminum by divalent ions, such as magnesium or iron. These changes lead to the formation of a charge on the lattice that is compensated by alkali metal or alkaline earth cations. The ions that are bonded to the sorbent skeleton by electrostatic forces are mobile and can be subject to an ion-exchange mechanism when in contact with a solution, similar to that for synthetic organic ion exchangers. The exchange is, however, greatly limited by the network effect, especially in three-dimensional zeolites. Various windows and channels are formed in the structure, depending on the Al:Si ratio in the zeolite and the steric arrangement of the tetrahedra on the surface of the zeolite particles. These openings can differ considerably among various materials, and are sterically selective. Ion exchange in these inorganic materials can be parallel to that on organic ion exchangers with a higher degree of cross-linking of the divinylbenzene type.

In addition to this ion-exchange mechanism, exchange can also occur on the surface of three-dimensional aluminosilicates. The crystal edges of these inorganic ion-exchange sorbents contain –OH groups that can be exchanged for anions or cations, depending on the pH. The exchange capacity for this type of surface exchange is, however, incomparably lower than that for ion capture in the internal lattice of the sorbent. The internal homogeneity of the three-dimensional aluminosilicate lattice cannot be compared with that of synthetic organic ion exchangers, as the steric ordering of the tetrahedral groups can form multidimensionally defined windows and various types of channels. For example, six different steric positions have been described for X-type zeolites.

It usually makes no difference to the sorption process on organic ion exchangers whether the ion exchanger is in the K^+ or Na^+ cycle, while the sorption properties of K-Y and hydrated Na-Y zeolites differ. The basic structural skeleton of organic ion exchangers practically does not change, depending on the ion form of the ion exchangers, while ion sorption in inorganic ion exchangers can result in structural changes. For example, the sorption of cesium on X-type zeolites leads to relocation of the Na^+ ions from the large α cavities into the smaller β cavities with simultaneous restructuring of the internal zeolite lattice. These changes lead to a change in the ion-exchange affinity of the sorbent for the sorbed ions, with corresponding changes in the corresponding affinity series.

On the other hand, purely physical sorption can occur on three-dimensional zeolites and layered aluminosilicates, where ions with a given charge are bonded by weak van der Waals forces to the surface layer of the sorbent, either to the charged surface or to the surface layers. The sorption of molecules and ion pairs in the cavities is termed molecular sorption, and is employed especially for gases that are sorbed on molecular sieves in the dry state.

It is obvious from this brief discussion of the possible sorption mechanisms for ions on inorganic aluminosilicates that the sorption process is complex and that a deeper understanding is necessary than for organic ion exchangers. Theoretical explanation is more difficult and cannot be compared with the classical concept of ion exchange on ion-exchange resins.

2. Acidic Salts of Polyvalent Metals

The mechanism of ion exchange on the acidic salts of polyvalent metals is also a very complex process. The exact work of Clearfield et al., who carried out structural studies and studies of the phase conversions during ion exchange to elucidate the sorption mechanism on these materials and correlated the degree of crystallinity with the sorption properties, led to an understanding of these sorption processes comparable to that for zeolites.

For example, differences in the structure, composition, and exchange behavior of amorphous zirconium phosphate can result from different preparation conditions. This substance is not usually completely amorphous, but rather, exhibits three or four very broad X-ray reflections of low intensity. The general formula of such a sorbent can be written as $Zr(OH)_x \cdot (HPO_4)_{2-2/x} \cdot yH_2O$, where x varies from 0 to above 2. It has been found that it is possible to obtain an almost continuous increase in the crystallinity until true crystals of α-zirconium phosphate are obtained. Evidence has been found for the presence of different-sized entrances into the cavities, and the disordered gel structure can give rise to hydrogen bonds with a range of bond energies. The sorption process on amorphous zirconium phosphates is a function of the loading. The exchangeable sites are not energetically equivalent. The incoming ions initially occupy the most favorable sites, i.e., the large cavities, so that the alkali ions are more or less hydrated when they enter the cavities. As the exchange process proceeds, smaller and smaller cavities are occupied and the less strongly bonded water is forced out by the larger cations. This process is accompanied by an increase in the interlayer spacing. Above 50% exchange, two cations must approach each other closely as they occupy the same cavity. Lattice disordering occurs and is reflected in the release of phosphate into the solution.

Semicrystalline zirconium phosphates are formed that have properties and behavior intermediate between those of gels and crystals; attempts have been made to relate their behavior to the degree of crystallinity. Consequently, zirconium phosphates cannot be considered as a single type of compound.

The α-modification of zirconium phosphate is most clearly defined, and has a layered structure with an interlayer distance of 7.56 Å. The metal atoms lie in a plane and the phosphate groups form bridges. The three O atoms of each phosphate group are bonded to three different zirconium atoms, and each zirconium atom is thus octahedrally coordinated by oxygens. The 12-membered rings consist of three P atoms, three metal atoms, and six oxygen atoms. Two 12-membered rings form a six-sided cavity containing a water molecule. One mole of cavities is formed per formula weight of α-zirconium phosphate and there are two exchange sites per cavity. During exchange, the layers expand or contract so that recording the first reflection in the X-ray powder pattern during exchange permits the reaction to be followed. The main interlayer spacings for various zirconium phosphate preparations are as follows:

Formula	**Proposed name**	**Interlayer spacing (Å)**
$Zr(HPO_4)_2$	ε-ZrP	5.59
$Zr(HPO_4)_2 \cdot {}^1/_2\ H_2O$	δ-ZrP	7.13
$Zr(HPO_4)_2 \cdot H_2O$	α-ZrP	7.56
$Zr(HPO_4)_2$	β-ZrP	9.40
$Zr(HPO_4)_2 \cdot 8H_2O$	θ-ZrP	10.40
$Zr(HPO_4)_2 \cdot 2H_2O$	γ-ZrP	12.20

The exchange reactions on crystalline zirconium phosphate are generally not reversible, and exchange on this sorbent is thus a much more complicated process than that on amorphous zirconium phosphate. These complications reflect the phase transitions occurring during exchange and also the nonuniformity of the exchange sites within each phase. The largest entrances into the cavities would allow a spherical cation 2.61 Å in diameter to diffuse unobstructed into the cavities. Thus, the ion-exchange behavior of crystalline zirconium phosphate is conditioned by two features of its structure, the size of the openings into the cavities and the weak forces holding the layers together. The crystals do not swell in water, so the size of the cavities determined for a dry crystal is the same as that encountered by exchanging cations. Thus, it should be possible for unhydrated Li^+, Na^+, and K^+ to enter the structure, but Rb^+ and Cs^+ would be excluded. Such a hydrated cation loses most of its water at the surface of the crystal, diffuses into the cavity as either an unhydrated or partially hydrated species, and is distributed throughout the exchanger as a solid solution without any phase change. However, the layered structure is not as rigid as that of zeolites and, as the pH is increased, the tendency to replace the internal protons becomes so great that the large cations can force the openings and enter the cavities. It should be pointed out that the forces holding the layers of the salt together are essentially the result of coulombic attraction between the negative fixed charges of adjacent planar macroions and the counterions distributed among them.

It was found that the forward and backward titration curves for the titration of α-zirconium phosphate with NaOH do not coincide. This hysteresis is a result of the formation of different solid phases during the titrations:

$$Zr(HPO_4)_2 \cdot H_2O + Na^+ \rightleftharpoons Zr(NaPO_4) \cdot (HPO_4) + H_3O^+$$

In this procedure, the interlayer spacing of the sorbent increases from 7.56 to 11.8 Å in the half-exchanged phase. The hysteresis between the forward and reverse curves is attributed to the limited miscibility of the phases.

A number of other inorganic sorbents have been found to behave similarly to crystalline zirconium phosphate. For example, the loading of α-titanium phosphate with Na^+ ions leads to the formation of three phases, α-Ti $(HPO_4)_2 \cdot H_2O$, α-$TiHNa(PO_4)_2 \cdot 4H_2O$, and α-Ti(Na $PO_4)_2 \cdot 3H_2O$, which then coexist in the solid sample. Hysteresis can thus be explained by the coexistence of these different hydrated phases. At a substantially higher pH, up to 75% exchange can be attained; however, other phase transitions can also be expected and inorganic phosphate can even be precipitated in the presence of some multivalent cations. This phenomenon has been described for divalent magnesium.

The sorption of metal cations into the half-exchanged phase is very slow, as is their elution procedure. This site binding in the back-titration procedure, even at pH values below 2, causes equilibrium conditions to not be achieved even after 4 d. Such a sorption process via selective trapping of ionic species may well be valid for a great many ion-exchange sorbents.

Zirconium phosphates with interlayer spacings of greater than 10 Å are often called expanded-layer zirconium phosphates. The half-exchanged sodium form is a pentahydrate with interlayer spacing of 11.8 Å and the hexahydrate (Θ-ZrP) and dihydrate (γ-ZrP) also exist with interlayer spacings of 10.4 and 12.2 Å, respectively. These substances can exchange large ions without appreciable barriers to diffusion and do not suffer from the restrictions resulting from small interlayer spacing; thus, complexes can also be exchanged under favorable conditions.

All these findings explaining the sorption mechanism on zirconium phosphates are described and were taken from the excellent book *Inorganic Ion-Exchange Materials* edited by Abraham Clearfield.

The above discussed facts indicate that the sorption mechanism for the exchange of ions on the acidic salts of multivalent metals is even more complex than that for aluminosilicates, as the sorption process is dependent on the degree of crystallinity of the sorbent, which can vary continuously from a highly amorphous substance to well-crystallized compounds; the irreversibility of the sorption process increases with increasing crystallinity.

3. Hydrous Oxides of Polyvalent Metals

The hydrous oxides of polyvalent metals occupy an important position among inorganic ion-exchange sorbents. The term hydrous oxide is used in its broadest sense here to refer to insoluble materials containing a metal oxide-water system. Metal oxides often combine nonstoichiometrically with water and in some cases, such as FeOOH and AlOOH, the terms oxyhydrate or oxide hydrate are mostly applicable. It sometimes happens (for example, for the hydrous oxides of tetravalent metals) that there is no evidence for the existence of a definite hydrate.

The adsorption mechanism on silica gel is mostly interpreted in terms of ion exchange. Conventional silica gel is an agglomeration of spheroids bounded by interparticle siloxane (≡Si–O–Si≡) links. Most hydrous oxides weakly exhibit acid and/or basic ion-exchange behavior. Some results have indicated that, in addition to this mechanism, other processes such as physical adsorption and ion inclusion play a role. Adsorbed cations are weakly bonded to the surface sites of SiO_2 and are readily exchangeable.

The mechanism of cation retention on stannic oxide is best described in terms of the hydrolysis process. Alkaline earth-metal cations are weakly adsorbed, whereas trivalent cations such as Al^{3+} and Cr^{3+} are strongly adsorbed. It appears that a coprecipitation process takes place in these systems.

The ion-exchange mechanism also seems probable for zirconium oxide sorbents. In the polymer of zirconium oxide prepared for this purpose, each zirconium atom is surrounded by eight hydroxyl groups in a distorted cubic arrangement. Oxygen-bridge formation between the layers leads to a three-dimensional structure with the general formula $[ZrO_b(OH)_{4-2b} \cdot xH_2O]_n$. Two types of hydroxyl groups are attached to the zirconium atoms: bridging and nonbridging. The latter lie on the surface of the crystallites. The reversibility of the sorption of alkali metal cations and of anions was determined for this substance, which indicates that the incoming ions are strongly dehydrated in the exchanger phase.

Crystalline antimonic acid has essentially a cage structure similar to that in three-dimensional zeolites. When small cations on crystalline antimonic acid are exchanged for large cations, the ion exchange becomes progressively more difficult as the proportion of one of the large cations in the solid phase is increased (as a result of steric effects). This sort of complication can occur in a rigid structure which undergoes relatively little swelling if there are large differences in the size of the two cations. If no site is available for the ingoing large cations within the intercrystalline volume, incomplete exchange results. When the cation is much larger than the size of the windows of the cage structure, an ion-sieve effect is operative. It has been found for ion exchange on crystalline antimonic acid that Ca^{2+}/H^+ and Sr^{2+}/H^+ ion exchange is "irreversible" because of the extremely slow rate of desorption and, in addition, the extremely high adsorption ability for Ca^{2+} and Sr^{2+} ions.

4. Uranyl Phosphates

To provide a complete picture of the exchange mechanism on inorganic sorbents, ion exchange on substances that are simple, defined inorganic compounds should be described. A good example is provided by uranyl phosphate, $(H_3O)^+ (UO_2PO_4)^- \cdot 3H_2O$. The UO_2^{2+} and PO_4^{3-} ions are bonded together in puckered, negatively charged layers that can be stacked in two ways, as the autenite or meta-autenite structure. Monovalent metal salts adopt the meta-autenite structure, while divalent metal salts exist in the autenite structure. The exchangeable cations and water molecules occupy positions between the layers. Under some conditions, it has been suggested that ion exchange occurs for both H^+ and $(H_3O)^+$ ions; a precipitation mechanism has also been described. It has been concluded on the basis of the sorption properties of uranyl hydrogen phosphate that sorption involves a process of reprecipitation and therefore should be controlled by the solubility products of the uranyl phosphates produced. However, it was found that the solubility products did parallel the selectivities. It is still not clear whether the reactions involve true ion exchange.

5. Conclusions

It follows from the above brief discussion of ion-exchange mechanisms on inorganic sorbents that the mechanisms differ considerably between organic and inorganic ion exchangers, although certain parallels can be found. The overall phase and chemical instability of inorganic sorbents differentiate them clearly from organic ion exchangers; this instability can lead to changes in the ion-exchange properties. These materials are highly varied, and their properties and ion-exchange mechanisms can vary widely.

F. ION-EXCHANGE SELECTIVITY

The ionic selectivity of organic ion exchangers is given primarily by the character of the functional groups incorporated in the polymer skeleton and by the degree of cross-linking. The selectivity of exchange on inorganic sorbents has been explained by a great many theories, but no unified hypothesis has been drawn up. Some authors have stated that the selectivity is given primarily by the magnitude of the hydration energy of the exchanged ion, charge density on the sorbent lattice, and its structure. For example, the greatest lattice charge density is exhibited by zeolites with low Si/Al ratios, low internal water contents, and higher affinity for small ions. In general, it can be stated that the selectivity of inorganic sorbents containing windows and internal channels into which the ions can diffuse is determined not only by the diameters of these windows, but also by their ability to adapt to the size of the entering ions. For example, zeolites with large windows exhibit the selectivity series Cs> Rb> K> Na> Li and Ba> Sr> Ca. Some ions, such as Cs^+ and Rb^+, pass through windows with a crystallographic diameter smaller than that of these hydrated ions, as the ions are dehydrated during sorption and are bonded very strongly inside the six-membered ring of the zeolite structure through resolvation of these ions in the sorbent phase. A similar mechanism of increased selectivity is valid in a more marked form for zeolites whose structure contains an eight-membered ring, e.g., chabazite, erionite, and clinoptilolite.

G. KINETICS OF ION EXCHANGE

The kinetics of ion exchange on inorganic ion exchangers is somewhat similar to that for organic ion exchangers with a higher degree of internal cross-linking. In general, the kinetics of the exchange mechanism at inorganic materials is diffusion controlled, so that the rate of ion exchange is dependent on the rate of diffusion in the liquid film surrounding the sorbent particles and on the rate of diffusion in the crystalline phase of the sorbent. Surface diffusion usually is not very dependent on the sorbent properties, while internal diffusion is highly dependent on the internal ordering of the studied substance. The sorption kinetics at inorganic materials is determined primarily by internal diffusion, as diffusion in the surface liquid film is usually faster by an order of magnitude. However, this situation can be reversed in sorption from very dilute solutions or of trace concentrations. Exact treatment of kinetic data for sorption on inorganic materials is often complicated by the fact that the diffusion rate can change with the degree of exchange, or phase changes can occur during ion exchange as a result of the fact that the internal phase of the studied substance is not homogeneous and cannot be prepared reproducibly. Thus, the results do not have exact character but, rather, tend to be qualitative.

H. THERMODYNAMICS OF ION EXCHANGE

The number of works dealing with the thermodynamics of ion exchange on inorganic sorbents is surprisingly large. Equilibrium constants, free energies, and enthalpy and entropy changes have been found for ion-exchange processes for both classical organic ion exchangers and for inorganic materials. Thermodynamic data were found by calculation from equilibrium sorption data or from primary calorimetric measurements. However, calculations methods require knowledge of the activity coefficients of the ions in the solid sorbent phase. These activity coefficients of the individual ions can be considered to be quantities characterizing the ion and exchanger with a given composition and can have values from 0 to 1; unity is considered to correspond to the sorbent in the standard state where the cycle of the sorbent contains only the ion for which the activity is considered. As the sorption sites on inorganic materials have various exchange capacities and the internal structure of the substance is usually insufficiently homogeneous, the definition of a standard state is usually rather complicated for these materials and is often not thermodynamically justified. In addition, calculation of the activity coefficients according to the Gibbs-Duhem equation assumes that the change in the activity of water in the aqueous and solid phase of the inorganic material is negligible; this is justifiable for highly dilute systems, but is no longer acceptable at higher concentrations. A relation of the calculated changes in the enthalpy and entropy in the exchange reactions to the number of exchangeable sites per gram of the sorbent is dubious, as they represent the integral value for all the exchangeable sites into which the exchanged ions can enter, while the selectivity of exchange for the individual sites can vary for inorganic materials by several orders of magnitude. Also, the determination of changes in the free enthalpy from the temperature dependence of the free energy is usually highly inaccurate.

Direct calorimetric measurement of the enthalpy change in ion-exchange reactions has been carried out with considerable precision, but it is still not clear to which quantity the measured value should be related, to which degree the studied material is thermodynamically defined and reproducibly prepared. Consequently, most determinations of thermodynamic data have limited general validity.

IV. CLASSIFICATION OF INORGANIC ION-EXCHANGE MATERIALS

Synthetic organic ion exchangers can be classified on the basis of exchangeable species (cation, anion, ampholyte, and multifunctional types), functional group (strong, weak, and acid/base types), skeleton types (polymers, copolymers, and polycondensates), and the rigidity of the polymeric structure [microporous gel, macroporous (macroreticular), isoporous].

A similar classification of inorganic ion-exchange materials is practically impossible. The chemical variety of these materials, their very characteristic properties for ion exchange, and the general inhomogeneity of the structure of sorbents with ion-exchanging sites with various selectivity all make it very difficult to find a single classification system. Consequently, classification of these materials on the basis of the chemical characteristics of the ion-exchanging species appears still useful, as proposed by Veselý et al. (1972):

1. Hydrous oxides
2. Acidic salts of multivalent metals
3. Salts of heteropolyacids
4. Insoluble ferrocyanides
5. Aluminosilicates

A. HYDROUS OXIDES

1. Hydrous Oxides of Bivalent Metals

This group of sorbents includes the hydroxides of Be^{2+}, Mg^{2+}, Zn^{2+}, and their mixtures with $Fe(OH)_3$ or $Al(OH)_3$. Hydrous BeO with composition $BeO \cdot 1.7H_2O$ acts as both a cation and anion-exchanging material. $Mg(OH)_2$ exhibits anion-exchanging properties. The sorption of Zn^{2+} on $Mg(OH)_2$ can be explained through a coprecipitation mechanism rather than ion exchange, which is also confirmed by the sigmoid character of the adsorption isotherm. A mixture of $Mg(OH)_2$ with $Fe(OH)_3$ or $Al(OH)_3$ exhibits interesting sorption properties, with irreversible sorption, and has thus been successfully used to adsorb radionuclides in trace concentrations.

2. Hydrous Oxides of Trivalent Metals

Pseudomorphic iron hydroxide acts as both an anion and cation-exchange material with an isoelectric point at pH 7.1 to 7.2. The adsorption isotherms more or less obey the Freundlich adsorption law, but clear evidence was not obtained for the exchange of OH^- ions. The cation-exchange mechanism was shown to be operative for alkali metal ions, but a coprecipitation mechanism was found for Ni^{2+} ions. Both hydrolytic and ion-exchange mechanisms were found for Cu^{2+} and some other bivalent cations.

Hydrous alumina also exhibits cation- and anion-exchanging properties with isoelectric points at pH 5.59, 6.78, and 6.80 for α-Al_2O_3, α-AlOOH, and α-$Al(OH)_3$, respectively.

These sorbents have been widely studied and a number of applications have been described, especially in thin-layer chromatography for ion separation.

The remaining sorbents of this group, such as MnOOH, $La(OH)_3$, $Sb_2O_3 \cdot 3H_2O$, $In_2O_3 \cdot 3.5H_2O$, Ga_2O_3, $Bi_2O_3 \cdot 3H_2O$, and $Cr(OH)_3$, seem less promising.

3. Hydrous Oxides of Tetravalent Metals

This group includes primarily the following oxides: SiO_2, SnO_2, TiO_2, ThO_2, ZrO_2, and MnO_2. They are usually described as hydrous oxides or hydrated hydroxy oxides and behave either as cation exchangers in alkaline solutions or as anion exchangers in acid solutions, depending on the basicity of the central atom and the strength of the M-O bond relative to that of the O-H bond in the hydroxyl group. The exchange sites are heterogeneous.

Hydrous silica in the form of hydrogels and xerogels has been studied intensively for many years. Although the ion-adsorption mechanism on silica gel has not yet been completely elucidated, it generally acts as a weakly acidic cation exchanger. It can be considered to be an agglomeration of spheroids cemented by interparticle siloxane links, and this structure is reflected in its porosity. The specific surface area and the porosity of silica gel are the decisive factors affecting its sorption behavior. The adsorption mechanism of nonhydrolyzing cations is mostly interpreted in terms of ion exchange. It has been suggested that physical adsorption and ion exclusion from small pores could also play a role, even in the sorption of alkali metal ions. Molecular sorption of the complexes of metals has been found on the SiO_2 surface and lattice during all the ion-exchange steps. Zr and Hf are probably hydrolyzed after diffusion into the silica gel phase. The mechanism of the sorption of hydrolyzed cations on SiO_2 is very complex and only qualitative conclusions can be drawn from the experimental results. In addition to the ion-exchange process, surface adsorption phenomena are probably also important for many hydrolysates. SiO_2 has been used widely in analytical chemistry, especially in thin-layer chromatography.

Hydrous stannic oxide, $SnO_2 \cdot xH_2O$ has cation- and anion-exchanging properties. Trivalent metal ions, Al^{3+} and Cr^{3+}, are sorbed by a coprecipitation mechanism.

Hydrous titanium oxide has the general formula $TiO_{2-x} \cdot (OH)_{2x} \cdot yH_2O$ ($x<1$) or $TiO(OH)_2 \cdot nH_2O$ and is insoluble in acids and alkaline solutions. It has suitable particle size stability for use in column applications. The titration curves indicate that the sorbent has polyfunctional character and can exchange three protons for every ten Ti atoms. The chemisorption mechanism has been described for Fe^{3+} adsorption on this sorbent. Its high chemical and radiation stability have made it useful for applications in nuclear technology. The adsorption of In^{3+}, SeO_3^{2-}, PO_4^{3-}, and Cu^{2+} ions occurs through coprecipitation.

The general formula of thorium oxide can be written as $ThO(OH)_n \cdot mH_2O$. Hydrous thorium oxide acts as both a cation- and an anion-exchanging material. The formation of thorium phosphate was found to occur in the coprecipitation of phosphates with hydrous thorium oxide. This sorbent was found to be useful for the preparation of inorganic ion-exchange membranes.

It has been suggested that hydrous zirconium oxide is a polymeric oxo-hydroxide with the general formula $[ZrO_b(OH)_{4-2b} \cdot xH_2O]_n$. However, the exact nature of the polymer species is not known. It acts as both a cation- and an anion-exchanging material and has found important applications in water desalination and in the preparation of membranes.

Manganese dioxide can form structures similar to β-, γ-, and δ-MnO_2 or a nonstoichiometric compound with the empirical formula $(MnO_x)_2OH$, where $x = 1.70$ to 1.75. The OH groups in hydrous manganese oxide have different acid-base properties. The sorption mechanism on this substance seems to be rather complex. The ion-exchange mechanism has been reported for alkali metal ions, the sorption of MCl_2 at higher concentrations, where M is an alkaline earth metal, is assumed to occur through molecular sorption, Cr^{3+} has been found to react chemically, a colloidal suspension is formed with Fe^{3+} ions in the sorbent solid phase and a precipitation mechanism has been proposed for WO_4^{2-}, MoO_4^{2-}, and AsO_4^{3-}.

4. Hydrous Oxides of Pentavalent and Hexavalent Metals

Antimony oxide, also called antimonic acid, is known to exist in crystalline, amorphous, and glassy forms, with different chemical compositions and ion-exchange properties. A tentative structure has been proposed for the crystalline species, with 14 exchangeable hydrogen atoms. The properties of the amorphous substance depend on the drying procedure, aging, and the preparation method. The glassy gel exhibits weak lines in the X-ray pattern and its properties are similar to those of the amorphous substance. Antimonic acid acts as a cation exchanger, but the ion-exchange isotherms are S-shaped for many ions. Crystalline antimonic acid and the glassy substance are more useful for column applications. A great variety of applications have been described.

Phosphoantimonic acid has a variable P/Sb ratio which is reflected in its variable selectivity for a number of cations. It has high thermal and radiation resistance and is not attacked by salt solutions or strong acids.

Other hydrous oxides in this group of penta- and hexavalent metals, such as hydrous Nb_2O_5 and Ta_2O_5, are of limited use and the oxides of Mo(VI), W(VI), etc. have not been found to be particularly useful.

B. ACIDIC SALTS OF POLYVALENT METALS

The ion-exchange materials in this group form a large number of compounds, as mentioned in the discussion of the history of the development of these substances. The renaissance of the practical application of these substances is closely connected with the development of nuclear science and also with the modern, more exact approach to the understanding and elucidation of sorption processes, based on structural studies, as first presented by Clearfield et al. (1982). Tetravalent metals were first studied most extensively, while the salts of Al^{3+}, Fe^{3+}, Cr^{3+}, and U (VI) were of lesser interest. The anions most extensively employed included antimonate, arsenate, molybdate, oxalate, phosphate, silicate, tellurate, tungstate, vanadate, etc.

1. Acidic Salts of Tetravalent Metals

Most of the research carried out in this area has dealt with zirconium phosphate ion exchangers. However, because these compounds can exist in a wide variety of crystalline forms, Clearfield et al. have suggested that they be classified into the following groups.

a. Acidic Salts with Layered Structure of the α Type

This group includes α-zirconium phosphate, α-titanium phosphate, α-titanium arsenate, α-zirconium arsenate, α-hafnium phosphate, α-tin phosphate and arsenate, organic-inorganic α-layered exchangers, and other α-layered materials.

Zirconium *bis* (monohydrogen orthophosphate)-monohydrate (α-ZrP), also termed the α modification, is one of the best-known crystalline compounds. It can be prepared by refluxing the amorphous gel in 10 to 15 *M* phosphoric acid. Larger crystals can be prepared by prolonged refluxing. As described in the section dealing with the ion-exchange properties of these materials, the structure is layered and the compound is monoclinic with unit cell dimensions of a = 9.076, b = 5.298, c = 16.22 Å, and β = 111.5°.

α-Ti $(HPO_4)_2 \cdot H_2O$ (α-TiP) is isomorphous with α-ZrP and can also be prepared by refluxing of amorphous species. The unit cell dimensions are a = 8.631, b = 5.002, c = 16.176 Å, and β = 110.2°. The entrance ways into the cavities of α-titanium phosphate are narrower than those in α-ZrP and thus ions with larger ionic radii, such as the K^+ ion, are excluded. This sorbent also exhibits unidimensional intercrystalline swelling when amines are incorporated into the structure.

α-Zirconium arsenate, $Zr(HAsO_4)_2 \cdot H_2O$ (α-ZrAs) and α-titanium arsenate (α-TiAs), $Ti(HAsO_4)_2 \cdot H_2O$, can be prepared by refluxing amorphous species in 8 *M* arsenic acid for 45 h. α-Zirconium arsenate is isomorphous with α-ZrP, with unit cell dimensions of a = 9.178, b = 5.378, and c = 16.55 Å, and β = 111.3°. Crystalline zirconium arsenate acts as an ion sieve. Alkali metal ions are not exchanged in acid solution; exchange in neutral or alkaline solutions is accompanied by extensive hydrolysis of the sorbent. Arsenate ions from α-titanium arsenate are released into solutions even in acid media, and thus the use of this material as an ion-exchange sorbent is dubious.

The α modifications of crystalline tin phosphate (α-SnP) and arsenate (α-SnAs) were also prepared by refluxing freshly prepared gels in concentrated H_3PO_4 or H_3AsO_4 solutions. $Sn(HPO_4)_2 \cdot H_2O$ is highly hygroscopic and adsorbs an additional mole of water at higher humidities without any change in the X-ray pattern. Both sorbents are readily hydrolyzed in alkaline solutions and exhibit ion-sieve properties, predominantly adsorbing Li^+ and Na^+ cations. Tin phosphate exchanges more Li^+ than corresponds to the proposed formula.

b. Acidic Salts with Layered Structures of the β and γ Types

These compounds include anhydrous compounds or dihydrates. In moist air or on contact with water, the β modification changes to the γ modification. As mentioned in the preceding section, these substances have much larger interlayer spacing than α-ZrP, but the atoms in their layers lie directly over one another. These sorbents can adsorb cations with larger dimensions, such as Cs^+.

c. Acidic Salts with Less Definite Structures

This group includes compounds with a fibrous structure. Their structures are mostly unknown and it is sometimes rather difficult to give a systematic account of their ion-exchange behavior. The phosphates of Ce(IV), Th(IV), and Ti(IV) and titanium arsenate have this type of structure.

A further group consists of acidic salts with a three-dimensional structure and with the general formula $HM_2^{4+}(XO_4)_3$, where M is Zr, Ti, or Ge and X is P, As, or Sb. Zirconium molybdate is hydrolyzed so readily that it is not useful in practical applications.

The last group consists of acidic salts with an unknown structure and can include thorium arsenate, cerium(IV) phosphate, cerium(IV) arsenate, and titanium tungstate.

As mentioned above, these substances can exhibit different degrees of microcrystallinity, from amorphous or gel-like substances to semicrystalline and exactly defined, crystalline substances. They contain nonhomogeneous functional groups that can participate in the exchange process, and are widely employed in analytical practice. These properties have already been described for zirconium phosphate and for other sorbents in this group.

Amorphous zirconium arsenate, antimonate, molybdate, and tungstate have limited applications because of their chemical instability. The titration curve of zirconium arsenate has no marked inflection point, reflecting the wide range of exchangeable acidic sites. Amorphous zirconium molybdate and tungstate also act as electron exchangers.

Attempts have been made to prepare inorganic ion-exchanging materials by combining the zirconium phosphate with other organic or inorganic compounds. Zirconium phthalophosphate, zirconium sulfosalicylophosphate, zirconium phosphate-silicate, zirconium phosphate-ammonium molybdophosphate, zirconium-titanium or hafnium or niobium phosphate, and zirconium phosphate-tungstophosphate have all been described.

Amorphous gels of titanium phosphate have lower exchange capacities than zirconium phosphate gels and are hydrolyzed even in acid solutions, which limits their use in practice. Glassy amorphous titanium arsenate with the composition $Ti(HAsO_4)_2 \cdot 2 \cdot 5H_2O$ and amorphous titanium antimonate with an Sb:Ti ratio of ~ 1, consisting of TiO_2 and Sb_2O_5 condensed in a polymeric structure, have been prepared for separations. However, their use is limited by hydrolysis.

The gels of cerium phosphate, with the tentative formula $Ce_3(OH)_8(H_2PO_4)_4$, are relatively strongly resistant to hydrolysis. The amorphous form exhibits both ion-exchange properties, with a maximum capacity of 2.9 meq/g, and electron-exchange properties. Microcrystalline cerium phosphate with a Ce:PO_4 ratio of 1.5, corresponding to the formula (Ce-O-Ce) · $(HPO_4)_3$ · H_2O, has a relatively low ion-exchange capacity, while fibrous crystalline cerium phosphate, with the formula $Ce(HPO_4)_2 \cdot H_2O$, has the relatively high ion-exchange capacity of ~ 5.2 meq/g with enhanced hydrolysis stability. Cerium phosphate-sulfates with the formula $Ce_2O(HPO_4)_{3-x}(SO_4)_x \cdot H_2O$, where $0<x<1$, have a zeolitic structure with ion-sieve properties that are very useful for separation purposes.

Thorium phosphate has been prepared and has variable composition, low stability toward hydrolysis, and a fibrous structure, with the composition $ThO_2 \cdot P_2O_5 \cdot 4H_2O$. Its fibers can be employed to prepare support-free flexible sheets.

Zirconium pyrophosphate, hypophosphate, and polyphosphates have been used in some analytical applications. In the form of glassy gels, these sorbents have a variable composition (P:Zr = 2 to 2.8) and their exchange properties depend greatly on the preparation conditions. However, the high degree of cross-linking in phosphates negatively affects the kinetics of the sorption process.

A great variety of inorganic ion-exchange materials have been developed from tetravalent cations and various anions, but no extensive applications have been found; they include zirconium tellurate, $Zr(H_2TeO_6) \cdot 4H_2O$, zirconium oxalate, $Zr(OH)C_2O_4H$, and zirconium silicate. A complete description is beyond the scope of this chapter.

2. Other Acidic Salts

Of the substances prepared from trivalent cations, chromium polyphosphate is of some interest, especially the product with the general formula $H_2CrP_3O_{10} \cdot 2H_2O$, which is highly selective for Ag^+. The anhydrous compound is formed in the sorption process. Chromium, aluminum, and ferric phosphates have limited applications. Aluminum vanadate can be used as an ion-exchange material at higher temperatures (900°C) in the form of membranes or sinters with a ceramic material. Of the divalent cations, uranyl phosphate should be mentioned; the sorption mechanism of this substance has already been discussed, but its potential applications seem very limited.

C. SALTS OF HETEROPOLYACIDS

The parent acids of these salts are the 12-heteropolyacids with the general formula $H_mXY_{12}O_{40} \cdot nH_2O$, where m = 3, 4, or 5, X can be phosphorus, arsenic, silicon, germanium, or boron, and Y one of the number of elements, such as molybdenum, tungsten, or vanadium. The prepared salts are usually impure and are contaminated during the preparation of the above acids. The heteroacids $H_3[PO_4(W_3O_9)_4]$ and $H_3[PO_4(Mo_3O_9)_4]$ with 29 water molecules usually crystallize out from solutions; they are dried to form the microcrystalline, stable, and well-defined pentahydrate. However, dehydration can be accompanied by partial decomposition. Both the acid hydrates have cubic crystal structures, and the unit cell dimension of the pentahydrate of 12-tungstophosphoric acid is a = 12.141 Å; the unit cell contains two formula units in the space group Pn3m. The 29-hydrate has dimension a = 23.281 Å and Z = 8 in the space group Fd3m, with a face-centered cubic crystal structure in which the atom packing is in a diamond-like arrangement. The more highly hydrated water-cation structure fills the larger and geometrically more intricate interstitial voids.

The salts of heteropolyacids with small cations are relatively soluble, whereas those with larger cations are far less soluble (e.g., K^+, NH_4^+, Rb^+, Cs^+, Sr^{2+}, Ba^{2+}, and alkylammonium). Hydrolytic degradation occurs in strongly alkaline solutions.

1. Molybdophosphates (MP)

12-Molybdophosphates form normal (X_3MP) and acidic (X_2HMP) salts. Ammonium molybdophosphates [AMP = $(NH_4)_3MP$ or $(NH_4)_2HMP$] are the most common salts and have satisfactory mechanical properties for column applications; however, asbestos supports, for example, are nonetheless used for separation applications. The dynamic exchange capacity equals 1.57 meq/g. A number of authors have proposed the ion-exchange mechanism during sorption, especially for the sorption of alkali metal cations. The exchange of cesium for the first equivalent of ammonium ions occurs more readily than for the second. Smaller alkali metal cations are exchanged as the partially hydrated ions. It is not clear why Ag^+ and Tl^+ are sorbed much more strongly than the alkali metal ions of the same size. The partial irreversibility of the Th(IV) adsorption and the relatively slow adsorption kinetics were ascribed to the probable formation of the Th^{4+} heteropolyanion complex. In addition to AMP, the following salts have been described: 8-hydroxyquinolinium 12-molybdophosphate, dimethylammonium 12-molybdophosphate, K_3MP, Rb_3MP, Cs_3MP, methylammonium 12-molybdophosphate, trimethylammonium 12-molybdophosphate, tetramethylammonium 12-molybdophosphate, Tl_3MP, Tl_2HMP, Cs_2HMP, K_2HMP, Rb_2HMP, Hg_3MP, and Ag_3MP.

2. Tungstophosphates (TP)

The tungstophosphates constitute the second-most-widely studied group of heteropolyacids. These substances exhibit strong base-exchange properties. Ammonium tungstophosphate (ATP) is a specific exchange material for Cs^+ with a sorption capacity of 0.66 meq/g, which is lower than that of AMP. Pyridinium tungstophosphate was found to be very specific for Cs^+. Ammonium tungstophosphate has better acid stability than AMP. The following mixed 12-heteropolyammonium molybdotungstophosphate salts have been prepared: AM_4T_8P, AM_6T_6P, and AM_8T_4P. The following mixed-alkali metal salts of the tungstophosphates are known: Cs_2NaTP, Cs_2KTP, Cs_2RbTP, Cs_2LiTP, $(NH_4)_2CsTP$, $(NH_4)_2RbTP$, and K_2CsTP. The following salts of 12-tungstophosphate acid have also been described: $(NH_4)_3TP$, Tl_3TP, Tl_2HTP, BaHTP, $Ba_3(TP)_2$, $Ce(Cs_2TP)_3$, $Al(K_2TP)_3$, $Al[(NH_4)_2TP]_3$, $UO_2(Cs_2TP)_2$, pyridinium-TP, picolinium-TP, lutidinium-TP, and collidinium-TP.

3. Other Heteropolyacids

The following heteropolyacids and their salts have been described in the literature: ammonium 12-molybdoarsenate (A_3MA), ammonium 12-molybdosilicate (A_4MS), ammonium 12-tungstoarsenate (A_3TA), ammonium 12-tungstosilicate (A_4TS), ammonium 12-molybdogermanate (A_4MG), 8-hydroxyquinolinium 12-molybdosilicate, pyridinium 12-molybdosilicate, 8-hydroxyquinolinium 12-molybdogermanate, pyridinium 12-molybdogermanate, and 6-molybdo-4-tungsto-2-vanadophosphoric acid.

D. INSOLUBLE FERROCYANIDES

When metal salt solutions are mixed with $H_4[Fe(CN)_6]$, $Na_4[Fe(CN)_6]$, or $K_4[Fe(CN)_6]$ solutions, precipitates with various compositions are formed, depending on the acidity, order of mixing, and the initial ratio of the reacting components. Pure compounds or mixtures with limiting compositions $M_4^+[Fe(CN)_6]$ (M^+ = Ag), $M_2^{2+}[Fe(CN)_6]$ (M^{2+} = Zn, Cd, Cu, Co, Ni, Mn), $M_2^+M^{2+}[Fe(CN)_6]$ (M^+M^{2+} = CsZn, HCu, KCo, KNi, HMn), $M_2^+M_3^{2+}[Fe(CN)_6]_2$ (M^+M^{2+} = HZn, NaZn, KZn, HCo), $M^+M3^+[Fe(CN)_6]$ (M^+M^{3+} = NaFe, KFe, RbFe, CsFe), and $M_4^+M_4^{2+}[Fe(CN)_6]_3$ (M^+M^{2+} = RbNi, KNi) with various amounts of water of crystallization (often not stated) have been reported. Some more complex ferrocyanides and ferricyanides have also been described, such as $5Cd_2[Fe(CN)_6] \cdot 4K_4[Fe(CN)_6]$, $M_6Ni[Fe(CN)_6]_2$, (M = Na^+, K^+, Rb^+, NH_4^+), $Cs_{16}Fe_4^{3+}[Fe(CN)_6]_7 \cdot nH_2O$, $[H_4[Fe(CN)_6](MoO_3(H_2O)_x)_{2.2-2.5}]_i$, and $[H_4[Fe(CN)_6](MoO_3(H_2O)_x)_{3.8-4.5}]_i$. Insoluble ferrocyanides with structures containing an amine have also been described in the literature, such as $AHCo_3[Fe(CN)_6]_2$ and $A_2Co_3[Fe(CN)_6]_2$, where A denotes the amine. Titanium can be incorporated in ferrocyanides as the hydrolytic form, e.g., $[(TiO_2)(OH)(HFe(CN)_6) \cdot 4H_2O]_n$. Ferrocyanides containing V(V), W(VI), and U(VI) are characterized by ratios of these elements to Fe of V:Fe = 3.86:1 or 1:1, W:Fe = 1.32:1 or 2:1 to 12:1, and U:Fe = 1.6:1.

The chemical stability of ferrocyanides is satisfactory in acid solutions up to a 2 *M* concentration. Nitric acid causes slight oxidation of Fe^{2+} to Fe^{3+}, while Co, Ni, and Fe^{3+} ferrocyanides have been stated to be resistant to this acid. Cu and Co ferrocyanides have been found to be radiation resistant.

The ion-exchange mechanism operative for ferrocyanides seems rather complex and has not yet been completely elucidated. For example a mixture of $Zn_2[Fe(CN)_6]$ and $Cs_2Zn[Fe(CN)_6]$ has been found to be formed in the sorption of Cs^+ on $Zn_2[Fe(CN)_6]$. The ion-exchange mechanism on Mo ferrocyanides is controlled by two processes, film and particle diffusion. We can point out two typical unit cells of FeMo. One of them contains two $[Mo_6O_{22}H_{16}]^{8+}$ groups, five $[Fe(CN)_6]^{4-}$ groups, and 36 H^+ ions, corresponding to an exchange capacity of 12 meq/g. The second one contains two $[Mo_6O_{22}H_{16}]^{8+}$ groups, one $[Mo_4O_{17}H_{12}]^{2+}$ group, and four $[Fe(CN)_6]^{4-}$ groups, providing a total of 44 H^+ ions. The structures of both forms are relatively open, permitting a high degree of exchange and very fast ionic transfer. The structure can swell on hydration and to some extent resembles clay minerals. The apparent diffusion coefficients thus differ by two to three orders of magnitude from those in the film surrounding the particles to those in the solid phase. There is also a difference of about two orders of magnitude between the reaction rates of the 1-1, 2-1, and 3-1 exchange reactions.

Insoluble ferrocyanides have found various applications in analytical chemistry and in technological practice because of their highly selective ion-exchange properties and satisfactory chemical and mechanical properties.

E. ALUMINOSILICATES

As ion exchange on zeolites is treated in a separate chapter and a brief description has already been given of the mechanism of ion exchange on zeolites, this section will be limited to classification of the very great number of these sorbents and a description of some of their properties associated with this classification. Their structure, sorption mechanism, and selectivity for ions during sorption, as related to the sorbent structure and analytical applications, will not be considered here.

Aluminosilicates can be divided into three main groups: amorphous substances, two-dimensional layered aluminosilicates as synthetic analogues of clay minerals, and three-dimensional structures (zeolites). The greatest attention will be paid to synthetic zeolites, as their molecular and ion-sieve properties are useful for analytical applications.

The chemical composition of zeolites is often expressed by the formula $M_{x/n}[(AlO_2)_x(SiO_2)_y]_{zH_2O}$, where M is a metal cation with valence n and y/x usually varies from 1 to 5.

Zeolites and their sorption properties are characterized not only by the free diameter of their cages (cavities), windows, and inner channels, but also by their unit cell compositions.

Synthetic aluminosilicates can be approximately related to their naturally occurring analogues. Thus, it is possible to classify all these materials into the following groups: faujasite-, mordenite-, heulandite-, chabazite-, analcite-, and phillipsite-like zeolites, and other zeolites.

The most important of these compounds are listed in the following table.

Zeolite	Idealized unit cell composition	Free diameter of the largest channel (Å)	Total exchange capacity (meq/g of water-free species)
Heulandite	$Ca_4[(AlO_2)_8(SiO_2)_{28}] \cdot 24H_2O$	2.4—6.1	3.45
Analcite	$Na_{16}[(AlO_2)_{16}(SiO_2)_{32}] \cdot 16H_2O$	2.8	4.95
Linde T	$Na_8[(AlO_2)_8(SiO_2)_{28}] \cdot 27H_2O$	3.5	3.43
Chabazite	$Ca_2[(AlO_2)_4(SiO_2)_8] \cdot 13H_2O$	3.7—4.2	4.95
ZK 5	$Na_{30}[(AlO_2)_{30}(SiO_2)_{66}] \cdot 98H_2O$	3.8	4.57
ZK 4	$Na_9[(AlO_2)_9(SiO_2)_{15}] \cdot 27H_2O$	4.1	5.50
Linde A	$Na_{96}[(AlO_2)_{96}(SiO_2)_{96}] \cdot 216H_2O$	4.1	7.04
Clinoptilolite	$(Ca,Na_2,K_2)_3[(AlO_2)_6(SiO_2)_{30}] \cdot 24H_2O$	4.1—6.2	2.64
Phillipsite	$(K,Na)_{10}[(AlO_2)_{10}(SiO_2)_{22}] \cdot 20H_2O$	4.2—4.4	4.67
Mordenite	$Na_8[(AlO_2)_8(SiO_2)_{40}] \cdot 24H_2O$	6.7	2.62
Linde X	$Na_{85}[(AlO_2)_{85}(SiO_2)_{107}] \cdot 256H_2O$	7.4	6.34
Linde Y	$Na_{52}[(AlO_2)_{52}(SiO_2)_{140}] \cdot 256H_2O$	7.4	4.10

1. Faujasite-Like Zeolites

Linde X and Linde Y are the most frequently studied materials. These zeolites are prepared by hydrothermal treatment of sodium silicate, aluminate, and hydroxide solutions at ~ 100°C or by evaporating this mixture until crystallization occurs. Depending on the experimental conditions, products with different pore size distributions and adsorption activity can be prepared. The sorbent structure contains sodalite cages, supercages (~ 12 Å), and windows (8 and ~ 2.5 Å). The Si:Al ratio is variable for Linde X and Y from 1.1 to 1.5 and from 1.5 to 3.0, respectively. Introduction of cations into the basic Na^+ forms of these zeolites produces changes in their lattice parameters, window diameters, pore size, and electrostatic field strength. Deformation or even destruction of the crystal lattice can occur. The different cationic forms of a single zeolite species can have different thermal, hydrolytic, and radiation stabilities. After thermal treatment of the NH_4^+ form of Linde X and Y at 100 to 150°C and 300°C, respectively, a maximum appears for adsorption of H_2O vapor.

2. Mordenite-Like Zeolites

Zeolite AW-300 and Zeolon are the synthetic counterparts of natural mordenite: these substances have a nearly rigid, three-dimensional framework that is thermally stable and has a varying degree of porosity. If this zeolite is synthesized in the presence of Sr^{2+} ions, the resulting structure of the zeolite in the Sr form conforms slightly to the spatial requirements of the incorporated cation. After removal of the strontium by ion exchange, the zeolite retains a "frozen-in" structural arrangement, and a strontium-selective mordenite-type zeolite can thus be prepared.

3. Heulandite-Like Zeolites

Heulandite-like zeolites are characterized by a framework that tends to be lamellar. The minerals heulandite, clinoptilolite, stilbite, and ferrierite all have well-known synthetic analogues. Synthetic heulandite-clinoptilolite has also been prepared. Mordenite, clinoptilolite, and ferrierite can be prepared with "frozen-in" structures that are selective for Sr^{2+}.

4. Chabazite-Like Zeolites

These zeolites have a rigid, three-dimensional framework. Synthetic counterparts of natural chabazite, levynite, and especially erionite have been described. The chabazite-like synthetic zeolite AW-500 contains ellipsoidal cavities with a free diameter of ~ 10 Å. Cations can enter these cavities through elliptical rings with maximum and minimum free diameters of 4.1 and 3.7 Å, respectively. The erionite-type synthetic zeolite is usually designated as AW-400, Zeolite T, or Linde T. ZK zeolite is isostructural with levynite. Synthetic erionites are stable up to 800°C, probably as a result of the higher Si:Al ratio.

5. Analcite- and Phillipsite-Like Zeolites

The analcite group contains two main minerals: analcite and wairakite. Like phillipsite, they are characterized by SiO_2: Al_2O_3 ratios that usually vary from 3.18 to 3.26 and have a rigid three-dimensional framework. Phillipsite is a zeolite that is closely related to zeolite B, which has the same composition as zeolite X. Three different structures have been found for phillipsite in the Na^+ form, with a primitive cell and two different body-centered cells.

6. Other Zeolites

The zeolite Linde A is usually included among faujasite-like zeolites and is one of the most extensively studied species. The preparation procedure and sensitivity to the preparation conditions are the same as for Linde Y and X. The Si:Al ratios of Linde A and zeolite ZK-4 are 1.0 and $1.0 < Si/Al < 1.7$, respectively. The Linde A structure consists of a three-dimensional network of large cavities formed by connecting supercages through shared rings of eight tetrahedra. Isolated sodalite cages open out into supercages through shared rings of six tetrahedra. After thermal treatment to 150°C, the NH_4^+ form of Linde A yields sorbents with maximum sorption capacity for H_2O vapor. The Na^+ form of zeolite A prefers cations of higher valency, atomic weight, and ability to form more strongly dissociated complexes.

Synthetic yugawaralite and synthetic zeolite Sr-Q are structurally very similar to the natural mineral. The synthetic sodalite-like zeolite is termed zeolite Zh and has the lowest SiO_2 content of the whole zeolite family. Synthetic zeolites with anion-exchanging properties have also been prepared. MCP II is an acid-resistant zeolite, and MCP III is a layer-type mineral of the montmorillonite type with Li^+ as the exchangeable ion. Zeolite L has a unique framework structure analogous to that of natural minerals. Zeolite of this type is thermally stable and exchanges trivalent ions.

V. ANALYTICAL APPLICATIONS AND CONTEMPORARY TRENDS IN INORGANIC ION-EXCHANGING MATERIALS

A. CHOICE OF SORBENT

For more than 30 years, inorganic ion-exchange materials have attracted the attention of analytical chemists. A great variety of suitable inorganic compounds are available that can be used as sorbents for ions from aqueous solutions (through various mechanisms).

Tables 1 and 2 list more than 100 different inorganic substances that have been prepared over the last 40 years and whose sorption and other properties have been studied. These compounds constitute the basic range of inorganic ion-exchange materials. Various crystal modifications of these substances, mixed sorbents, etc. have also been prepared. In addition to these synthetic sorbents, a large group of natural and synthetic aluminosilicates are available which are utilized primarily in the analysis of organic or biochemical substances and in gas analysis.

In spite of the wide variety of compounds that can be used as inorganic ion-exchanging materials, they have a number of more or less common properties that assist the chemist-analyst in orientation in this very large number of compounds and help him to select the most suitable for a given analytical problem.

Inorganic ion-exchange materials can be employed as cation or anion exchangers, and some act as amphoteric exchangers; the properties of some are similar to those of organic redoxites.

The pH value of the solution has a marked effect on the character of amphoteric exchangers, which consist primarily of the hydrated oxides. Typical suitable sorbents are SiO_2, SnO_2, Al_2O_3, and ZrO_2. Under suitable conditions, these sorbents can act as cation or anion exchangers, and under certain conditions both cation and anion exchange can occur simultaneously.

TABLE 1
Schematic Formulas of Various Inorganic Ion-Exchange Materials Suitable for Analytical Application

Metal								
Ti	**Zr**	**Fe(III)**	**Sn(IV)**	**Sb(V)**	**Th**	**Cr**	**Ce**	**Hf**
TiO_2	ZrO_2	Fe_2O_3	SnO_2	Sb_2O_5	ThO_2	$Cr(OH)_3$	CeO_2	HfO_2
TiP	ZrP	FeP	SnP	SbP	ThP	CrP	CeP	HfP
TiAs	ZrAs	FeAs	SnAs	SbAs	ThAs	CrAs	CeAs	HfAs
TiSb	ZrSb	FeSb	SnSb		ThSb	CrSb	CeSb	
TiMo	ZrMo	FeMo	SnMo		ThMo	CrMo		
TiV								
TiT	ZrT	FeT	SnT	SbT	ThT		CeT	
TiSe	ZrSe	FeSe	SnSe					
TiTe				SbTe				
TiPSi	ZrPSi			SbPSi			Ce-tripoly-P	
	ZrSi							
	Zr-oxalate						Ce-oxalate	
	Zr-pyro-P							
	Zr-hypo-P							
	Zr-phtalo-P							
$[Na(Ti_2O_3H)]$	$[Na(Zr_2O_5H)]$							
Ta	**Nb**	**U**	**La**	**Bi**	**Al**	**Mn**	**Si Ge**	**W**
Ta_2O_5	Nb_2O_5	UO_3	La_2O_3	Bi_2O_3	Al_2O_3	MnO_2	SiO_2	WO_3
TaP	NbP	UP					GeP	
TaAs	NbAs		LaAs					
TaSb	NbSb				AlSb			
TaMo	NbMo							
					AlV			
		UT	LaT	BiT				
TaSe	NbSe							
			LaTe	BiTe				
$[Na(Ta_2O_6H)]$	$[Na(Nb_2O_6H)]$							

Note: O_x—hydrous oxide; SbP—phosphoantimonic acid; P—phosphate; SbAs—arsenoantimonic acid; As—arsenate; Sb—antimonate; SbT—tungstatoantimonic acid; SbTe—telluratoantimonic acid; Se—selenide; Si—silicate; Mo—molybdate; V—vanadate; *Te*—tellurate; T—tungstate.

Their dissociation under various conditions can be depicted schematically as:

$$MOH \rightleftharpoons M^+ + OH^- \tag{1}$$

$$MOH \rightleftharpoons MO^- + H^+ \tag{2}$$

where M is the central atom.

Scheme 1 is favored in acid solution, where the substance can act as an anion exchanger. In alkaline solutions (scheme 2), the substance can act as a cation exchanger. According to both schemes, (hydrous oxide) dissociation can take place near the isoelectric point of the substance and both types of exchange can occur simultaneously.

The following two examples illustrate this phenomenon:

TABLE 2
Schematic Formulas of Various Inorganic Ion-Exchange Materials Suitable for Analytical Application

Heteropolyacids and their salts			Hexacyanoferrates (Fo)	Hexacyanoferrites (Fi)	Other exchanging materials
NH_4^+ salts of 12-hetero polyacids[a]		MoP MoSi TWAs TWSi	ZnFo, CuFo, CoFo, NiFo, Fe(II)Fo, Sn(II)Fo, MnFo, PbFo, CdFo, Fe(III)Fo, Cr(III)Fo, AlFo, TiFo, ZrFo, ThFo, Sn(IV)Fo, SbFo, VFo, MoFo, U(VI)Fo, WFo	CrFi CoFi ZnFi	$CaCO_3$, NiSb, CoSb, $PbSO_4$, $Cd(OH)_2$, PbO_2, MgP, NiP, SrP, PbF_2, CuSi, ZnSi, various sulfides (S) (Ni, Co, Pb), various mixed types, synthetic zeolites
Zr salts		AsP PSi MoSi			
Sn salts	MoSi MoAs VP TWAs AsP SeWT	BMo BWT BP BAs BS			

Note: See Table 1 for list of symbols. Fo: hexacyanoferrates, Fi: hexacyanoferrites, S: sulfides.

[a] In addition to NH_4^+, various organic ammonium salts were also used.

Hydrous Zirconium Oxide

	Cation exchange					Anion exchange				
pH	**1**	**3**	**5**	**7**	**9**	**1**	**3**	**5**	**7**	**9**
Capacity (mmol/g)	0.05	0.35	0.5	0.7	0.8	0.5	0.4	0.3	0.2	0.0

Hydrated SnO_2

	Cation exchange				Anion exchange				
pH	**3**	**5**	**7**	**9**	**1**	**3**	**5**	**7**	**9**
Capacity (mmol/g)	0.25	0.4	0.6	0.9	1.0	0.9	0.8	0.6	0.0

A large number of examples of this type of exchange material have been studied in recent years, some more extensively than others. Abe and Ito (1965) studied the titration curves and sorption capacities, and suggested the following approximate acidity series for various hydrous oxides: for divalent metals, $MgO<BeO$; for trivalent metals, $La_2O_3<Bi_2O_3<Al_2O_3<In_2O_3<Ga_2O_3<Mn_2O_3$; for tetravalent metals, $ThO_2<CeO_2<ZrO_2<TiO_2<SnO_2<SiO_2<MnO_2$; and for pentavalent metals, $Nb_2O_5<Ta_2O_5<Sb_2O_5$.

The selection and usefulness of a suitable sorbent also depends on its stability in the medium employed. The concentration of free acid or base in the solution can be a decisive factor. Sorption on some substances occurs in a very wide range of acid concentrations; on the other hand, some inorganic ion-exchange materials are sensitive to an increase in the concentration of ammonium or alkaline hydroxides in the analyzed solution. Exchange on some types of sorbents is possible in only a very narrow pH range.

Suitable sorbents can be chosen for applications in the presence of high concentrations of strong oxidizing or complexing agents; on the other hand, other sorbents are very sensitive to these agents.

The concentration of H^+ or OH^- ions in solution or the presence of various complexing agents can dramatically affect the solubilities of some sorbents. The properties of some inorganic ion-exchange materials can also be affected by the time elapsed from their preparation: long-term storage in the air can result in a number of changes connected with the sorption properties of the material, its crystal structure, stability, selectivity, and exchange capacity (e.g., a decrease in the exchange capacity of the sorbent as a result of condensation or destruction of the –OH groups of the sorbent).

Inorganic sorbents are used most often in ion-exchange columns. Provided that the sorbent does not form large species and the hydrodynamic resistance of the column is too high, suitable ion-exchanger particles must be produced by agglomeration, incorporation into various organic or inorganic polymers, or precipitation on porous inert materials, or suitable particles must be prepared by some other method (sol-gel procedure, freezing out, etc.). The hydrodynamic resistance of the sorbent in the column can also be decreased by mechanically mixing the exchange material with a suitable inert packing.

Precipitation of the sorbent on paper as a support or mixing with cellulose fibers can yield a suitable form of the sorbent for the separation of ions in paper chromatography. Similarly, the sorbent can be mixed with a suitable inert support to prepare plates for thin-layer chromatography. Inorganic ion-exchange materials can also be fixed on various supports to prepare membranes. These materials are especially useful in the preparation of ion-selective electrodes.

A further important property of inorganic sorbents that plays an important role in the selection of a suitable exchange material is its selectivity for the analyzed element or for a small group of elements to be analyzed. Very high selectivity of the sorbent for a given element permits it to be separated from a number of other elements, but it is usually rather difficult to elute this element from the sorbent. When radioisotopes emitting gamma radiation are employed, this disadvantage can be avoided by quantitatively determining the test element directly on the exchanger (without elution). A number of analytical methods are based on this phenomenon and are used in the analysis of various biological materials or admixtures in substances of high purity using neutron activation analysis.

B. SELECTIVITY OF INORGANIC ION-EXCHANGE MATERIALS

Compared with classical high-molecular-weight ion exchangers, inorganic ion-exchange materials are of interest to the chemist-analyst not only for their high stability in various media, but also for their selectivity for groups of elements or individual elements. This selectivity is sometimes extremely high.

An example is the separation of a mixture of ions of the alkali metals, which is difficult using organic ion exchangers. If this mixture is separated using crystalline antimonic acid, the separation factors between the light alkali metals are two to three times higher under similar conditions than on Amberlite IR-120. Similarly, the difference in the separation factors for the transition metals is tenfold for the Ni-Mn pair, 20-fold for Ni-Cr, and 240-fold for Cu-Cd, when crystalline antimonic acid is used rather than the organic cation exchanger AG-50W, under identical conditions. A great many similar examples could be given for a wide range of inorganic sorbents.

In spite of this optimistic view of the properties of inorganic ion-exchange materials and the consequent capabilities for the selective separation of mixtures of inorganic ions, a number of difficulties are encountered in the utilization of the selectivities of these materials, their manufacture, and application.

It has been found in a number of experiments and careful analysis of the results obtained that the selectivity of these exchanging materials depends, in general, on the conditions of their preparation, e.g., on the pH value at which the precipitate is formed, drying, aging, surface and crystalline structure, and other conditions. For example, a slight variation in the preparation conditions of thorium arsenate resulted in new crystalline forms of this material, and the compounds $Th(HAsO_4)_2 \cdot 2.5H_2O$ and $Th(HAsO_4)_2 \cdot 4H_2O$ were obtained. The interlayer distances in these new crystal forms were greater (8.59 and 8.23 Å, respectively). than those for the material reported in the literature-$Th(HAsO_4)_2 \cdot H_2O$ (7.05 Å). As a result of this increased interlayer distance, both of these new sorbents are capable of exchanging not only Li^+, for which the original sorbent was selective, but also larger ions, such as Na^+ and K^+.

Similarly, amorphous and glassy forms of antimonic acid form an affinity series: Li < Na < K < Rb < Cs. However, the selectivity on crystalline antimonic acid is very different: Li < K < Rb < Na.

A change in the conditions during the preparation of a given ion-exchange material can yield sorbents with different selectivities for a given ion.

The selectivities of inorganic ion-exchange materials is also greatly affected by the loading of the exchanger with exchanging ions. This effect can be illustrated by the following data for the exchange of the alkali metals on amorphous zirconium phosphate:

Observed selectivity	**Loading range**
Cs> Rb> K> Na> Li	$\bar{x}_M < 0.19$
K> Rb> Cs> Na> Li	$0.19<\bar{x}_M<0.34$
K> Na> Rb> Cs> Li	$0.42<\bar{x}_M<0.44$
Na>K>Li>Rb>Cs	$0.52<\bar{x}_M<0.54$
Na>Li>K>Rb>Cs	$0.54<\bar{x}_M<0.62$
Li> Na> K> Rb> Cs	$\bar{x}_M > 0.62$

It can be seen that the selectivity of the sorbent can change drastically as a result of the loading of the sorbent. In the sorption of the alkaline earth metal cations by the crystalline form of antimonic acid, the following selectivity series was found at low cation concentrations ($<10^{-3}$ *M*): Mg < Ba < Ca < Sr. As the concentration increases, this series changes (at concentrations >0.1 *M*): Mg < Ba < Sr < Ca.

It is important in analytical applications that the selectivity of the sorbent for a given element can often be substantially affected by varying the composition of the sorption solution. For inorganic exchange materials, this usually involves a change in the pH or free-acid concentration. A good example is the sorption on crystalline antimonic acid (hydrated antimonic oxide), where the following selectivity series were found in dilute nitric acid:

Li < K < Cs < Rb < Na
Mg < Ba < Ca < Sr
Ni < Mn < Zn < Co < Cu < Cd < Pb
Al < Tm < Ga = Yb < Y < Er < Dy < Fe < Eu < Sm < Sc < La < In

The selectivity changes dramatically when this sorbent is used in HCl medium (Girardi, 1968). At 6 to 12 *M* HCl, only sodium and tantalum were completely adsorbed out of a total of 60 elements (in microconcentrations); fluorine was also partly adsorbed. The remaining elements were not sorbed on the material under these conditions. Similarly, sodium was adsorbed selectively and quantitatively in 7 *M* HNO_3. As the acid concentration decreased in the sorption solution, the selectivity of the sorption of various ions decreased and Na, K, Rb, Cs, Sr, Cu, Zn, Fe, Co, and Br were completely sorbed in about 1 *M* HNO_3. The alkali metals and alkaline earth metals were sorbed especially strongly.

This example of the effect on sorbent selectivity of a change in the acid concentration in the sorption solution has been used to develop an important analytical method for the removal of Na^+ ions from biological materials or high-purity substances for the determination of trace elements by neutron activation analysis.

A number of procedures for separating ions with the same valence and similar separation coefficients have successfully employed the possibility of changing the valence of one of the ions to increase the value of the separation factor sufficiently to enable separation. Separation of the Am^{3+}-Cm^{3+} mixture is carried out on amorphous zirconium phosphate, where Am^{3+} is converted to Am^{5+}. Similarly, a mixture of Np, Pu, and Pr is separated more readily when these elements are converted to the forms Np^{5+}, Pu^{4+}, and Pr^{3+}.

C. ANALYTICAL APPLICATIONS OF INORGANIC ION-EXCHANGE MATERIALS

Inorganic ion-exchange materials have not been used as extensively in analytical chemistry as classical high-molecular-weight ion exchangers, especially for the analysis of inorganic ions. Nonetheless, their use for separation and concentration, especially of radioisotopes, has greatly increased in the last 10 to 15 years. The loading of the exchanger is almost always low in the separation of radioactive isotopes, so that the assumed selectivity of the sorbent is retained.

The use of a suitable inorganic sorbent under specific conditions can sometimes approach the ideal of a "specific sorbent" that sorbs only a single element from a very complex mixture of analyzed elements.

Inorganic ion-exchange materials can be used in a range of interesting separations that are far more difficult to carry out using organic ion exchangers.

Most inorganic ion-exchange materials can be used for the separation of alkali metals. The selectivities for the alkali metals on synthetic inorganic ion-exchange materials (e.g., zirconium phosphate, tungstate, molybdate, various ferrocyanides, hydrated oxides, antimonates, arsenates, and heteropolyacids) are generally more favorable than those on organic ion-exchange resins. Separations of mixtures of the alkali metals have been carried out using relatively short columns.

Table 3 compares the separation factor values (Dg) for alkali metals on various exchangers. It follows from these values that a suitable inorganic sorbent is often far more useful for the separation of the individual alkali metals than a strongly acidic organic cation exchanger.

The separation will occur on the basis of the high selectivity of the inorganic sorbent only at a low loading of the exchanging material. This fact is illustrated by the data in Table 4, giving the values of the distribution coefficient (Dg) for various loadings. It can be seen that the Dg value increases with decreasing concentration of the sorbed element, with simultaneous improvement of the separation properties of the sorbent. This dependence of the selectivity (and thus the efficiency of the separation) on the concentrations of the separated elements is of general validity.

A number of inorganic separation materials have sufficient selectivity for the separation of mixtures of alkali metals, either binary or multicomponent. There are also sorbents with very high selectivity for the individual alkali metals. For example, crystalline antimonic acid is highly selective for Na^+ ions in 6 to 12 *M* HCl; lithium can be separated selectively on sorbents such as Sn(IV) antimonate, thorium arsenate, or Sn(IV) arsenate. Because of its unusual selectivity for Li^+ ions, Ti(IV) antimonate can be successfully employed for the recovery of this element from sea water or various mineral waters.

TABLE 3
Comparison of the Separation Factors for Alkali Metals on Various Ion Exchangers

Ion exchanger	Solution	Separation factor			
		Na/Li	K/Na	Rb/K	Cs/Rb
Zirconium phosphate	0.1 *M* HNO_3	2.7	3.4	2.8	2.3
Ammonium molybdophosphate	0.1 *M* NH_4NO_3	—	—	39	28
Phosphoantimonic acid	0.05 *M* NH_4NO_3	1.2	1.8	2.3	3.7
Amberlite IR-120	0.2 *M* HNO_3	1.5	2.1	1.3	1.2
Dowex 50W-x8	0.1 *M* NH_4NO_3	—	1.5	1.1	1.2
Heteropolyacids	0.01 *M* HNO_3 + NH_4NO_3				
$(NH_4)_3PMo$		—	—	—	23.4
$(NH_4)_3PW$		—	—	—	21.3
$(NH_4)_2HPMo$		—	—	—	18.2
$(NH_4)_2HPW$		—	—	—	14.5
Tl_2HPMo		—	—	—	33.3
$T1_2HPW$		—	—	—	28.0
Amorphous antimonic acid	0.1 *M* HNO_3	4.3	4.4	1.6	1.2
		Li/K	Cs/K	Rb/Cs	Na/Rb
Crystalline antimonic acid	0.1 *M* HNO_3	1.4×10^3	3.1	5.8	10.2

Cesium can be successfully separated using the ferrocyanides of a number of heavy metals, such as Ti, Fe, Ni, Cr, Mo, or W. On the other hand, Sb(V) ferrocyanide is selective for Sr^{2+} ions. Sb(V) silicate has been found to be slective for Rb^+ ions (Rb can be separated from the other alkali metals above pH 1). Similarly, other inorganic ion-exchange materials have been found to be selective for other elements under suitable conditions, and can be used for the selective sorption of these ions.

Zirconium antimonate is highly selective for Sr^{2+} ions. The high selectivity of Ti(IV) molybdate for Pb^{2+}, Ba^{2+}, and Tl^+ ions has been used for their separation from binary and ternary mixtures of various ions. On the other hand, the high selectivity of Ti(IV) tunsgtate for Ca^{2+} ions permits them to be separated from the other alkaline earth elements. The increased sensitivity of thorium (IV) molybdate for Fe^{3+} ions has been similarly utilized for the separation of Fe^{3+} from various binary mixtures.

Zirconium (IV) molybdate and tungstate also exhibit electron-exchange properties. They are reduced by a $SnCl_2$ solution to products that probably contain trivalent tungsten. The reduced form of the sorbent produced can be reoxidized, e.g., using a V(V), Fe(III), Ce(IV), or H_2O_2 solution. Its electron exchange capacity is about 0.2 meq/g. Some other inorganic ion-exchange materials have similar properties.

A number of other sorbents of the group of acidic salts of multivalent metals exhibit increased selectivity for various ions, especially the heavy metals; ferric phosphate, for example, has been used analytically for the separation of thorium or lead from mercury, zinc, strontium, manganese, and nickel. Similarly, iron (III) selenite is sufficiently selective for the separation of Pb^{2+} from a number of other metals.

Various salts of heteropolyacids can be used for the selective separation of cesium. Under suitable conditions, their selectivity can be modified for other elements: Zr(IV) and Ti(IV) arsenophosphates or tungstophosphates have been used for the separation of thorium, and Sn(IV) tungstoarsenate has a high affinity for Cu^{2+} and La^{3+} ions.

It is beyond the scope of this general discussion to describe the utilization of the selective properties of inorganic sorbents in analytical chemistry. However, it should be noted that these sorbents are used extensively in neutron activation analysis. Inorganic sorbents have been found to be especially useful in this field because of their high radiation and thermal resistance and especially high selectivity for various elements. The acidic salts of multivalent

TABLE 4
Values of the Distribution Coefficient Dg for Alkali Metals in Various Concentrations on Crystalline Titanium Phosphate

Metal concentration (molarity)	Dg				
	Li	Na	K	Rb	Cs
10^{-1}	6	41	1	1	1
10^{-2}	1	4	1	1	1
10^{-3}	11.5	15	19	22.5	45
10^{-4}	—	55	80	115	205
10^{-5}	—	85	—	370	780

elements (molybdates, phosphates, antimonates, and tungstates), ferrocyanides, sulfides, and the hydrated oxides of four-, five-, and six-valent elements (SnO_2, TiO_2, MnO_2, ZrO_2, Nb_2O_5, Ta_2O_5, and Sb_2O_5) have been extensively used in this field. A number of interesting separations can be carried out using these sorbents by varying the composition of the sorption solution.

Both individual sorbents and mixtures have been used to attain the required separation and concentration effects. They can be complemented by various organic ion exchangers. Attempts have been made to automate the separation on sorbents to facilitate routine analyses.

The wider application of inorganic ion-exchange materials in analytical practice is prevented by a lack of suitable materials. Although a great many kinds exist, large manufacturers produce and market only a small number of types.

The analyst who wishes to employ these materials can use a commercial product if a suitable material is available or can prepare the material in the laboratory. In the former case, the quality of the sorbent is ensured (provided that changes in the exchange properties do not occur on aging). In the latter case, if the synthesis of the sorbent is simple, reliable results can be obtained. However, the preparation of a highly selective material requires considerable experience and the preparation procedure must be followed carefully. It often happens that the analyst is disappointed with the properties of the ion-exchange material prepared in the laboratory, which often differ from those in the literature.

This dilemma is hard to solve at the present; consequently, the number of publications dealing with the analytical applications of inorganic sorbents will not correspond to their use in analytical practice.

D. CONTEMPORARY TRENDS IN THE APPLICATION OF INORGANIC ION-EXCHANGE MATERIALS IN CHEMICAL ANALYSIS

It follows from the number and purposes of the published works on the use of inorganic ion-exchange materials in analytical chemistry that their main application lies in their use for the selective separation and concentration of elements in various mixtures. They are and will be used intensively in combination with neutron activation analysis of biological and very pure materials.

The greatest interest centers on exchange materials based on hydrated oxides and the acidic salts of multivalent metals. The exchange properties of a number of sorbents is being verified, as data obtained in earlier publications, especially in the 1950s, were obtained using incompletely defined sorbents. The published data on their properties, especially their selectivity, are not always in agreement with new data obtained using precisely defined exchange materials.

The use of inorganic ion-exchange materials for the separation of transplutonium elements from acid and salt solutions seems promising, especially for the separation of berkellium. The sorbent must have high radiation and thermal stability and it must also be highly stable in strongly acidic and oxidizing media.

Interesting separations have been carried out on mixtures of ions using paper chromatography, electrochromatography, and thin-layer chromatography using inorganic sorbents. Work in this field is in its infancy.

It would be useful to carry out systematic studies and to use new exchange materials characterized by large surface areas. It can be assumed that these new sorbents, bonded to a suitable support, will improve the separation capabilities in both TLC and paper chromatography.

So far, little work has been carried out on the use of inorganic ion exchange materials in gas chromatography and for the separation of organic substances. However, some acid salts of multivalent elements have been used for the separation of mixtures of alkaloids.

The protection of the environment and health protection require strict limits on the concentrations of the heavy metals and radioisotopes in the hydrosphere. The analyst must often determine the studied element that is highly diluted in a large volume of water or air. It is often necessary to treat a large amount of sea or river water or complex industrial effluents with high salt contents. The use of suitable inorganic ion-exchange materials with high selectivity for the studied elements often facilitates their concentration and separation, which are necessary for successful determination. It is thus necessary to employ materials with suitable sorption properties, high selectivity, and high chemical and mechanical stability. Inorganic ion-exchange materials can be especially useful in this field.

E. SELECTED MONOGRAPHS AND REVIEWS

Almost half a century of research on inorganic ion-exchange materials and their great development in the 1960s and 1970s is naturally reflected in the number of monographs and reviews dealing with this subject.

The first important monograph that is also of historical importance was written by one of the first research workers in the development of modern inorganic sorbents, C. B. Amphlett in 1964, and describes the beginning of the rapid development of this subject.

Barrer wrote an excellent monograph on contemporary zeolite and clay minerals, and collected information on the latter representative and independent group of inorganic sorbents.

In the 1980s, the monograph of Clearfield et al. made a great contribution to the understanding of the structure and mechanism of sorption processes on inorganic sorbents consisting of the acidic salts of multivalent metals and hydrous oxides. The monograph by Nikolsky et al. describing the work of Soviet authors was published at about the same time.

However, monographs necessarily provide a long-term picture of the given field, while reviews give new information on the state of the art in a much shorter time interval. From this point of view, the reviews by Fuller (1971), Qureshi et al. (1972), Veselý and Pekárek (1972), Clearfield et al. (1972), Abe et al. (1974), and Alberti et al. (1983) have played an important role.

REFERENCES

1. **Amphlett, C. B.,** *Inorganic Ion Exchangers,* Elsevier, Amsterdam, 1964.
2. **Aripov, E. A.,** Prirodnye Mineralnye Sorbenty, ikh Aktivirovanie i Modificirovanie, Izd FAN, Tashkent, 1970 (in Russian).
3. **Barrer, R. M.,** *Zeolites and Clay Minerals as Sorbents and Molecular Sieves,* Academic Press, London, 1978.

4. **Breck, D. W.**, *Zeolite Molecular Sieves: Structure, Chemistry and Use*, John Wiley & Sons, New York, 1974.
5. **Celishev, Yu. P.** *Ionoobmennye Svoystva Mineralov*, Nauka, Moscow, 1971 (in Russian).
6. **Cicishvili, G. V., Andronikashvili, T. G., Kirov, G. N., and Filizova, L. D.**, *Prirodnye Ceolity*, Khimiya, Moscow, 1985 (in Russian).
7. **Clearfield, A., Eds.**, *Inorganic Ion Exchange Materials*, CRC Press, Boca Raton, FL, 1982.
8. **Flank, W. H., Eds.**, *Adsorption and Ion Exchange with Synthetic Zeolites*, ACS Symp. Ser. No. 135, American Chemical Society, Washington, D.C., 1980.
9. **Gould, R. F., Ed.**, *Molecular Sieve Zeolites*, Adv. Chem. Ser. No. 101, American Chemical Society, Washington, D.C., 1971.
10. **Greenland, D. J. and Hayes, M. H. B., Eds.**, *The Chemistry of Soil Processes*, John Wiley & Sons, Chichester, 1981.
11. **Kepák, F.**, *Sorpce a Koloidní Vlastnosti Radionuklidů ve Vodných Roztocích*, Academia, Prague, 1985 (in Czech).
12. **Laskorin, B. N.**, *Sorbenty na Osnove Silikagelya v Radiokhimii, Khimicheskie Svoystva, Primeneniye*, Atomizdat, Moscow, 1977 (in Russian).
13. **Marhol, M.**, Chelating Resins and Inorganic Ion Exchangers, in *Wilson and Wilson's Comprehensive Analytical Chemistry*, Vol. 14, Ion Exchangers in Analytical Chemistry, Elsevier, Amsterdam, 1982.
14. **Marhol, M. and Vaňura, P.**, Anorganické měniče iontů a sorbenty, *Academia*, Prague, in press (in Czech).
15. **Meier, W. M. and Uytterhoeven, J. B., Eds.**, *Molecular Sieves*, Adv. Chem. Ser. No. 121, American Chemical Society, Washington, D.C., 1973.
16. **Murakami, Y., Iijima, A., and Ward, J. W., Eds.**, New developments in zeolite science and technology, in *Proc. 7th Int. Zeolite Conf., Kodasha-Tokyo*, Elsevier, Amsterdam, 1986.
17. **Nikolskyi, B. P., Ed.**, *Neorganicheskie Ionoobmennye Materialy*, Izdat. Leningrad. University, Vol. 1 and 2, Leningrad, 1974, 1980 (in Russian).
18. **Olson, D. and Bisio, A., Eds.**, *Proc. Int. Zeolite Conf.*, Butterworths, Guilford, 1984.
19. **Pushkarev, V. V. and Nikiforov, A. F.**, *Sorbciya Radionuklidov Geteroholikiclotami*, Energoizdat, Moscow, 1982 (in Russian).
20. **Rees, L. V. C.**, *Proc. 5th Int. Conf. Zeolites*, Heyden, London, 1980.
21. **Sand, L. B. and Mumpton, F. A., Eds.**, *Natural Zeolites: Occurrence, Properties, Use*, Pergamon Press, Oxford, 1978.
22. **Sukharev, Yu. I. and Yegorov, Yu. V.**, *Neorganicheskie Ionity Tipa Fosfata Cirkoniya*, Energoatomizdat, Moscow, 1983 (in Russian).
23. **Tarasevich, Yu. I.**, *Prirodnye Sorbenty v Processakh Ochiski Vody*, Naukova Dumka, Kiev, 1981 (in Russian).
24. **Towsend, R. P.**, *The Properties and Applications of Zeolites*, Spec. Publ. No. 33, Royal Society of Chemistry, 1980.
25. **Volkhina, V. V., Ed.**, *Neorganicheskie Ionoobmeniki: Sintez, Struktura, Svoystva*, Vydav. Permsk. Polit. Inst., 1977 (in Russian).
26. **Yegorov, Yu. V.**, *Statika Sorbcii Mikrokomponentov Oksigidratami*, Atomizdat, Moscow, 1975 (in Russian).
27. **Zhdanov, S. P., Khvoshchev, S. S., and Samulevich, N. N.**, *Sinteticheskie Ceolity*, Khimiya, Moscow, 1981 (in Russian).
28. Inorganic Ion Exchangers and Adsorbents for Chemical Processing in the Nuclear Fuel Cycle, IAEA-TECDOC-337, Vienna, 1985.
29. Use of Local Minerals in the Treatment of Radioactive Wastes, IAEA Tech. Rep. Ser. No. 136, Vienna, 1972.
30. **Abe, M.**, *Jpn. Anal.*, 23, 1254, 1974 (in Japanese).
31. **Abe, M.**, *Jpn. Anal.*, 23, 1561, 1974 (in Japanese).
32. **Alberti, G., Casciola, M., and Constantino, U.**, *J. Membr. Sci.*, 16, 137, 1983.
33. **De, A. K. and Sen, A., K.**, *Sep. Sci. Technol.*, 13, 517, 1978.
34. **Barsukova, K. V. and Myasoedov, B. F.**, *Radiokhimiya*, 23, 489, 1981 (in Russian).
35. **Belinskaya, F. A. and Militsina, E. A.**, *Usp. Khim.*, 49, 1904, 1980 (in Russian).
36. **Clearfield, A., Nancollas, G. H., and Blessing, R. H.**, *Ion Exch. Solvent Extr.*, 5, 1, 1973.
37. **Ermolenko, I. F. and Efros, M. D.**, *Vesc. Acad. Nauk Belarus SSSR, Ser. Khim. Nauk*, 1, 21, 1966 (in Russian).
38. **Fuller, M. J.**, *Chromatogr. Rev.*, 14, 45, 1971.
39. **Inczedy, J.**, *Rev. Anal. Chem.*, 1, 157, 1972.
40. **Pekárek, V. and Veselý, V.**, *Talanta*, 19, 1245, 1972.
41. **Qureshi, M., Nabi, S. Z., Gupta, J. P., and Rathore, H. S.**, *Sep. Sci.*, 7, 615, 1972.
42. **Veselý, V. and Pekárek, V.**, *Talanta*, 19, 219, 1972.

Chapter 2

MODERN THEORIES OF ION EXCHANGE AND ION EXCHANGE SELECTIVITY WITH PARTICULAR REFERENCE TO ZEOLITES

A. Dyer

TABLE OF CONTENTS

I. INTRODUCTION

Interest in the ion-exchange processes taking place in inorganic materials has undergone a recent revival. Historically, all early investigations were on inorganic substances, such as those of Way[1] on soils and Eichhorn[2] on natural zeolites. This interest waned with the discovery of organic ion-exchange resins in 1935[3] and, until comparatively recently, the majority of theoretical and experimental studies on ion exchange were devoted to resins.

The revival of studies relating to inorganic exchangers can be ascribed to a variety of reasons. Prominent among these has been the discovery of novel synthetic zeolites[4] because the ion-exchange properties of these aluminosilicates are critical to (1) their current economic importance as detergent builders, cracking catalysts, and molecular sieves, and (2) their potential applications in waste-water treatment, agriculture, and horticulture.[5]

A further major reason stems from studies of the role of clay minerals in soil chemistry, as well as those industrial applications of clays which depend upon their ion-exchange capability. Finally, the growth in nuclear installations has created the need to investigate inorganic exchange media as a means of treating nuclear wastes. Clays, zeolites, and phosphates (e.g., zirconium phosphate and ammonium molybdophosphate) have been widely investigated and used for this purpose.[6] Linked to these areas of research is the necessary concern with the ability of clays to "fix" radionuclides. This arises from the need to acquire knowledge relevant to the safe storage and containment of intermediate and low-level nuclear wastes.

Clays are favored "backfill" materials for nuclear waste repositories, and clearly there is a vital safety requirement to understand the course of radionuclides from waste forms should a leak occur to the surrounding environment.[7]

Throughout all these studies, there has been an emphasis on the unique selectivities shown by inorganic materials, and this has prompted theoretical studies designed to understand the general phenomenon of ion exchange through the use of inorganic models. It is salutary to think that arguably the oldest physical property known to man[8] still is very far from a complete understanding. In most instances, the model substances chosen have been the synthetic zeolites, and it is with these materials that this chapter will be concerned.

Before embarking on a more detailed discourse of the theories of inorganic ion exchangers, the point should be made that *all* the theory to be considered relates to *cation* exchange. No equivalent theory exists to describe *anion* selectivities; in fact, it is likely that true anion exchange in an inorganic matrix is very rare. Most of the processes described in the literature relate to either an irreversible uptake of an anionic species or to surface sorption phenomena. Even when ion-exchange isotherms can be constructed, they often show severe hysteresis.[9]

No consideration will be given to nonsilicate exchangers, as these have been described in detail in a recent book of this series.[10]

II. ZEOLITES AS MODELS FOR THE CATION EXCHANGE PROCESS

A. CLASSIFICATION OF ZEOLITES AND THEIR STRUCTURES

By definition, both natural and synthetic zeolites have three-dimensional aluminosilicate framework structures. This means that the fundamental structural building units are the $[SiO_4]^{4-}$ and $[AlO_4]^{5-}$ tetrahedra (Figure 1). These complex ions link to each other by sharing all their corners to build a three-dimensional array (Figure 2).

It is convenient to describe the complex aluminosilicate skeletons of zeolite structures by defining a series of secondary building units (SBU's), each of which is a specific assemblage of the primary tetrahedral units of structure. The simplest example of an SBU is

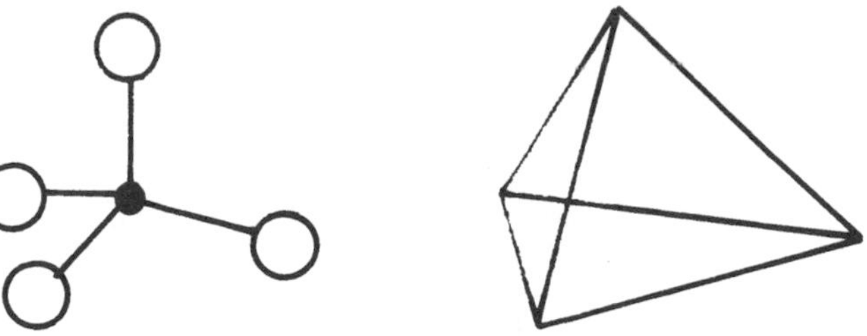

FIGURE 1. $[SiO_4]^{4-}$, $[AlO_4]^{3-}$ tetrahedra.

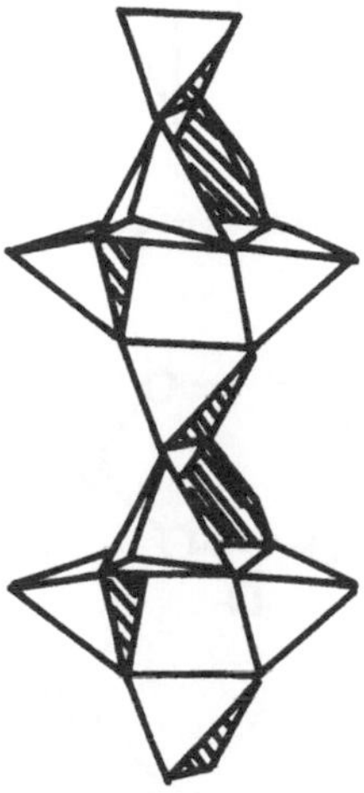

FIGURE 2. Linkages of tetrahedra to produce a zeolite structure (natrolite).

that of a ring created by linking four tetrahedra (Figure 3). This is known as a "single four ring" (S4R).

Diagrammatically, it is most convenient to show SBUs by using a straight line to represent the oxygen atom and a point to show the tetrahedral atom (Si or Al). This shorthand is illustrated in Figure 3, and all the SBUs used to illustrate zeolitic frameworks are in Figure 4.

This approach enables a classification to be drawn up, which is shown in full in Tables 1 through 5. This includes varieties of sodalite and cancrinite, which belong to the mineralogical group of felspathoids. This is because these substances contain elements of structure also found in zeolites. An example of this is the sodalite unit represented in Figure 5.

This is an easily recognized building "cage" present in the structures of the faujasitic group of zeolites. These particular zeolites are the most well-investigated zeolites, especially for their ion-exchange properties, and virtually all theoretical studies have been completed on these zeolites.

The structure of zeolites X and A can be described as follows. (1) In zeolite X, the sodalite units are joined by one half of their hexagonal faces to create a tetrahedral array in which each sodalite cage occupies a position in space like that of a carbon atom in the diamond lattice. (2) In zeolite A, the sodalite units are linked by every square face to form a cubic array in which the units are spatially similar to the positions of the ions in a sodium chloride lattice.

The structures of A and X are shown in Figures 6 and 7, and it can be seen that they are composed of large cavities (β) cages and other smaller sodalite (or α) cages.

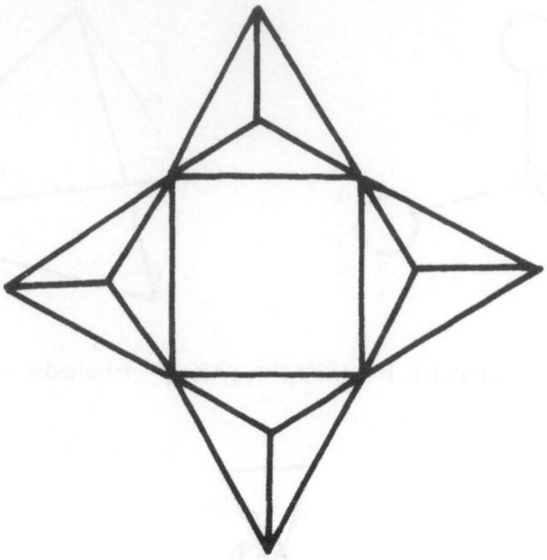

FIGURE 3. Linkage of tetrahedra in an SBU (S4R).

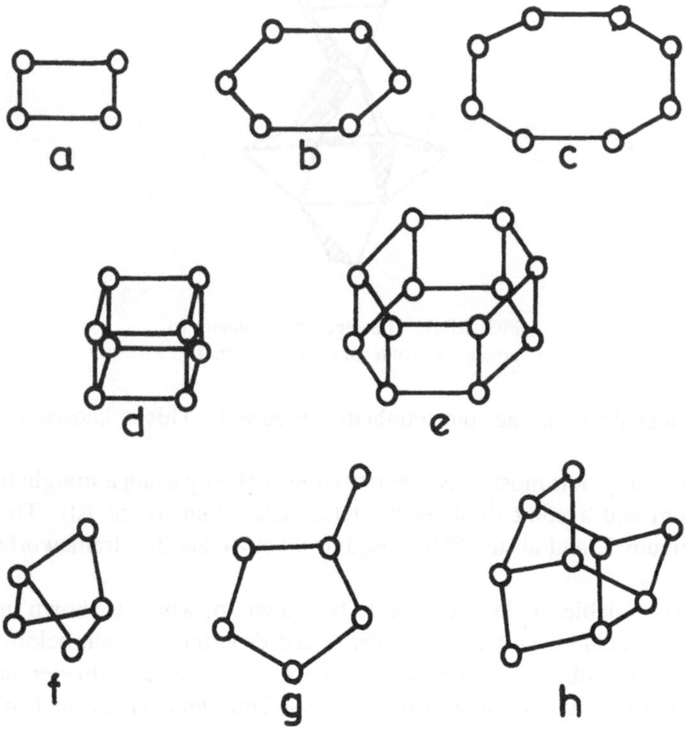

FIGURE 4. Zeolite SBU's (a) S4R, (b) S6R, (c) S8R, (d) D4R, (e) D6R, (f) 4-1 unit, and (g) 4-4-1 unit.

The structures of other zeolites of most relevance to this discourse are those of chabazite and mordenite. Chabazite is built from linkages of the double six-ring (D6R) SBU, as illustrated in Figure 8, while the mordenite structure is based upon the 5-1 SBU (Figure 9).

B. CATION POSITIONS IN ZEOLITES

The presence of aluminum isomorphously substituted in silicon tetrahedral positions confers a net negative charge on the zeolite aluminosilicate framework. This charge is

TABLE 1
Zeolite Structures Based Upon Single Oxygen Ring SBU's

Secondary building Unit (SBU)	Structure type	Name	Typical unit cell content
S4R	ANA	Analcime	$Na_{16}Al_{16}Si_{32}O_{96}10H_2O$
	ANA	Wairakite	$Ca_8Al_{16}Si_{32}O_{96}16H_2O$
	GIS	Gismondine	$Ca_4Al_8Si_8O_{32}16H_2O$
	GIS	Amicite	$K_4Na_4Al_8Si_8O_{32}10H_2O$
	GIS	Garronite	$NaCa_{2.5}Al_6Si_{10}O_{32}14H_2O$
	GIS	Gobbinsite	$Na_5Al_5Si_{11}O_{32}11H_2O$
	GIS	Zeolite NaP-1	$Na_6Al_6Si_{10}O_{32}12H_2O$
	LAU	Laumontite	$Ca_4Al_8Si_{16}O_{46}16H_2O$
	MER	Merlionite	$K_5Ca_2Al_9Si_{23}O_{64}24H_2O$
	PAU	Paulingite	$(K_2Na_2Ca,Ba)_{76}Al_{152}Si_{520}O_{1344}700H_2O$
	PHI	Phillipsite	$K_2Ca_{1.5}NaAl_6Si_{10}O_{32}12H_2O$
	PHI	Harmotome	$Ba_2Ca_{0.5}Al_5Si_{11}O_{32}12H_2O$
	YUG	Yugawaralite	$Ca_2Al_4Si_{12}O_{32}8H_2O$
S6R	CAN	Cancrinite-hydrate	$Na_6Al_6Si_6O_{24}8H_2O$
	ERI	Erionite	$Na_2K_2Mg_{0.5}Ca_2Al_9Si_{27}O_{72}27H_2O$
	LEV	Levynite (Levyne)	$NaCa_3Al_7Si_{11}O_{36}18H_2O$
	LTL	Zeolite L	$K_6Na_3Al_9Si_{27}O_{72}21H_2O$
	LOS	Zeolite Losod	$Na_{12}Al_{12}Si_{12}O_{48}19H_2O$
	MAZ	Mazzite (Zeolite Omega)	$Mg_2K_3Ca_{1.5}Al_{10}Si_{26}O_{72}28H_2O$
	OFF	Offretite	$KCa_2Al_5Si_{13}O_{36}15H_2O$
	SOD	Sodalite-hydrate (HS)	$Na_6Al_6Si_6O_{24}8H_2O$
S8R	Occurs in many structures but with other SBU's (see structure of Zeolite A, Chabazite, etc.).		

TABLE 2
Zeolite Structures Based upon Double Oxygen Ring SBU's

Secondary building unit (SBU)	Structure type	Name	Typical unit cell content
D4R	LTA	Zeolite A	$Na_{12}Al_{12}Si_{12}O_{48}27H_2O$
D6R	CHA	Chabazite	$Ca_2Al_4Si_8O_{24}13H_2O$
	CHA	Wilhendersonite	$K_2Ca_2Al_6Si_6O_{24}10H_2O$
	FAU	Faujasite	$Na_{12}Ca_{12}Mg_{11}Al_{58}Si_{134}O_{384}235H_2O$
	FAU	Zeolite X	$Na_{88}Al_{88}Si_{104}O_{384}220H_2O$
	GME	Gmelinite	$Na_8Al_8Si_{16}O_{48}24H_2O$
	KFI	Zeolite ZK-5	$Na_{30}Al_{30}Si_{66}O_{192}98H_2O$
	RHO	Zeolite rho	$(Na,Cs)_{12}Al_{12}Si_{36}O_{96}46H_2O$

TABLE 3
Zeolite Structures Based upon the 4-1 SBU

Secondary building unit (SBU)	Structure type	Name	Typical unit cell content
4-1	EDI	Edingtonite	$Ba_2Al_4Si_6O_{20}8H_2O$
	NAT	Natrolite	$Na_{16}Al_{16}Si_{24}O_{80}16H_2O$
	NAT	Tetranatrolite	$Na_{16}Al_{16}Si_{24}O_{80}16H_2O$
	NAT	Paranatrolite	$Na_{16}Al_{16}Si_{24}O_{80}24H_2O$
	NAT	Mesolite	$Na_{16}Ca_{16}Al_{48}Si_{72}O_{240}64H_2O$
	NAT	Scolecite	$Ca_8Al_{16}Si_{24}O_{80}24H_2O$
	THO	Thomsonite	$Na_4Ca_8Al_{20}Si_{20}O_{80}24H_2O$
	THO	Gonnardite	$Na_5Ca_2Al_9Si_{11}O_{40}14H_2O$

TABLE 4
Zeolite Structures Based upon the 5-1 SBU

Secondary building unit (sbu)	Structure type	Name	Typical unit cell content
5-1	BIK	Bikitaite	$Li_2Al_2Si_4O_{12}2H_2O$
	DAC	Dachiardite	$Na_5Al_5Si_{19}O_{48}12H_2O$
	EPI	Epistilbite	$Ca_3Al_6Si_{18}O_{48}16H_2O$
	FER	Ferrierite	$NaCa_{0.5}Mg_2Al_6Si_{30}O_{72}20H_2O$
	MFI	Zeolite ZSM-5	$Na_nAl_nSi_{96-n}O_{192}{\sim}16H_2O(n \sim 3)$
	MOR	Mordenite	$Na_8Al_8Si_{40}O_{96}24H_2O$

TABLE 5
Zeolite Structures Based upon the 4-4-1 SBU

Secondary building unit (sbu)	Structure type	Name	Typical unit cell content
4-4-1	BRE	Brewsterite	$Sr_2Al_4Si_{12}O_{32}10H_2O$
	HEU	Heulandite	$Ca_4Al_8Si_{28}O_{72}24H_2O$
	HEU	Clinoptilolite	$Na_6Al_6Si_{30}O_{72}24H_2O$
	STI	Stilbite	$Na_2Ca_4Al_{10}Si_{26}O_{72}34H_2O$
	STI	Stellerite	$Ca_4Al_8Si_{28}O_{72}28H_2O$
	STI	Barrerite	$Na_8Al_8Si_{28}O_{72}26H_2O$

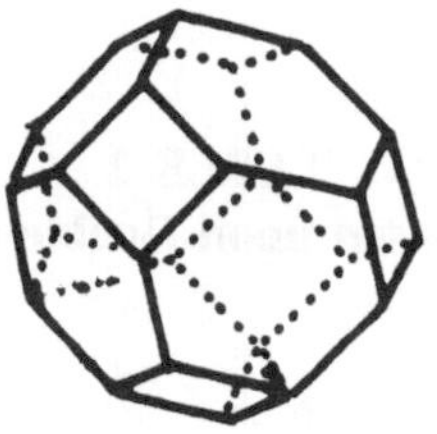

FIGURE 5. The sodalite cage.

compensated by the presence of cations inside the structure, and one of the reasons for the use of zeolites A, X, and mordenite as ion-exchange models is that the positions occupied by cations within their structures are known in detail.

The principal cation sites in the X framework are shown diagrammatically in Figure 10, and summarized with those in zeolite A in Table 6. In mordenite, the cations occupy specific positions within its channel structure — again listed in Table 6.

C. ZEOLITE ION EXCHANGE

From the foregoing structural descriptions, it can be seen that, in the open zeolites A and X, the ions are sited within the framework cavities and those in the large cavities are coordinated to water molecules. It is these ions which undergo the facile ion exchange for which zeolite materials are well known.

This statement, however, requires qualification in that not all zeolites readily exchange their ions. Natrolite, for instance, requires extreme conditions of external salt concentration and elevated temperature to achieve ion replacements. Most data on this zeolite has been obtained by using molten salt conditions.[11] Furthermore, it should be noted that the three-dimensional nature of the A and X structures is not always that of other zeolites. Mordenite provides a convenient example of a zeolite whose structure is of a two-dimensional nature

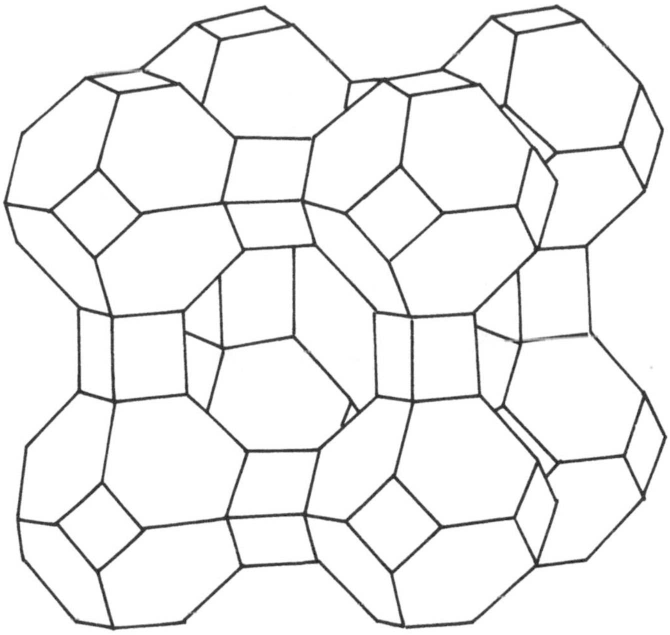

FIGURE 6. Structure of zeolite A.

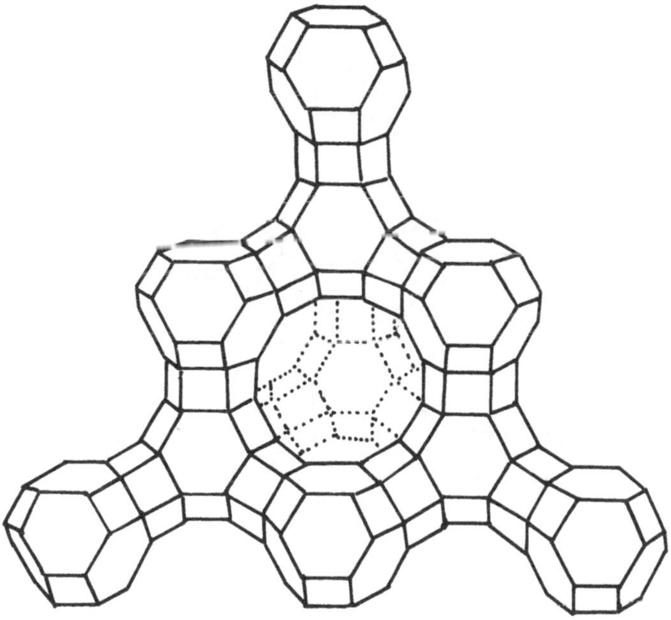

FIGURE 7. Structure of zeolite X (faujasite framework).

in that the openings between successive ''layers'' of the structure are small and likely to restrict the passage of an ion. Another zeolite (L) has a structure composed of one-dimensional channels,[12] so that exchange ions have to pass each other in the channel.

This isotropy of exchange has not yet formed part of a zeolite ion-exchange theory, although it might well be yet another way in which the structural uniqueness of zeolite frameworks could contribute to a more general understanding of the phenomenon of ion exchange.

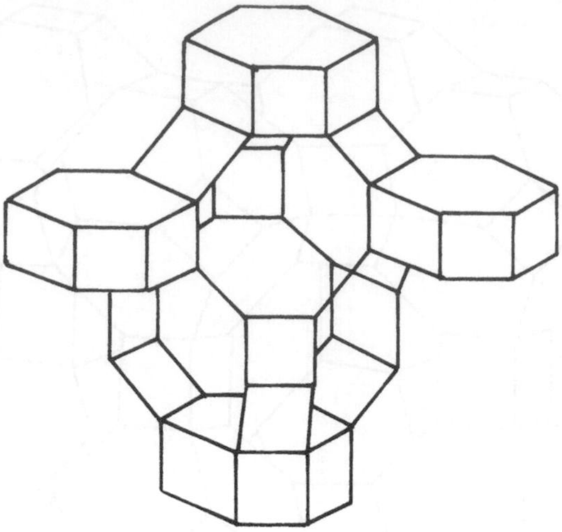

FIGURE 8. Structure of chabazite.

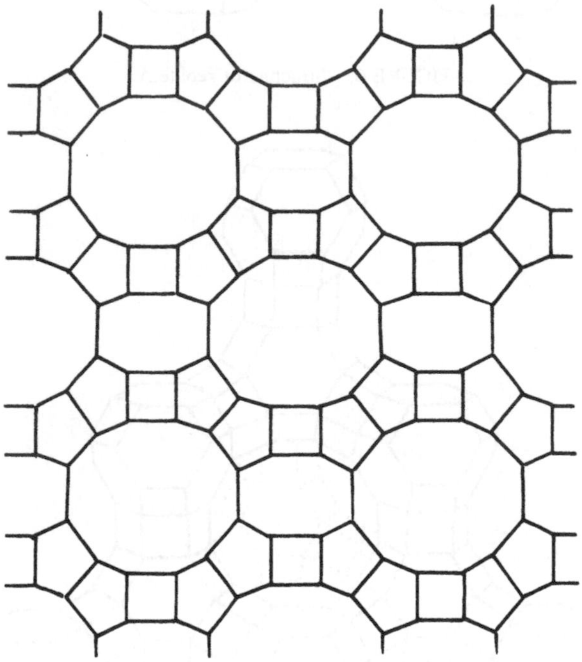

FIGURE 9. Structure of mordenite.

Aside from these qualifications, zeolites have very attractive features to encourage their use as models for ion exchange. These can be summarized:

1. Unlike organic resins, zeolites do not swell appreciably when placed in water.
2. Zeolite unit cell dimensions do not significantly alter as a function of the size or charge of the ion present in the exchange.
3. Zeolite structures are the best characterized of all exchange materials (including cation and water sites), so interpretations of the mechanisms of ion replacements can be formulated.

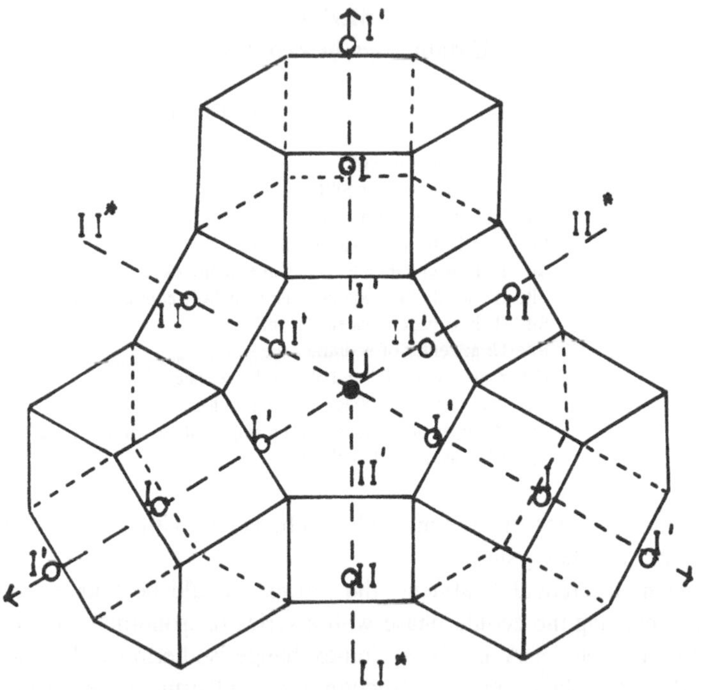

FIGURE 10. Cation exchange sites in zeolite X.

4. Some zeolites offer an isostructural series, so the effect of framework charge can be considered. The best known examples of this are the synthetic faujasites, where the only difference betwen zeolite X and zeolite Y is in the relative proportion of Si/Al isomorphously substituted in the framework (and the consequent change in total cation content).
5. The assemblage of SBU's into a framework creates a number of well-defined constraints to the passage of ions through the structure (e.g., entry to the large cage in A is through an eight-oxygen ring [S8R], while to gain entry to the sodalite cage, an ion must pass through a S6R. This gives rise to an ion sieve effect in some instances.).
6. Zeolites also lend themselves well to the study of ion exchange in nonaqueous and mixed-solvent systems.[13]

These attractions have created a large literature, and a variety of helpful reviews[14-16] summarize the earlier work, which will not be considered here, except to observe that the theoretical aspects of this early work depend upon the usual twin approaches of kinetic and equilibrium analyses. To help expedite this chapter, it is necessary to remind the reader of the basics of ion-exchange theory prior to a discussion of modern advances.

III. ION EXCHANGE EQUILIBRIA, WITH PARTICULAR REFERENCE TO ZEOLITES

A. ION-EXCHANGE ISOTHERM

For convenience, this will be related to zeolite properties so that an exchange between a cation A^{Z_A}, initially in solution, and a cation B^{Z_B}, initially in a zeolite, can be written as:

$$Z_B A^{Z_A} + Z_A \overline{B}^{Z_B} = Z_B \overline{A}^{Z_A} + Z_A B^{Z_B} \quad (1)$$

TABLE 6
Cation Sites in Zeolites

Zeolite	Cation sites
A	Site I: Adjacent to S6Rs
	Site II: near center of S8Rs
	Site III: near center of α cage
X, Y	Site I: in D6Rs (hexagonal prism)
	Site I′: near entrance to D6Rs in sodalite cage
	Site II′: inside the sodalite cage near S6R entrances to α cage
	Site II: in α cage adjacent to S6Rs
	Site U: at center of sodalite cage
	Other sites (e.g., IV and V): in the α cage
Mordenite (Major sites)	Site I: in centers of S8Rs forming a channel
	Site II: adjacent to the 5 oxygen rings linking the larger 12 "O"-ring channels
	Site III: coordinated to S4Rs

where $Z_{A,B}$ are the valencies of the ions and the characters with a bar as superscript refer to a cation inside a zeolite crystal.

For subsequent theoretical treatment, the process should be both stoichiometric and reversible. By contacting the zeolite phase with a series of isonormal solutions constituted from differing proportions of A and B, an ion-exchange isotherm can be constructed.

The methodology of the correct construction of an isotherm has been fully described by Dyer et al.[17] and reinforced by Townsend[18] who, more recently, has emphasized the importance of carrying out complete analyses on both zeolite and solution phases.

From analyses made when ion exchange equilibrium has been reached (checked by an independent kinetic experiment), the distributions of A and B between the phases can be determined. This enables an isotherm to be plotted which records the equivalent fraction (A_S) of the entering ion (A) in solution against its equivalent fraction ($\bar{A}_Z$) in the zeolite.

These quantitites are defined by

$$A_S = Z_A m_A/(Z_A m_A + Z_B m_B) \tag{2}$$

and

$$\overline{A}_Z = Z_A M_A/(Z_A M_A + Z_B M_A) \tag{3}$$

where m_A, m_B and M_A, M_B are the ion concentrations (mol dm^{-3}) in solution and solid phases, respectively.

Idealized isotherm shapes are displayed in Figure 11, and give pictorial indications of the relative preference of the ions for solution and solid phases. If the solid phase has an equal affinity for A and B, then the isotherm would be a straight line joining A_S, $\bar{A}_Z = 0$ to A_S, $\bar{A}_Z = 1$ (dotted line in Figure 11). It can then be seen that isotherm 2 in Figure 11 illustrates the case where A is selectively exchanged by the zeolite and isotherm 3 is that where A remains in solution, while B is preferred by the zeolite.

The selectivity of a zeolite for an ion (A) can be expressed quantitatively as a separation factor (α), viz:

$$\alpha = \overline{A}_Z m_B/\overline{B}_Z m_A \tag{4}$$

Where, by definition

$$\overline{B}_Z = 1 - \overline{A}_Z \tag{5}$$

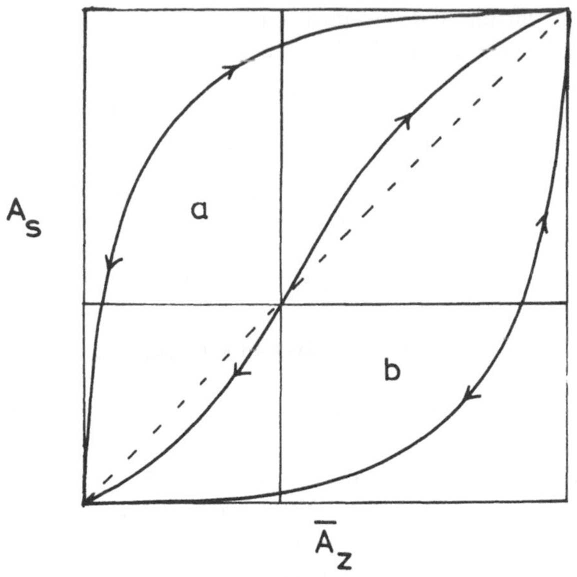

FIGURE 11. Ion-exchange isotherms.

In Figure 11, this is represented by area "a" divided by area "b".

B. ANALYSIS OF ISOTHERMS TO PROVIDE THERMODYNAMIC DATA

Where the isotherm has been shown to be reversible, a mass action quotient (K_m) can be defined thus:

$$K_m = A_Z^{Z_B} m_B^{Z_A} / B_Z^{Z_A} m_A^{Z_B} \tag{6}$$

From this, the thermodynamic constant (K_a) can be measured, where

$$K_a = K_m \, \Gamma \, (f_A^{Z_B}/f_B^{Z_A}) \tag{7}$$

with

$$\Gamma = \gamma_B^{Z_A}/\gamma_A^{Z_B} \tag{8}$$

γ_A and γ_B represent the single ion activity coefficients of A^{Z_A} and B^{Z_B} in solution, respectively, and f_A and f_B are the activity coefficients of the same ions in the solid phase.

K_a can be determined by graphical integration of a plot of ln $K_m\Gamma$ against $\bar{A}_Z$ (or by an analytical integration of the polynomial, which gives the computed best fit to the experimental data).

The quantity $K_m\Gamma$ often is described by

$$K_c = K_m\Gamma \tag{9}$$

where K_c is the Kielland coefficient related to K_a by the simplified Gaines and Thomas equation,[21] i.e.,

$$\ln K_a = (Z_B - Z_A) + \int_0^1 \ln K_c \, dA_Z \tag{10}$$

Values for γ_A and γ_B cannot be determined, but Γ can be obtained from the mean stoichiometric activity coefficients in mixed salt solution via

$$\Gamma = \gamma_B^{Z_A}/\gamma_A^{Z_B} = \left([\gamma_{\pm BX}^{(AX)}]^{Z_A(Z_B + Z_X)}/[\gamma_{\pm AX}^{(BX)}]^{Z_B(Z_A + Z_X)}\right)^{1/Z_X} \qquad (11)$$

Here, Z_X is the charge on the common anion $\gamma_{\pm BX}^{(AX)}$, and $\gamma_{\pm AX}^{(BX)}$ can be calculated from $\gamma_{\pm BX}$ and $\gamma_{\pm AX}$ using the method of Glueckauf;[19] f_A and f_B are available from the Gibbs-Duhem equation.

Having obtained K_a, a value of ΔG^{θ} can be gained from

$$\Delta G^{\theta} = -(RT \ln k_a)/Z_A Z_B \qquad (12)$$

where R and T have their usual meaining and ΔG^{θ} now is the standard free energy per equivalent of exchange.

The standard states in the zeolite relate to the respective homoionic forms of the exchanger immersed in an infinitely dilute solution of the corresponding ion, so the water activity in the zeolite in each standard state is equal to the water activity in the ideal solution, and the standard states in the solution phase are defined as the hypothetical ideal molar ($mol \cdot dm^{-3}$) solutions of the pure salts according to the Henry law definition of an ideal solution.[20]

Before progressing further, it must be pointed out that this summary is based upon the simplified Gaines and Thomas treatment.[21] In the complete version of Equation 10, the LHS should read $\ln K_a - \Delta$, where Δ is a water activity term given by[23]

$$\Delta = -Z_A Z_B \int_{\ln a_w = 0\ (E_A = 1,\ I = 0)}^{\ln a_w = 0\ (E_B = 1,\ I = 0)} \nu_{w,\,AB}\, d \ln a_w \qquad (13)$$

In Equation 13, $\nu_{w,AB}$ represents the number of moles of water per equivalent of exchanger, I is the ionic strength in solution, and a_w is the activity of water in the exchanger. E_A and E_B are cationic equivalent fractions based upon the Gapon[22] method of expressing an ion-exchange process, viz:

$$Z_B \cdot A^{Z_A+} \text{(solution)} + Z_A Z_B \cdot B_{(1/Z_B)} L = Z_A \cdot B^{Z_B+} \text{(solution)} + Z_A Z_B \cdot A_{(1/Z_A)} L \qquad (14)$$

where L is defined as a portion of the zeolite framework holding unit negative charge. In practice, Δ has been shown to be negligible.[23,24]

Another important point arises at this juncture: a detailed comparison of the approach of Gapon leads to an apparent paradox in the application of the Gaines and Thomas equation to the ion-exchange process. This lead to the suggestion by Sposito[25] that such an application was incorrect. However, Barrer and Townsend[26] have provided an elegant elucidation to show that there is nothing wrong with the Gaines and Thomas approach. An addendum to the use of the Gaines and Thomas method is that it does provide a means of allowing for the ingress of salt molecules into the zeolite phase during ion exchange. This process has been described as "salt imbibition" and certainly occurs in open zeolites.[27]

ΔG^{θ} now provides a more accurate way of comparing zeolite selectivities and has been used to define selectivity series for a number of zeolites. Those measured for monovalent ions in the faujasitic zeolites can be used for the purposes of a general illustration. For the alkali metals, the observed selectivities (at 50% exchange levels) based upon measured ΔG^{θ} values are[15]

1. Zeolites Y (Si/Al ~ 2.5) Cs > Rb > K > Na > Li
2. Zeolite X (Si/Al ~ 1) Na > K > Rb > Cs > Li

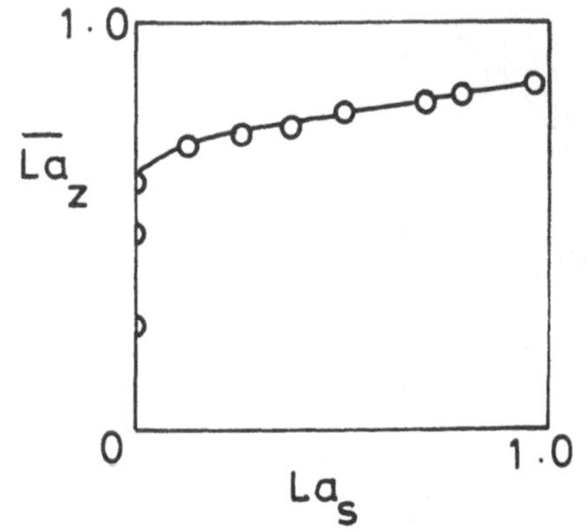

FIGURE 12. The $Na^+ \rightleftharpoons {}^1/_3La^{3+}$ isotherm in zeolite X. (From Sherry, H. S., in ACS Symp. Ser. 101, American Chemical Society, Washington, D.C., 1971, 350. With permission.)

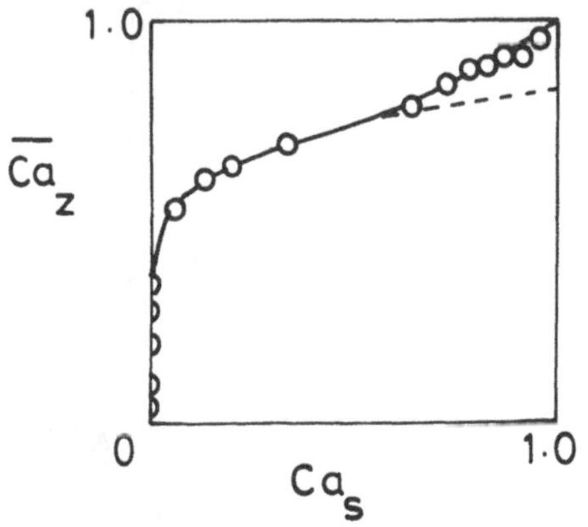

FIGURE 13. The $Na^+ \rightleftharpoons {}^1/_2Ca^{2+}$ isotherm in zeolite X. (From Sherry, H. S., in ACS Symp. Ser. 101, American Chemical Society, Washington, D.C., 1971, 350. With permission.)

Three general conclusions can be made from these selectivity series:

1. Zeolite X, with a high anion framework charge, prefers the univalent metals in the order of decreasing size.
2. Zeolite Y, with a low framework charge, prefers the alkali metals in the order of increasing size.
3. The Li^+ ion is an exception in Y.

Unfortunately, when similar analyses are attempted for divalent ion selectivities, no such simple pattern emerges. This lack of correlation can be ascribed to hydration effects, which vary with cations and the temperature of exchange. Because the cation sites are known in X and Y, it is possible to make detailed explanations of the selectivities observed, as explained by Sherry.[15] These are specific to the synthetic faujasite structure and cannot be extrapolated to other ion-exchange systems.

Observed ΔG^θ values can be used to generate values for ΔH^θ and ΔS^θ, the changes in enthalpy and entropy, respectively, for the ion-exchange process being measured (values are, of course, the standard changes per equivalent of exchange as before).

Evaluation of ΔH^θ requires measurements at more than one temperature. This may mean that more than one set of exchange sites is involved and, hence, interpretation should be used with caution.

Another important theoretical point may be discussed at this juncture which, in fact, relates to the philosophy behind the calculation of both ΔG^θ and ΔH^θ from isotherm data. It arises from the observed experimental case where, at equilibrium under isothermal conditions, an ion replaces only a fraction of the ions present in the exchanger. This is a well-known situation in zeolite X where, for example, La^{3+} is seen to replace only those Na^+ originally present in the large cavities of the zeolite at ambient temperature, unlike Ca^{2+}, which removes virtually all the Na^+ present. This is represented in Figure 12 and 13. The explanation offered is that an ion-sieve mechanism operates such that the larger hydrated La^{3+} cannot (at RT) pass through the S6R into the sodalite cages of the X structure.

The question arises as to whether or not a normalization procedure should be applied to the $Na^+ \rightleftharpoons {}^1/_3\ La^{3+}$ isotherm before calculating ΔG^θ. As yet, no comment can be found in the literature to support the calculation of ΔG^θ values from nonnormalized isotherm data when the object of the calculation is to generate a selectivity series for a particular zeolite.

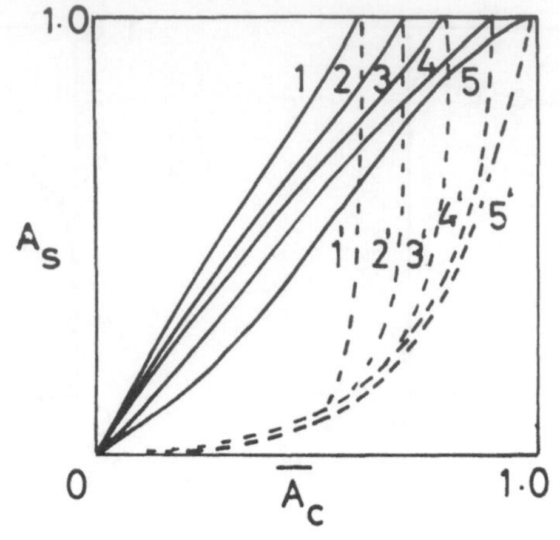

FIGURE 14. Exchange isotherms calculated for a zeolite with two site groups, 1 and 2. Group 1 is available to univalent ions A and B; 2 is available only to B. Isotherms with solid lines have K = 1; those with dashed lines, K = 10. $N/(N_A + N_B)$ = 1. (From Barrer, R. M. and Klinowski, J., *Philos. Trans.*, 285, 637, 1977. With permission.)

Such an aim may be of value when the ion exchanges being compared are binary processes. This usually was thought to be so in the early work on zeolite ion exchange, but modern applications require that more complex ternary and even quaternary cases be resolved. The simplest example of this is in the use of zeolite A as a detergent "builder" when a detailed knowledge of the Na/Ca/Mg A system is an advantage. This will be returned to later in this chapter, but suffice it to say at this juncture that a normalization procedure is not relevant when *more* than two ions are competing for exchange sites within a zeolite framework.[28]

C. MORE RECENT CONSIDERATIONS OF EXCHANGE ISOTHERMS

1. Shapes of Isotherms and Their Kielland Plots

To be complete, a theoretical treatment should be able to explain the observed shapes of an isotherm and its derived Kielland curve. It also should cope with the presence of a miscibility gap caused by ions in an exchanger clustering to nucleate a new phase. An example of this is that observed in the $Na^+ \rightleftharpoons K^+$ exchange in analcime.[29]

The progress made toward these aims has been reviewed by Barrer[30] and Townsend.[18,31] The approach described is a statistical thermodynamic one which models (1) miscibility gaps, (2) exchanges with a single exchange site, (3) ideal exchanges when there are n site groups available and (4) the effect of isomorphous framework replacement (e.g., a change in the Si/Al ratio). It is convenient to discuss a particular case when a zeolite (exchanger) has more than one site ($n > 1$).

Barrer and Klinowski[32] describe the case when n = 2, with one set of sites being available to ions A *and* B and another only to B. Isotherms calculated show that, as hypothetical equilibrium constants increase, A_Z^{max} increases, as shown in Figure 14. Barrer and Townsend liken curves 1′ and 2′ to those measured when transitional metal ammine ions (ions A) exchange with NH_4^+ (ions B) in mordenite.[33] In this competitive exchange, the ammine ions are thought to be too large to enter the side pockets which line the major channels in mordenite, in contrast to the NH_4^+ ions, which can.

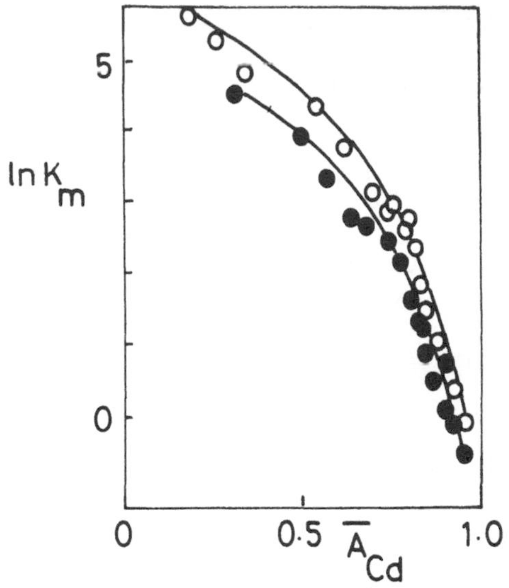

FIGURE 15. Plots of experimentally measured $\ln^{Cd}_{Na}K_m$ data, where nitrate (●) or an equivalent mixture of nitrate and chloride (○) were present as co-anions. Curves are fits to predicted K_m values. (From Townsend, R. P., Fletcher, P., and Loizidou, M., in *Proc. 6th Int. Conf. Zeolites,* Olson, D. and Bisio, A., Eds., Butterworth, Guildford, 1984, 110. With permission.)

2. Prediction of Exchange Equilibrium

The requirements for the prediction of an exchange equilibrium have been clearly outlined[23] and can be related back to the discussion of standard states earlier in this chapter. One requirement is that, during exchange, the ratio of activity coefficients for the exchanging component within the exchanger should remain sensibly constant for a given zeolite composition as the total concentration of salt solution external to the zeolite is varied (at equilibrium conditions). This constancy of ratio requires that there is no observable effect due to changes in water activity during the exchange, which has been demonstrated to be true in practice.

Another requirement is that undissociated salt molecules do not enter the zeolite. This imbibition of salts has been known for a long time,[27] as mentioned earlier, and will occur even at quite modest molarities ($\sim$ 0.1 *M*).[35] Given that these conditions exist, then for a given composition $\bar{E}_A$, $\bar{E}_B$, the appropriate corrected selectivity coefficient should be invariant with a change in total solution concentration. This means that selectivities of an exchanger depend *solely* on the nonideal behavior in the solution phase.

Examples of the application of this, with the major principles involved, have been fully described by Townsend et al.[31] for binary exchanges in zeolites X, mordenite, and ferrierite. Figure 15 shows ln K_m values for a $Na^+ \rightleftharpoons {}^1/_2Cd^{2+}$ exchange in mordenite. Note that the predictions take into account different anion compositions.

3. Multicomponent Exchangers

Using the principles of binary predictions, Townsend and co-workers have provided a basis for the study of exchanges in which more than two ions compete for sites in a zeolite framework. Townsend et al.[31] conveniently summarize their earlier work,[20,34,36] and it is instructive to take the case of a ternary exchange as an example. The reaction can be stated as:

$$2z_v z_w M_u^{z_u+} + z_u z_w \overline{M}_v^{z_v+} z_u z_w \overline{M}_w^{z_w+} = 2z_v z_w \overline{M}_u^{z_u+} + z_u z_w M_v^{z_v+} + z_u z_v M_w^{z_w+} \quad (15)$$

giving equilibrium constants described by

$$\binom{u}{v,w} K_a = \left[\binom{u}{v,w} K_m \gamma_v{}^{z_u z_w} \gamma_w{}^{z_u z_v} / \gamma_u{}^{2z_v z_w}\right]\left[g_{u,T}^{2z_v z_w} / g_{v,T}^{z_u z_w} g_{w,T}^{z_u z_v}\right] \quad (16)$$

$$\text{for} \quad \boxed{\begin{matrix} u \\ v \\ w \end{matrix}} = \boxed{\begin{matrix} A, B, C \\ B, C, A \\ C, A, B \end{matrix}}$$

$\binom{u}{v,w}K_m$ is the appropriate mass-action quotient and $g_{u,T}$ (for example) is the activity coefficient for the ion $M_u^{z_u+}$ in the zeolite phase in the presence of ions $M_v^{z_v+}$ and $M_w^{z_w+}$ and for a given combination $\bar{E}_u$, $\bar{E}_v$ ($\bar{E}_w$ is then defined, since $\Sigma_{i=u}^{w} \bar{E}_i = 1$). $E_{u,v,w}$ are equivalent fractions. When corrections for solution-phase nonideality have been made, the corrected selectivity coefficient, $\binom{u}{v,w}K_G$, is obtained and $\ln\binom{u}{v,w}K_a$ is calculated by integrating across the ternary surface, which describes the variation of $\ln\binom{u}{v,w}K_G$ with $\bar{E}_u$, $\bar{E}_v$.

The corrections for nonideality in the solution phase are not simple, and to achieve this, Fletcher and Townsend[36] have extended the original theories of Guggenheim[37] and Gluekauf[19] to apply to any number of different cations (and anions) in solution.

A further improvement lies in the use of a more elegant iteration procedure[28] which takes into account ionic charge and changes in total solution normality (T_N). It also is useful to closely define the thermodynamic formulation for the multicomponent system. This has been approached in two ways. Chu and Sposito[38] use a formulation based upon a pseudo-binary, "Vanselow-type" corrected selectivity coefficient, whereas Fletcher and Townsend[20,34] use a function based upon the Gapon method (see back). These approaches have been shown to be compatible[39] and comparisons to other models have been provided.[40]

To date, these ternary ion-exchange systems have been examined in detail, namely:

1. Na/Ca/Mg — A[28,41,42]
2. Na/K/Cd — X[43]
3. Na/NH_4/Mg — Y[44]

and examples of the success of the predictive approaches are shown in Figure 16.

This is an appropriate point to mention that some *apparently* binary exchanges can be ternary in zeolite systems. In synthetic zeolites, extensive hydrolysis can occur; for example, it has been well documented in the NH_4/Na/Y system.[45] Exchanges in natural zeolites may well involve the displacement of more than one cation from similar (or dissimilar) sites by an incoming cation. This problem has not yet been the subject of a detailed theoretical approach, but can cause unusual selectivity reversals.[46] In all cases where ion exchange is being studied in zeolites (natural or synthetic), great care must be taken in the experimental methods used. The latter have been listed by Townsend in Reference 18 and will not be discussed here.

Deviations occur from predictions even outside obvious experimental errors. In the Na/K/Cd X system (see Figure 16), variations which appear at higher concentrations of salt may well be due to salt imbibition. Similar deviations have been observed in zeolite Y when Mg^{2+} is an exchanging ion, and these have been ascribed to $MgCl^+$ species being taken up by the zeolite.[43]

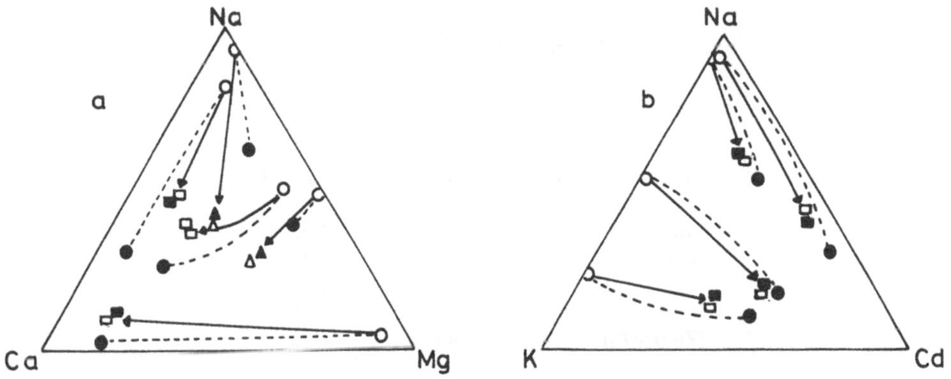

FIGURE 16. Examples of predicted composition for (a) Na/Ca/Mg-A and (b) Na/K/Cd-X systems. Solution phase (○) and corresponding compositions (●) measured at 0.1 equiv. dm^{-3} (points jointed by a tie-line -----); predicted (□) and measured (■) compositions at 1.4 equiv. dm^{-3}; predicted (△) and measured (▲) compositions at 0.025 equiv. dm^{-3}. (From Townsend, R. P., *Pure Appl. Chem.*, 58, 1359, 1986. With permission.)

Application of the predictive theories outlined is time consuming. A recent paper by Franklin and Townsend[47] seeks to simplify their own approaches by taking cognizance of work in which multicomponent exchanges in resins have been examined.[48,49] Franklin and Townsend provide a summary of the possibilities offered via four circumstances:

1. Data are needed for a ternary system at a unique concentration (at a given temperature and pressure).
2. Data are needed for a range of concentrations (at a given temperature and pressure).
3. The *binary* isotherms needed to help define ternary systems exhibit partial exchange.
4. The systems involve *more* than three ions.

If only circumstance 1 is involved, then either of the methods due to Bajpai et al.[48] or Brignal et al.[49] is applicable, provided adequate *binary* isotherm data are obtained under the defined experimental conditions and none of the binary isotherms are too selective. When circumstances 1 and 2 are needed, only the Brignal approach is applicable. Alternatively, when circumstances 1 and 4 are the criteria, only the Bajpai method is appropriate. The rigorous approach of Townsend and co-workers can be applied to any combination of the four criteria.

4. Effect of Temperature Variation

The implication of predictive ion-exchange equilibria theory to cases where a change of temperature is investigated awaits a solution.

Zeolite equilibria are illustrative of a general problem when more than one exchange site is available for ion occupancy. It is well known that cation site populations change in zeolites as a function of temperature and are conditioned by cation size and charge and the presence of water. Despite the vast literature on zeolite cation sites,[50] very few cases have been examined in detail sufficient for the stringent boundaries of ion-exchange theory.

5. Other Advances

Several authors have used extensions of dielectric theory to calculate the ΔG^{θ} for zeolite ion exchanges.

For ion A exchanging with ion B, the following general equation can be used:

$$\Delta G^{\theta} = Ne^2/8\pi \, (Z_A/r_A - Z_B/r_B)(1/\epsilon_c - 1/\epsilon_s) \qquad (17)$$

TABLE 7
ΔG^{θ} Values for Exchanges in Mordenite

Exchanging ion	ΔG^{θ} kJ equiv.$^{-1}$ Observed	Calculated
$[Mn(H_2O)_6]^{2+}$	3.86	2.97
$[Co(H_2O)_6]^{2+}$	4.42	[4.42][a]
$[Ni(H_2O)_6]^{2+}$	4.51	4.58
$[Cu(H_2O)_6]^{2+}$	4.02	4.85
$[Zn(H_2O)_4]^{2+}$	4.46	4.42

[a] Values calculated by reference to $[Co(H_2O_6)]^{2+}$.

From Barrer, R. M. and Klinowski, J., *J. Chem. Soc. Faraday Trans. 1*, 70, 2080, 1974. With permission.

where N is the Avogadro constant, e the charge on the electron, $r_{A,B}$ the ionic radii of the incoming ion (A^{Z_A}) and outgoing ion (B^{Z_B}). $\epsilon_{c,s}$ are the permittivities of the crystal (zeolite) and solution phases.[33] Calculated values agree well with experimental ones for transitional elements in mordenite (Table 7).

The simple theory has been extended to exchanges in the presence of ammine ligands[51,52] and to take into account changes in zeolite framework charge.[53]

Site selectivities for the Na^+ and Ag^+ ions in zeolite Y have been explained in terms of calculated electron densities at specific framework environments together with the crystallographic radii, hydration energies, and electronegativities of the competing cations.[54,55]

Another numerical approach calculates cation-lattice interaction energies to evaluate selectivities. An example of this is the calculated preferences of Mn^{2+}, Co^{2+}, Fe^{2+}, and Ni^{2+} for 6SR sites and Zn^{2+} for 8SR sites in zeolite A.[56] These studies (Nitta and coworkers) consider the dehydrated structures; more recently, Koh et al.[57,58] have carried out a theoretical calculation of selectivities on hydrated A. Their studies are based upon semi-empirical potential energy functions. Similar work has considered selectivities in hydrated chabazite.[59,60]

6. Statistical Thermodynamics of Ion Exchange

Barrer and Falconer[61] first considered the use of a statistical thermodynamic approach to exchange in felspathoids. Barrer and Klinowski[32] have extended this to (1) cation exchange on a homogeneous group of exchange sites, (2) cation exchange when more than one kind of exchange site is present, and (3) replacement of silicon by aluminium. The relevance of 1 and 2 to zeolite exchange was mentioned earlier in this chapter. The third extension covers cases in tectosilicates to explain why some tectosilicates have nearly fixed Al/Si ratios, while other have ratios which vary over a nearly infinite range. It also explains the experimental observation that Al/Si does not exceed unity. To date, no publication has considered framework isomorphous replacement by elements other than Si or Al.

The same authors[62] have applied order-disorder theory to cover the case where ion-exchange equilibrium occurs for two ions present in crystalline exchanger with more equivalent cation sites than the number of ions needed to neutralize the total framework charge.

IV. ION-EXCHANGE KINETICS IN ZEOLITES

A. ION-EXCHANGE DIFFUSIVITIES

Progress in the theory of zeolite ion-exchange kinetics has been much less evident than that in ion-exchange equilibria.

The work by Rees and Brooke[63-65] in 1968 to 1970 sought to extend the application of the Helfferich-Plesset equation[66] to nonideal systems. This still remains incomplete, despite the advances made by Rees and Brooke, due to the requirement to take into account the variation in the ionic diffusion coefficient with concentration and to consider the effect of water transport during the exchange process.

This and other earlier work has been reviewed by Sherry[15] and Barrer.[30] It includes an emphasis on the estimation of ionic self-diffusion coefficients and ion diffusivities, and recent work has continued in this theme.

Barrer et al.[67] used an equation developed by Ash[68] and co-workers of the form

$$I_o = \int_o^\infty (1 - Q_t/Q_\infty)dt = r_o^2/15D \tag{18}$$

where $O_{t,\infty}$ are the extents of exchange at times $t = t$ and $t = \infty$, r_0 is the radius of the zeolite particles, and D is the diffusivity. The estimation of D requires the construction of a plot of Q_t/Q_∞ v t for an ion exchange, and I_0 is then defined as the area above the curve between it and the horizontal asymptote, measured when t is large.

Barrer et al.[67] studied the exchanges $Na^+ \rightleftharpoons {}^1/_2Ca^{2+}$ and $Na^+ \rightleftharpoons {}^1/_2Ba^{2+}$ in the synthetic zeolite RHO, and obtained improved fits to experimental data by taking account of particle size distribution.

Ion-exchange kinetics in zeolite A have been measured[69] by using Paterson[70] solutions for limited volume diffusion after the manner described by Danes and Wolf.[71] Rates of complete replacement of Na^+ by Ca^{2+} have been obtained as well as the rates of small incremental replacements over the range 100% $Na^+ \rightarrow$ 90% Ca^{2+}. Kinetics of reverse exchanges and some values in the ternary system Na/Ca/Mg were obtained.

The interest in clinoptilolite for waste-water treatment and animal feedstuff amendation has prompted work on the kinetics of ion exchange in this natural zeolite. As with zeolite A, there have been differing mathematical analyses of the rate plots to estimate ion-exchange diffusivities. Neveu et al.[72] use a semiempirical equation

$$U_{(t)} = [1 - \exp(\beta\ Dt\pi^2/r_o^2)]^{1/2} \tag{19}$$

where

$$\beta = [1 - U_{(t)}]^{2/3} \tag{20}$$

$U_{(t)}$ is the exchanged fraction at time t, D the diffusivity, and r_0 the particle radius.

Similar parameters have been measured by White,[73] who has compared ion-exchange diffusivities in clinoptilolite computed by the methods of Ash et al.[68] and Paterson[70] to those calculated using the Carman and Haul[74] approach. Strictly, the latter equation is applicable to self-diffusion phenomena, but White demonstrates that good fits are obtained to experimental data. The diffusion parameters generally agree with those calculated from the Ash equation, but less well with results generated from the Paterson solution.

Araya and Dyer[75] earlier used the Carman and Haul analysis to measure cation self-diffusion coefficients in clinoptilolite, and the results of White are compatible with their conclusions.

B. CATION SELF-DIFFUSION

Earlier work concerned with the quantification of cation self-diffusion processes in zeolites has been collated by Barrer.[30] More recently, Todorovic et al.[76] have used the equation of Brown et al.[77] to elucidate Cd self-exchange in zeolite X, whereas Dyer and Yusof[78,79] have preferred the Carman and Haul approach to follow sodium, potassium, and water movements in synthetic analcimes. They were able to conclude that the diffusion mechanisms observed corresponded to isotropic pathways in the three nonintersecting channels in the zeolite framework.

In several cases where self-exchange processes are being followed, the phenomenon observed is that of a two-stage process.[75,80-92] Thompson and Tassopoulos[83] have considered this by a semianalytical interpretation and shown good fits between model simulations and experimental data. This agreement is seen only when the concept of a threshold concentration is invoked, which must be overcome in the larger cages of the zeolite before transport between the larger and smaller cages begins. This means that transport from the bulk fluid to the large cage and then to the smaller cages cannot be simply diffusion or kinetically controlled.

REFERENCES

1. **Way, J. T.,** On the power of soils to absorb manure, *J. R. Agric. Soc. Engl.*, 13, 123, 1852.
2. **Eichhorn, C. H.,** Ueber die Einwirking verdunnter Salzlosungen auf Silicate, *Poggendorf Ann. Phys. Chem.*, 105, 126, 1958.
3. **Adams, B. A. and Holmes, E. L.,** Adsorptive properties of synthetic resins, *J. Soc. Chem. Ind.*, 54, 1T, 1935.
4. **Breck, D. W., Eversole, W. G., Milton, R. M., Reed, T. B., and Thomas, T. L.,** Crystalline zeolites. I. The properties of a new synthetic zeolite, type A, *J. Am. Chem. Soc.*, 78, 5936, 1956.
5. **Dyer, A.,** *An Introduction to Zeolite Molecular Sieves,* John Wiley & Sons, Chichester, 1988.
6. Operation and Control of Ion Exchange Processes for the Treatment of Radioactive Wastes, Tech. Rep. Ser. No. 78, I.A.E.A., Vienna, 1967.
7. **Nowak, E. J.,** Composite backfill materials for radioactive waste isolation by deep burial in salt, *Sci. Basis Nucl. Waste Manage.*, 3, 545, 1981.
8. **Thompson, H. S.,** On the absorbent power of soils, *J. R. Agric. Soc. Engl.*, 11, 68, 1950.
9. **Dyer, A. and Malik, S. A.,** Studies on inorganic anion exchangers, *J. Inorg. Nucl. Chem.*, 43, 2975, 1981.
10. **Clearfield, A.,** *Inorganic Ion Exchange Materials,* CRC Press, Boca Raton, FL, 1982.
11. **Belitskii, I. A. and Gabuda, S. P.,** Water diffusion in ion-exchanged forms of some natural zeolites, *Chem. Erde,* 27, 79, 1968.
12. **Barrer, R. M. and Villiger, J.,** Structure of the synthetic zeolite L, *Z. Kristallogr.*, 128, 353, 1969.
13. **Huang, P. C., Mizany, A., and Pauley, J. L.,** Ion exchange of synthetic zeolites in various alcohols, *J. Phys. Chem.*, 68, 2575, 1964.
14. **Rees, L. V. C.,** Ion exchange in zeolites, *Ann. Rep.*, Chemical Society, London, 1970, 191.
15. **Sherry, H. S.,** Cation exchange on zeolites, in *Molecular Sieve Zeolites—I,* Adv. Chem. Ser. No. 101, American Chemical Society, Washington, D.C., 1971, 350.
16. **Barrer, R. M.,** *Cation Exchange Equilibria in Zeolites and Feldspathoids in Natural Zeolites, Occurence, Properties,* Sand, L. B. and Mumpton, F., Eds., Pergamon Press, New York, 1978, 385.
17. **Dyer, A., Enamy, H., and Townsend, R. P.,** The plotting and interpretation of ion-exchange isotherms, *Sep. Sci. Tech.*, 16, 173, 1981.
18. **Townsend, R. P.,** Ion exchange in zeolites: some recent developments in theory and practice, in *New Developments in Zeolite Science Technology, Proc. 7th Int. Zeolite Conf.*, Murakami, Y., Iijima, A., and Ward, J. W., Eds., Kodansha, Tokyo, 1986, 273.
19. **Glueckauf, E.,** Activity coefficients in concentrated solution containing several electrolytes, *Nature,* 163, 414, 1949.
20. **Fletcher, P. and Townsend, R. P.,** Ternary ion exchange in zeolites. I. Problem of predicting equilibrium compositions, *J. Chem. Soc. Faraday Trans.* 2, 77, 955, 1981.

21. **Gaines, G. L. and Thomas, H. C.,** Adsorption studies on clay minerals. II, *J. Chem. Phys.*, 21, 714, 1953.
22. **Gapon, E. N.,** Theory of exchange adsorption in soils, *J. Gen. Chem. U.S.S.R.*, 3, 144, 1933.
23. **Barrer, R. M. and Klinovski, J.,** Ion exchange selectivity and electrolyte concentration, *J. Chem. Soc. Faraday Trans. 1*, 70, 2080, 1974.
24. **Laudelout, H. and Thomas, H. C.,** The effect of water activity on ion-exchange selectivity, *J. Phys. Chem.*, 69, 339, 1965.
25. **Sposito, G.,** *Thermodynamics of Soil Solutions*, Clarendon Press, Oxford, 1981, 150.
26. **Barrer, R. M. and Townsend, R. P.,** Concentration scales and ion exchange thermodynamics, *Zeolites*, 5, 287, 1985.
27. **Barrer, R. M. and Walker, A. J.,** Imbibition of electrolytes by porous crystals, *Trans. Faraday Soc.*, 66, 171, 1964.
28. **Franklin, K. R. and Townsend, R. P.,** Multicomponent ion exchange in zeolites. II. Prediction of exchange equilibria over a range of solution concentrations, *J. Chem. Soc. Faraday Trans. 1*, 81, 3127, 1985.
29. **Barrer, R. M. and Hinds, L.,** Ion exchange in crystals of analcite and leucite, *J. Chem. Soc.*, 1879, 1953.
30. **Barrer, R. M.,** Zeolite exchangers — some equilibrium and kinetic aspects, in *Proc. 5th Int. Conf. Zeolites*, Rees, L. V. C., Ed., Heyden, London, 1980, 273.
31. **Townsend, R. P., Fletcher, P., and Loizidou, M.,** Studies on the prediction of multicomponent, ion exchange equilibria in natural and synthetic zeolites, in *Proc. 6th Int. Conf. Zeolites*, Olson, D. and Bisio, A., Eds., Butterworths, Guildford, 1984, 110.
32. **Barrer, R. M. and Klinowski, J.,** Theory of isomorphous replacement in aluminosilicates, *Philos. Trans. R Soc. London, A*, 285, 637, 1977.
33. **Barrer, R. M. and Townsend, R. P.,** Transitional metal ion exchange in zeolites. II. Ammines of Co^{3+}, Cu^{2+} and Zn^{2+} in clinoptilolite, mordenite and phillipsite, *J. Chem. Soc. Faraday Trans. 1*, 72, 2650, 1976.
34. **Fletcher, P. and Townsend, R. P.,** Ternary ion exchange in zeolites. II. A thermodynamic formulation, *J. Chem. Soc. Faraday Trans. 2*, 77, 965, 1981.
35. **Dyer, A., Harjula, R., and Townsend, R. P.,** unpublished work.
36. **Fletcher, P. and Townsend, R. P.,** Ternary ion exchange in zeolites. III. Activity coefficients in multi-component electrolyte solutions, *J. Chem. Soc. Faraday Trans. 2*, 77, 2077, 1981.
37. **Guggenheim, E. A.,** Specific thermodynamic properties of aqueous solutions of strong electrolytes, *Philos. Mag.*, 19, 588, 1935.
38. **Chu, S. YH. and Sposito, G.,** The thermodynamics of ternary cation exchange systems and the subregular models, *Soil Sci. Soc. Am. J.*, 45, 1084, 1981.
39. **Townsend, R. P.,** Thermodynamics of ion exchange in clays, *Philos. Trans. R. Soc. London, A*, 311, 301, 1984.
40. **Fletcher, P., Franklin, K. R., and Townsend, R. P.,** Thermodynamics of binary and ternary ion exchange in zeolites, the exchange of sodium, ammonium and potassium ions, *Philos. Trans. R. Soc. London A*, 312, 441, 1984.
41. **Franklin, K. R. and Townsend, R. P.,** Multicomponent ion exchangers in zeolites. I. Equilibrium properties of the sodium/calcium/magnesium zeolite A system, *J. Chem. Soc. Faraday Trans. 1*, 81, 1071, 1985.
42. **Barri, S. A. I. and Rees, L. V. C.,** Kinetics and equilibria of Na;Ca;Mg exchange in zeolite A, *J. Chromatogr.*, 201, 21, 1980.
43. **Franklin, K. R. and Townsend, R. P.,** Multicomponent ion exchange in zeolites. III. Equilibrium properties of the sodium/potassium/cadmium-zeolite X system, *J. Chem. Soc. Faraday Trans. 1*, 84, 687, 1988.
44. **Franklin, K. R. and Townsend, R. P.,** Multicomponent ion exchanges in zeolites. IV. The exchange of magnesium ions in zeolites X and Y, *J. Chem. Soc. Faraday Trans. 1*, 84, 2755, 1988.
45. **Franklin, K. R., Townsend, R. P., and Whelan, S. J.,** Ternary exchange equilibria involving H_3O^+, NH_4^+ and Na^+ ions in synthetic zeolites of the faujasite structure, in *New Developments in Zeolite Science Technology, Proc. 7th Int. Conf. Zeolites*, Murakami, Y., Iijima, A., and Ward, J. W., Eds., Kodansha, Tokyo, 1986, 289.
46. **Ahmad, Z. B. and Dyer, A.,** Ion exchange in ferrierite, a natural zeolite, in *Ion Exchange Technology*, Naden, D. and Streat, M., Eds., Ellis Horwood, Chichester, 1984, 519.
47. **Franklin, K. R. and Townsend, R. P.,** Prediction of multicomponent ion exchange equilibria in zeolites: a comparison of procedures, *Zeolites*,
48. **Bajpa, R. K., Gupta, A. K., and Gopala-Rao, M. J.,** Binary and ternary ion-exchange equilibria sodium-cesium-manganese-Dowex 50W-X8 and cesium-manganese-strontium-Dowex 50W-X8 systems, *J. Phys. Chem.*, 77, 1288, 1973.

49. **Brignal, W. J., Gupta, A. K., and Streat, M.**, in Theory and Practice of Ion Exchange, Soc. Chem. Ind. Conf., Cambridge, 1976.
50. **Mortier, W. J.**, *Compilation of Extra Framework Sites in Zeolites*, Butterworths, Guildford, 1982.
51. **Fletcher, P. and Townsend, R. P.**, Exchange of hydrated and amminated silver(I) ions in synthetic zeolites X, Y and mordenite, *J. Chromatogr.*, 201, 93, 1980.
52. **Fletcher, P. and Townsend, R. P.**, Transitional metal ion-exchanges in zeolites. IV. Exchange of hydrated and amminated silver in sodium X and Y zeolites and mordenite, *J. Chem. Soc. Faraday Trans. 1* 77, 497, 1981.
53. **Fletcher, P. and Townsend, R. P.**, Exchange of ammonium and sodium ions in synthetic faujasites, *J. Chem. Soc. Faraday Trans. 1*, 78, 1741, 1982.
54. **Costenoble, M. and Maes, A.**, Site group interaction effects in zeolite Y. I. Structural examination of the first stages of the Ag ion exchange, *J. Chem. Soc. Faraday Trans. 1*, 74, 1978.
55. **Maes, A. and Cremers, A.**, Site/group interaction effects in zeolite Y. II. Na-Ag selectivity in different site groups, *J. Chem. Soc. Faraday Trans. 1*, 74, 136, 1978.
56. **Nitta, M., Ogawa, K., and Aomura, K.**, A theoretical study of the site selectivity of transitional metal ions in dehydrated zeolite A, in *Proc. 5th Int. Conf. Zeolites*, Rees, L. V. C., Ed., Heyden, London, 1980, 291.
57. **Koh, K. O. and Jhon, M. S.**, Theoretical study of ionic selectivities and adsorption of water in zeolite A, *Zeolites*, 5, 313, 1985.
58. **Koh, K. O., Chon, H., and Jhon, M. S.**, Site selectivity of alkaline earth metal cations in zeolite A., *J. Catal.*, 98, 126, 1986.
59. **Barreto Perez, M. C.**, Interpretation of the theory of the selectivity of zeolite in ion exchange, *Folia Chim. Theor. Lat.*, 14, 101, 1986.
60. **Barreto, M. C., Ciambelli, P., Del Re, G., and Peluso, A.**, A theoretical study of ion selectivity of zeolites — application to chabazite, *J. Phys. Chem. Solids*, 48, 1, 1987.
61. **Barrer, R. M. and Falconer, J. D.**, Ion exchange in felspathoids as a solid state reaction, *Proc. R. Soc. London A*, 236, 227, 1956.
62. **Barrer, R. M. and Klinowski, J.**, Order-disorder model of cation exchange in silicates, *Geochim. Cosmochim. Acta*, 43, 755, 1979.
63. **Brooke, N. M. and Rees, L. V. C.**, Kinetics of ion exchange. I, *Trans. Faraday Soc.*, 64, 3383, 1968.
64. **Brooke, N. M. and Rees, L. V. C.**, Kinetics of ion exchange. II, *Trans. Faraday Soc.*, 65, 2728, 1969.
65. **Rees, L. V. C. and Brooke, N. M.**, Kinetics of ion exchange: comparison of experimental and computed kinetics of ion exchange, in *Proc. Int. Conf. Ion Exchange in the Process Industries, London, Soc. Chem. Ind. London*, 1970, 352.
66. **Helfferich, F. and Plesset, M. S.**, Ion exchange kinetics, a nonlinear diffusion problem, *J. Chem. Phys.*, 28, 418, 1958.
67. **Barrer, R. M., Barri, S., and Klinowski, J.**, Zeolite RHO. II. Cation exchange equilibria and kinetics, *J. Chem. Soc. Faraday Trans. 1*, 76, 1038, 1980.
68. **Ash, R., Barrer, R. M., and Craven, R. J. B.**, Sorption kinetics and Time-Lag Theory, *J. Chem. Soc. Faraday Trans. 2*, 74, 40, 1978.
69. **Drummond, D., De Jonge, A., and Rees, L. V. C.**, Ion-exchange kinetics in zeolite A, *J. Phys. Chem.*, 87, 1967, 1983.
70. **Paterson, S.**, The heating or cooling of a solid sphere in a well-stirred fluid, *Proc. Phys. Soc. London*, 59, 50, 1947.
71. **Danes, F. and Wolf, F.**, Ion exchange of sodium ions with divalent ions in zeolite molecular sieve Type A. III. Kinetics of ion exchange — calculation of diffusion coefficients, *Z. Phys. Chem. Leipzig*, 252, 15, 1973.
72. **Neveu, A., Gaspard, M., Blanchard, G., and Martin, G.**, Intracrystalline self-diffusion of ions in clinoptilolite, ammonia and sodium cation studies, *Water Res.*, 19, 611, 1985.
73. **White, K. J.**, Ion-Exchanges in Clinoptilolite, Ph.D. thesis, University of Salford, U.K., 1988.
74. **Carman, P. C. and Haul, R. A. W.**, Measurement of diffusion coefficients, *Proc. R. Soc. London Ser. A*, 222, 109, 1954.
75. **Araya, A. and Dyer, A.**, Studies on natural clinoptilolites. II. Cation mobilities in near homoionic clinoptilolites, *J. Inorg. Nucl. Chem.*, 43, 594, 1981.
76. **Todorovic, M. B., Gal, I. J., Paligoric, I. D., and Radak, V. M.**, Self-diffusion of cadmium(2^+)ions in hydrated Cd X zeolite, *Croat. Chem. Acta*, 55, 287, 1982.
77. **Brown, L. M., Sherry, H. S., and Krambeck, F. J.**, Mechanisms and kinetics of isotopic exchange in zeolites. I. Theory, *J. Phys. Chem.*, 75, 3846, 1971.
78. **Dyer, A. and Yusof, A. M.**, Diffusion in heteroionic analcimes. I. Sodium-potassium-water system, *Zeolites*, 7, 191, 1987.
79. **Dyer, A. and Yusof, A. M.**, Diffusion in heteroionic analcimes. II. Diffusion of water in sodium/thallium, sodium/lithium, and sodium/ammonium analcimes, *Zeolites*, 9, 129, 1989.

80. **Dyer, A. and Townsend, R. P.,** The mobility of cations in synthetic zeolites with the faujasite framework. V. The self-diffusion of zinc into X and Y zeolites, *J. Inorg. Nucl. Chem.*, 35, 3001, 1973.
81. **Dyer, A. and Enamy, H.,** Mobility of magnesium ions in the synthetic zeolites A, X and 2.62-Y, *Zeolites,* 1, 7, 1981.
82. **Dyer, A. and Enamy, H.,** Self-diffusion of zinc and cadmium cations in synthetic zeolite A, *Zeolites,* 4, 319, 1984.
83. **Thompson, R. W. and Tassopoulos, M.,** A phenomenological interpretation of two-step uptake behaviour by zeolites, *Zeolites,* 6, 9, 1986.

Chapter 3

INSOLUBLE HETEROPOLYACID SALTS

J. van R. Smit

TABLE OF CONTENTS

I. INTRODUCTION

The ion-exchange properties of the insoluble 12-heteropolyacid (HPA) salts have been known qualitatively,[1] even semiquantatively,[2,3] for several decades, at least for its best-known member, ammonium 12-molybdophosphate (AMP), of formula $(NH_4)_3PMo_{12}O_{40}$. These authors also realized that the reaction was reversible: if an AMP recipitate is washed with a potassium nitrate solution, up to about two of the three NH_4^+ ions of AMP may be replaced by K^+ ions;[4] subsequent washing with an ammonium nitrate solution restores the AMP to the triammonium salt (for comments on these observations, see later).

It was after the present author, in 1958, showed[5] that AMP could be used for the column chromatographic separation of the alkali metal ions, with quite remarkable selectivity, that the insoluble inorganic HPA salts were recognized as representing a new type of cation exchange material of note.

Research on the HPA salts has been confined largely to AMP and, to a lesser extent, to the corresponding 12-tungstophosphate (AWP), for these are the only two ammonium salts of sufficiently low solubility. AWP has received less attention, largely because of the extreme fineness of the microcrystalline particles obtained with the most favorable preparation procedures.

The unusual properties of this class of ion exchangers may be gauged from the following summary of the results reported for AMP:

1. AMP is an entirely anhydrous cation exchanger showing unique selectivity in that, in acid and weakly acid medium, only those large monovalent cations forming insoluble 12-molybdophosphates, viz., K^+, Rb^+, Cs^+, Tl^+, and Ag^+ (which we may call LMCs, for convenience), are adsorbed strongly and exchange to a significant extent, and unusually large selectivities are shown for an alkali metal ion over that above it in Group I of the Periodic Table. Thus, in 0.1 *M* NH_4NO_3 solution, for example, the selectivity coefficient α_{Rb}^{Cs} for AMP is about 26, compared to only about 1.2 for an 8% cross-linked sulfonic acid resin with a polystyrene-DVB matrix,[5] and about 3 to 8 for zirconium phosphate gel exchangers.[6] Since the ions not listed above do not significantly compete with the LMCs for the exchange sites, Cs^+, for example, is adsorbed essentially as well from acidified sea water[7] as from distilled water, and the column exchange capacity of AMP for Cs^+ realized when a cesium-spiked simulated fission product waste solution (~3 *M* in HNO_3 and about 3 *N* in other metal cations) is passed through a bed of AMP is only a few percent lower than for a pure, dilute cesium nitrate solution.[8]
2. Not all of the NH_4^+ ions on AMP are exchangeable, despite the fact that all three NH_4^+ ions in the formula occupy structurally equivalent positions in the crystal lattice.[9,10] The maximum exchange capacity of AMP for the individual LMCs is not a simple function of ionic radius or any other thermodynamic property of the incoming ion.[9,11]
3. Because of the anhydrous nature of AMP, cations adsorbed onto this exchanger are stripped of their hydration water upon entering the lattice. Relative rates of diffusion in the AMP are therefore determined by bare ionic radii. Rb^+ ions thus diffuse faster than Cs^+ ions,[12,13] unlike the situation with other ion exchangers, where relative diffusion rates are determined by hydrated ionic radii.
4. From neutral (pH 2 to 5) solutions, the smaller alkali ions (Li^+ and Na^+) and multivalent ions are also significantly adsorbed[14] onto AMP, particularly if the solution is buffered at a pH of about 4 to 5. This type of exchange, which is still not properly understood, permits excellent group separations between bi- and trivalent ions, but separation of two or more trivalent ions from one another is not possible because of a pH-dependent hysteresis effect preventing dynamic reversibility of the exchange.

A. EARLIER REVIEWS AND SCOPE OF THE PRESENT REVIEW

Excellent reviews of the earlier literature are given by Amphlett[15] and Churms.[16] Clearfield et al.[17] reviewed the literature up to 1973. In recent years, most monographs on ion exchange have devoted small sections to ion exchange on the insoluble HPA salts. In many cases, these treatments have not been altogether satisfactory, the authors often finding it difficult to present a balanced and coherent view in the absence of an understanding of the general principles governing selectivity and mechanism in ion exchange on HPA salts.

The present author will attempt to provide a review of the essential features of ion exchange on HPA salts and its applications in chemical analysis based on what is hoped to be a more complete understanding of the fundamentals of this ion-exchange system that was possible up to now. In this, he will have the advantage of drawing on the recently published results of the major study of McDougall.[9-11] No effort will be made to catalog all the papers published in this field, and the review will deliberately be selective.

II. PREPARATION AND GENERAL CHARACTERISTICS OF INSOLUBLE 12-HETEROPOLYACID SALTS

A. PREPARATION METHODS

The insoluble salts of the HPAs are generally characterized by poor cohesion in their crystals, resulting, in general, in the formation of very fine microcrystalline powders even within the most favorable preparation procedures, including precipitation from homogeneous solution. Preparation procedures, which are generally designed to give a material of acceptable particle size, may be classified into three approaches: (1) direct precipitation of the crystals by the mixing of solutions containing, respectively, the ammonium or other cation salt and the appropriate HPA, usually prepared by the classical solvent extraction methods of Wu,[18] or adaptations thereof, (2) precipitation from homogeneous solution, and (3) preparation of agglomerates of the desired overall particle size by a metathesis reaction of crystals of the parent HPA in an ammonium or other LMC salt solution.

Products obtained by mixing solutions of the HPA and salt of the desired cation generally have particle sizes varying from about 1 μm to colloidal dimensions. These particles tend to peptize further during use in columns (for example, together with a suitable carrier such as Gooch-quality asbestos or filter paper pulp) and are not recommended, although in a fair proportion of published separations the materials used were prepared by such direct-precipitation methods.

Precipitation from homogeneous solutions takes place where the anion of the HPA is slowly formed in aqueous solution in the presence of the desired cation, such as in classical analytical methods for the precipitation of orthophosphate as AMP.[19] These methods for AMP are generally slow, the product is frequently contaminated with molybdenum oxide, and since they require a large excess of molybdate reagent, the yield, based on the quantity of molybdate used, is relatively low. A better method of routine preparation of AMP which conserves molybdate is the citromolybdate procedure of Kassner et al.,[20] as modified by Smit et al.[21] Since AMP is generally the most useful HPA salt ion exchanger, the procedure is given here in full:

> Dissolve ammonium nitrate (81 g), citric acid monohydrate (81 g), and ammonium paramolybdate, $(NH_4)_6Mo_7O_{24} \cdot 4H_2O$ (102 g), in 2140 ml of water without heating. Pour this solution slowly, with stirring, into a dilute nitric acid solution, obtained by mixing 60% nitric acid (456 ml) and water (390 ml). To the mixture add 10 ml of a 5% solution of diammonium hydrogen phosphate, with stirring. *Heat to boiling and maintain at the boiling point for two min. Allow the solid to settle. Decant the mother liquor and collect the precipitate for further treatment. Allow the mother liquor to cool for at least 30 min, slowly add, with stirring, 50 ml of a 16% ammonium paramolybdate solution, 2.6 ml of

60% nitric acid, and 10 ml of 5% diammonium hydrogen phosphate solution. Repeat the cycle several times from the asterisk, to produce the desired quantity of AMP. The yield per cycle is about 8 g. Collect the different crops of AMP in 1 *M* ammonium nitrate solution, mix to ensure homogeneity, and filter with suction. Wash repeatedly with 1 *M* ammonium nitrate solution. Suck well, dry, and expose the precipitate to the air until a fine, dry yellow powder is obtained. Heat in an oven at 280°C for 4 h to volatilize the residual ammonium nitrate. The product may now be exposed to the air or in a hydrostat over saturated ammonium nitrate to constant weight, and thereafter stored in an air-tight container, preferably in the dark.

The composition of the AMP thus prepared corresponds closely to the formula $(NH_4)_3PMo_{12}O_{40} \cdot xH_2O$, compared to AMP precipitated from acid solution which has a variable part of the ammonium ions isomorphously replaced by hydronium ions.[22]

The third method of preparing insoluble HPA salts makes use of the fact that the parent HPAs are highly hydrated, containing up to 30 moles of water per mole of HPA, and readily crystallize from solution as large, well-developed crystals. When these crystals are immersed into a relatively concentrated solution (say $\gtrsim$ 1 *M*) of an ammonium salt (or salt of an LMC, as appropriate), the cations rapidly diffuse into the HPA lattice and convert the HPA into the ammonium or LMC salt by a metathesis reaction. The latter is anhydrous or contains much less water of crystallization. As a result, the metathesis reaction is accompanied by release of the excess water molecules. The product comprises an agglomerate of fine crystallites[23] (of particles size of the order $\lesssim$ 1 μm) held together rather weakly, with retention of the morphology of the original crystals.

Particles prepared in this way may be graded by decantation in a salt solution or by wet screening, preferably using nylon mesh screens. The use of metal screens leads to a small degree of reduction of the HPA salt. AMP, for example, tends to become green. Reoxidation with a solution of the LMC salt containing some bromine will restore the original color.

The agglomerate particles peptize very readily if brought into contact with pure water or an electrolyte solution of concentration ;$\lesssim$ 0.1 *M*. If allowed to dry, the particles collapse into a very fine powder. HPA salts prepared in this way should therefore be stored in an electrolyte solution, preferably of concentration at least 0.2 *M*, preferably of a salt of the parent LMC.

It is also possible to precipitate insoluble HPA salts within the pores of silica gel, by imbibing the solution obtained when dissolving a sample of the salt in ammonia or alkali solution, into oven-dried silica gel, and thereafter immersing the imbibed silica gel particles in an acid solution of an appropriate LMC salt. The crystallites of HPA salt thus obtained by *in situ* precipitation in the silica gel pores are, however, extremely fine, and tend to further peptise and get washed out when these granules are subsequently used in column separations, even if the eluting solutions are pre-saturated with the insoluble HPA salt.

B. GENERAL PROPERTIES

1. Chemical Composition and Insolubility

Insoluble ammonium salts of the HPAs prepared by any one of the methods discussed in the previous section are obtained in the first instance as nonstoichiometric acid salts, in which part of the ammonium ions are replaced by hydronium ions and a further part by other ions, such as Na^+ and K^+, if present in the solution during the preparation. For AMP, in the absence of other LMCs, about 2.3 moles of NH_4^+ ions per mole of AMP are required to ensure insolubility. The remaining 0.7 moles of NH_4^+ ions may be reversibly exchanged for other ions, e.g., H_3O^+, Li^+, Na^+, Ca^{2+}, etc., which do not form insoluble salts with HPAs. To avoid variability of the starting material in, for example, column separations, it is desirable to convert the HPA salt into the tri-LMC form. In the case of ammonium salts, this is best done by washing the salt with ammonium nitrate solution[24] and volatilizing the residual ammonium nitrate by heating to about 250°C.

While nominally insoluble, the HPA salt exchangers have, of course, finite solubilities, the solubility of the salt analogs of a particular HPA invariably decreasing in the sequence $K^+ > NH_4^+ > Rb^+ > Cs^+ \sim Tl^+$. Thus, Thistlethwaite[22] reports the following solubilities (in g/100 g of water) for the tri-LMC 12-molybdophosphate (MP) salts (presumably at 25°C): KMP: 1.2, AMP: 0.3, RbMP: 0.07, and CsMP: 0.005. Healy and Ingham[25] report the following solubilities at 20°C for the tri-LMC 12-tungstophosphate (WP) salts (converted here to g/100 ml of water): KWP: 0.15, AWP: 0.015, RbWP: 0.070, and CsWP: 0.080. Interestingly, they report[25]; higher solubilities for the trirubidium and -cesium salts than for the corresponding acid di-cation salts Rb_2HPW and Cs_2HPW, indicating more favorable packing in the lattice of the acid salts than in the tri-cation salts. The same does not, however, apply in the case of the salts of ammonium and potassium.

2. "Water of Crystallization"

The fact that the insoluble salts of the HPAs, almost invariably precipitated from acid medium, have some of their LMCs replaced by hydrogen ions (or more likely, hydronium ions) was reflected in older textbooks[19] by the composition for AMP being given as $(NH_4)_3PMo_{12}O_{40} \cdot 2HNO_3 \cdot H_2O$. As indicated above, the hydronium ions may be completely displaced by ion exchange, e.g., by washing the crystals with a salt of the relevant LMC.

Air-dried LMC salts of the HPAs loose several molecules of water upon oven drying. The number of moles of water per mole of AMP (for example) thus lost depend upon the size and degree of perfection of the crystals, and thus upon the method of preparation.[26] For AMP, the water lost upon heating (or sorbed by heat-dried AMP or AMP dried over phosphorus pentoxide[27]) may vary between 4 and 8 (or even more[9]) moles of water per mole of AMP. This water does not, however, represent true water of hydration or "structural" water taken up in the AMP interstices. Instead, they are taken up within the relatively large channels between crystallites found in all AMP preparations, thus corresponding to inter-crystalline and intragranular adsorption.[27]

In fact, the AMP lattice is entirely anhydrous, as are probably the majority of phases found in the LMC salts of the HPAs. Probably the only unambiguous way to identify water in the LMC salts of HPAs as being accommodated within the lattice is by single-crystal X-ray crystallography (see Section III).

III. STRUCTURAL STUDIES ON AMMONIUM 12-MOLYBDOPHOSPHATE AND OTHER INSOLUBLE HETEROPOLYACID SALTS

A. EARLIER DATA BASED ON X-RAY POWDER PATTERNS

The structures of the 12-HPAs[28,29] and their salts[30-32] were established in the early 1930s by X-ray powder diffraction techniques. Keggin[28,30] studied and elaborated the structure of the pentahydrate of 12-tungstophosphoric acid (WPA), obtained by partial dehydration of crystals of the 29-hydrate. It was subsequently established that this so-called "pentahydrate" or "Keggin" structure was isomorphous with those of the corresponding 12-MPA and LMC salts of both of these acids. Heteropolyacid chemistry abounds with other examples of 12-heteropoly salts with the Keggin structure, but the LMC salts of these compounds are generally more soluble than AMP or AWP and/or offer no advantages over these two members as ion exchangers. The discussion below will therefore be confined to the ammonium and other LMC salts of MPA and WPA.

The anion in a Keggin structure consists of a large polyhedral closed basket formed by 12 MO_6 octahedra (where M = Mo or W) sharing corners with the tetrahedrally coordinated PO_4^{3-} hetero group in the center, the PO_4^{3-} group sharing each of its oxygens with three MO_6 octahedra. The Keggin anion (Figure 1) therefore comprises four sets of three MO_6

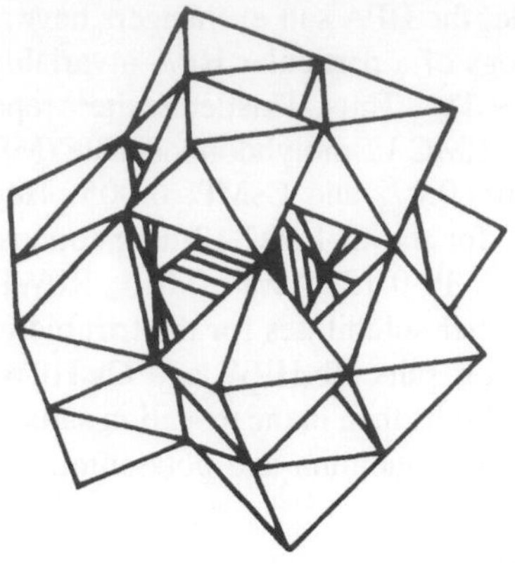

FIGURE 1. Model of the Keggin 12-heteropoly anion. The octahedra and shaded tetrahedron represent MoO_6 and PO_4, respectively.

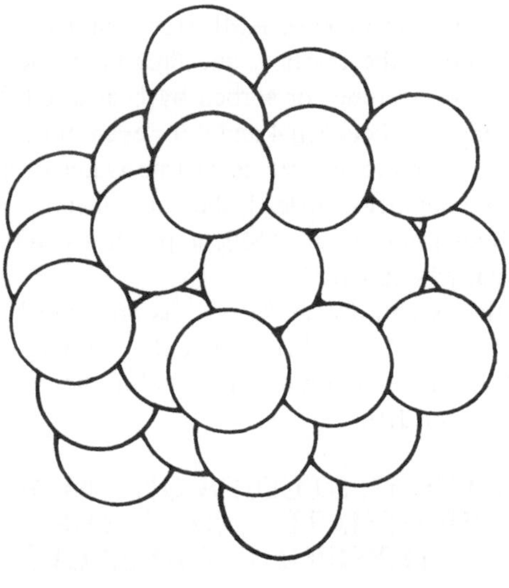

FIGURE 2. A space-filling model of $PMo_{12}O_{40}^{3-}$, a typical Keggin anion, showing oxygen atoms only.

octahedra, with edge sharing within each set and corner sharing between them. The highly symmetrical cubo-octahedral anions (Figure 2) are packed in the salts in a pseudo body-centered cubic lattice (Figure 3), of space group Pn3m. There are two formula units per unit cell.

According to Keggin,[28] any crystalline structure formed by the packing together of PW anions would be expected to contain a relatively large amount of water of crystallization, fitted in a relatively large space between the large, roughly spherical anionic units. Thus, until fairly recently, although the general structural type and shape of 12-heteropolyanions were known in fair detail from powder work, the distribution of the total water content between water of crystallization (if any) and capillary condensed water had been uncertain.

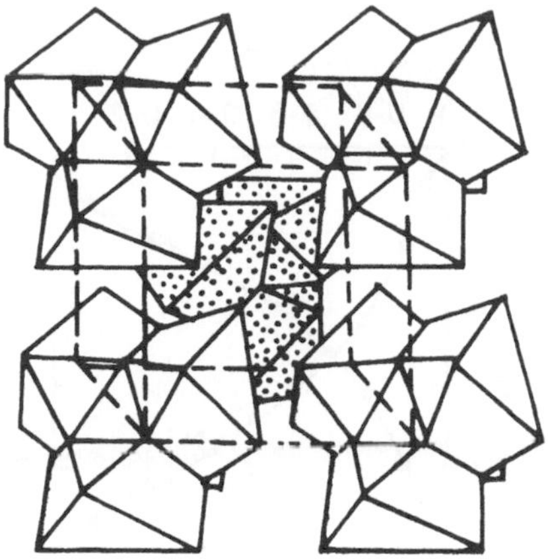

FIGURE 3. The pseudo body-centered arrangement of anions in the Keggin structure (space group Pn3m, on the left), and the true body-centered arrangement (space group 143m, on the right). The structure on the left is obtained from that on the right by rotation through 90° of the anions of one layer relative to those in the preceding layer. (From *Proc. R. Soc.*, A144, 75, 1934. With permission.)

B. SINGLE-CRYSTAL X-RAY CRYSTALLOGRAPHY OF AMP AND OTHER 12-MOLYBDOPHOSPHATE SALTS

A more complete understanding of the crystal structure of AMP and appreciation of the role of its crystallography in determining its ion-exchange properties emerged from the single-crystal X-ray crystallographic studies of McDougall.[9,10] This work was undertaken in an effort to explain the incomplete exchangeability (see Section IV) of the three crystallographically equivalent (according to the Keggin model) NH_4^+ ions referred to in Section I.

1. "Precipitation Isotherms" for $M^+/NH_4^+/MP$ Systems

Earlier, McDougall had shown,[9-11] from "precipitation isotherms" (Figure 4), that the achievement of a postulated minimum lattice energy state at maximum exchange of M^+ ions with the NH_4^+ ions on AMP could not explain the limited exchangeability. Precipitation isotherms obtained by adding MPA solutions to solutions containing variable mixtures of ammonium nitrate and MNO_3 (where M = Cs, Rb, K, Ag, or Tl) for the Rb^+/NH_4^+, Cs^+/NH_4^+, and Tl^+/NH_4^+ systems resembled ion-exchange isotherms that would have been expected had all the ammonium ions in the AMP been exchangeable. Moreover, they simply reflected the lower solubility of RbMP, CsMP, and TlMP over that of AMP, and the smoothness of the curves showed that the mixed crystals had nonstoichiometric compositions determined solely by the concentration ratios of M^+ and NH_4^+ in the precipitating solutions.

The precipitation isotherms for the Ag^+/NH_4^+ and K^+/NH_4^+ systems (which reflect the selectivity of the MPA anion for M^+ over NH_4^+ as being closer to 1) are anomalous. The former is made up of two S-shaped curves, the first part (at low relative NH_4^+ concentration) being for Ag^+-rich crystals having the non-Keggin structure of AgMP, with the second part reverting to the NH_4^+-rich Keggin phase of AMP. (For the relevance of these observations, see Section IV.)

The sigmoidal shape of the precipitation isotherm for the K^+/NH_4^+ system (Figure 4) markedly resembles some ion-exchange isotherms for zeolites, in which the S-shape is

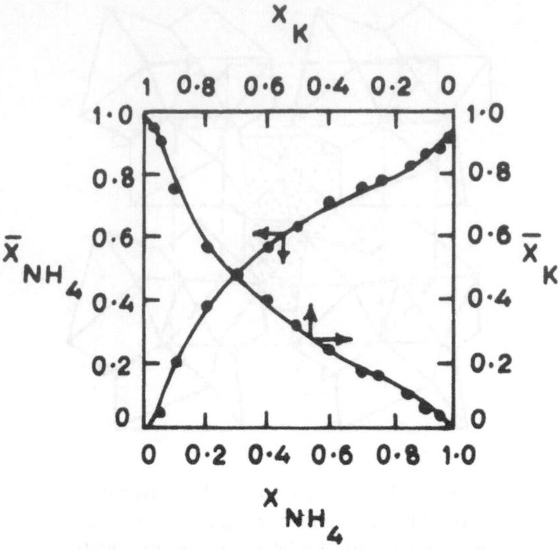

FIGURE 4. Precipitation isotherm for the K^+/NH_4^+ 12-molybdophosphate system. X is the ionic fraction in the aqueous phase and X is that in the solid phase. Initially, the total nitrate concentration was 1 *M*, and precipitation was brought about by adding an MPA solution dropwise to a large excess of the nitrate solution. (From *J. Solid State Chem.*, 18, 191, 1976. With permission.)

ascribed to selectivity reversals occurring as X_M traverses the range 0 to 1. Of the three possible causes given[33,34] for such selectivity reversals, the only one which could have a bearing on the shape of the precipitation isotherms for these two systems is a discontinuous rearrangement of the lattice, i.e., a phase transition, accompanying the ion-exchange reaction, i.e., traversing the range of $X_{NH4}{}^+$ from 0 to 1 involves, as for the $Ag^+/NH_4{}^+$ system, a change from a KMP phase to a different AMP phase.[9,10]

The above tentative conclusion has been confirmed by single-crystal X-ray crystallography.[9-11]

2. The McDougall Structure, a Refinement of the Keggin Structure for 12-Heteropoly Compounds

McDougall[9,10] for the first time carried out a single-crystal X-ray crystallographic study of KMP and AMP. Morphologically featureless crystals were obtained by allowing solutions of either ammonium nitrate or potassium nitrate and MPA to diffuse together over 2 d into a gelatin plug separating the solutions in a U-tube. Since the crystals were grown in acid medium, part of the potassium and ammonium ions were replaced by hydronium ions.

The results may be summarized as follows:

The crystal structure was shown to be cubic, of space group Pn3m, with two formula units per unit cell, thus confirming the powder work of Illingworth and Keggin.[30] The gross features of the Keggin structure[28,30] (see Section III.A) were confirmed as more or less correct. Close packing of the anionic units leaves sufficient space for the cations in orthogonally intersecting channels, but not for any water of crystallization, as suggested for Keggin's "pentahydrate" structure,[28] i.e., for $(H_3O)_3PW_{12}O_{40} \cdot 2H_2O$.

The four different types of oxygen in the anion are identified in Figure 5. The atoms designated O(1) and O(3) are shared between adjacent MoO_6 octahedra and act as bridges, with an average Mo-O bond length of 1.92 Å, a figure typical for Mo-O-Mo bridge bonds.[35] The average length of the bond between the molybdenum atoms and nonbridging peripheral O(2) atoms is 1.69 Å, compared to a typical double-bond length of 1.70 Å (Reference 35).

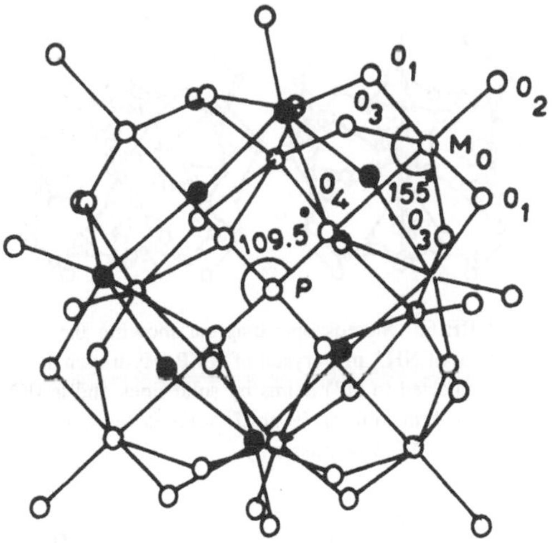

FIGURE 5. The 12-molybdophosphate anionic unit, showing the atomic numbering scheme. The P atom is at a site of symmetry $\overline{4}3m$.

The average Mo-O(4) distance of 2.45 Å (i.e., between a Mo atom and an oxygen atom attached to the phosphorus hetero atom) is considerably longer than any known true Mo-O bond length. The authors[9,10] therefore conclude that the MP anion is really a clathrated phosphate ion, entrapped in a poly molybdenum-oxygen shell, as originally proposed by Pauling.[36] According to the rules of nomenclature for clathrates,[37] AMP should therefore be named more appropriately as ammonium phosphate-12-polymolybdate(1/1), of formula $(NH_4)_3PO_4(MoO_3)_{12}$.

The cations in AMP and KMP (as well as in the exchanged salts, see next section) occupy only the body-centered 42m sixfold set, and are thus crystallographically equivalent, again in agreement with Keggin.[28]

3. Crystallographic Study of Ion Exchange with AMP and KMP

The reasons for the incompleteness of exchange of the ammonium ions in AMP, despite their crystallographic equivalence, emerged from studies by McDougall[9,10] of single crystals subjected to ion exchange. Diffraction data were collected on the parent single crystals before and after dipping the crystals, mounted on goniometer heads, for 20 min into 3 *M* solutions of either ammonium or potassium nitrate.

The following are details of the crystals examined, the cationic composition having been obtained from the diffraction data by occupancy refinement (The values in parentheses are the unit cell lengths in Å):

1. $K_{2.0}(H_3O)_{1.0}$MP: directly as obtained (11.62)
2. $K_{2.7}(H_3O)_{0.3}$MP: by exchanging 1 in KNO_3 solution (11.60)
3. $K_{0.9}(NH_4)_{2.1}$MP: by exchanging 1 in NH_4NO_3 solution (11.62)
4. $(NH_4)_{2.6}(H_3O)_{0.4}$MP: directly as obtained (11.70)
5. $(NH_4)_{1.8}K_{1.2}$MP: by exchanging 4 in KNO_3 solution (11.65)

The first feature to be noted from the above compositions is that, while only about 40% exchange occurred when dipping AMP into potassium nitrate solution (compared to a maximum exchange of about 53% in column experiments — see Section IV), the reverse process of NH_4^+ exchange on KMP had proceeded to 70% of completion. This indicates that while the K^+ ions in the KMP single crystal are exchangeable to a greater degree than in the

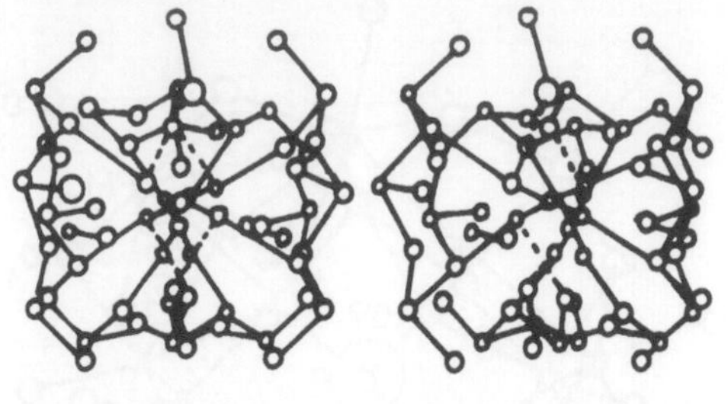

FIGURE 6. Stereoscopic diagram showing the environment of NH_4^+ in a crystal of AMP. Hydrogen atoms are connected to 0(3) atoms by solid lines and to 0(2) atoms by broken lines. (From *J. Solid State Chem.*, 18, 191, 1976. With permission.)

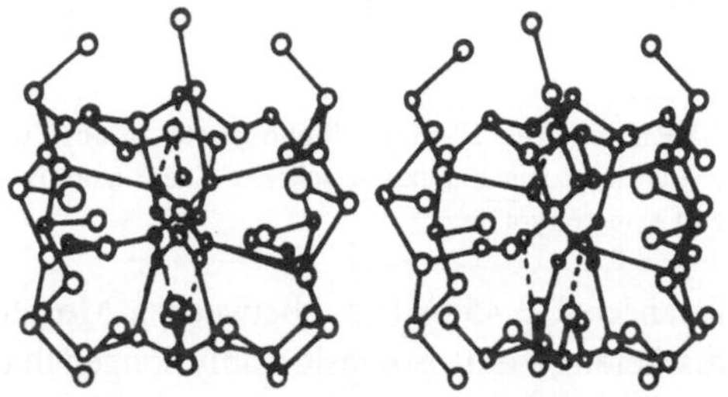

FIGURE 7. Stereoscopic diagram showing the environment of NH_4^+ in a crystal of AMP after subjection to K^+ ion exchange. Hydrogen atoms are connected to 0(1) atoms by solid lines and to 0(2) atoms by broken lines. (From *J. Solid State Chem.*, 18, 191, 1976. With permission.)

citromolybdate AMP, the exchange will not proceed to completion. More importantly, the exchange reactions AMP + $K^+ \rightarrow K^+$(AMP) and KMP + $NH_4^+ \rightarrow$ (NH_4^+)KMP lead not only to compositionally different, but also to structurally different partially exchanged products.

The exchange reaction AMP + $K^+ \rightarrow$ is, in fact, accompanied by a phase transition. In the AMP starting material, the cation-O(1) and cation-O(3) interatomic distances are 3.30 and 3.25 Å, respectively, but in the partially exchanged (K)AMP, these have interchanged to 3.26 and 3.31 Å, respectively. This interchange is accompanied by interchanges of the separations between neighboring O(1)-O(1) and O(3)-O(3) atoms, from 2.60 and 2.68 Å, respectively, in AMP to 2.64 and 2.55 Å, respectively, in (K)AMP. However, the O(1)-O(3) distance remains unaffected.

These changes are caused by a reorientation of the ammonium ion and its interaction with the cagework oxygen atoms during the exchange. The tetrahedral ammonium ion cannot be hydrogen bonded to the first-shell O(2) nearest neighbors in the cage, since these occur in square-planar arrangement around the cationic site. Instead, hydrogen bonding is to the second-nearest shell containing O(1) and O(3) sets, which are both tetrahedrally disposed with respect to the cation. In AMP, the O(3) set is significantly closer to the nitrogen atom than the O(1) set, and in (K)AMP, the order is reversed. McDougall[9,10] concludes that the observed N–O distance of 3.25 Å probably represents a bifurcated type of hydrogen bond (similar to the N-H–O bonds of 3.33 Å observed by neutron diffraction[38] in ammonium sulfate) and, depending on the composition, is directed preferentially toward either the O(1) or O(3) atoms. The "before" and "after" positions regarding the hydrogen-bonded systems are shown stereoscopically in Figures 6 and 7. In both diagrams, the N-H bonds are seen not to be directed straight toward the oxygen atoms, but, rather, toward the region between either the O(3) or O(1) and a neighboring O(2) atom, giving hydrogen bonds of the bifurcated type. In AMP, the NH_4^+ ion is thus H bonded to both O(2) and O(3) atoms, while, in (K)AMP, the bifurcated bonds involve O(2) and O(1) atoms.

This reorientation of the NH_4^+ ions occurs progressively as ion exchange for K^+ ions proceeds, with the exchange being arrested when only reoriented NH_4^+ ions remain in the crystal. Rupture of the H bonds during the exchange of NH_4^+ for K^+ produces an environment in an excited state, and relaxation occurs through the strengthening (and contraction) of bonds in the anionic unit. Thus, in going from AMP to (K)AMP, the Mo-O(2), Mo-O(3), and P-O(4) bonds are shortened from 1.69 to 1.62, from 1.93 to 1.91, and from 1.54 to 1.51 Å, respectively. The P-O(3) separation also decreases from 3.98 to 3.94 Å, which is expectedly accompanied by a weakening of the Mo-O(4) interaction from 2.46 to 2.48 Å.

The net effect of these changes taking place as the exchange of NH_4^+ for K^+ proceeds is a progressive elimination of H bonds, leading to a reorganization of the bond pattern in adjacent anions. "With increased restructuring of the anion, the orientation of the remaining NH_4^+ ions becomes less favorable for bifurcated H bonding to atoms O(3) and O(2), and they change their orientation to form new H bonds which involve the more favorably disposed atoms O(1) and O(2). At about 50% exchange, the situation is reached where all remaining NH_4^+ ions have reoriented to form new H bonds. Up to this point it has been possible to compensate for the rupture of H bonds by a general increase in bond strength within the anionic unit. Beyond this point, no further compensation for H-bond rupture in the new phase is possible. Therefore, the exchange of NH_4^+ for K^+ ions becomes energetically unfavorable, and ceases."[10]

In the exchange of K^+ for NH_4^+ ions on KMP, the exchanged structure, (NH_4)KMP, will apparently eventually approach the structure of AMP, but McDougall[9,10] observed no evidence of preferential direction of the H bonds toward any particular set of cagework oxygen atoms.

"The NH_4^+ ions appear to be disordered over two possible orientations, although an examination of the bond lengths shows that the AMP structure is favored. The more extensive exchange observed in this case is not surprising, since there is no energy barrier such as the rupture of H bonds to be overcome.

"Thus, while exchange of an NH_4^+ ion for a K^+ ion in AMP involves the rupture of H bonds of probably all or most of the NH_4^+ ions along the channel, and the consequent relaxation of the lattice (accompanied by a distortion of the molybdophosphate cages) leading to a lower energy state, the limited convertibility of KMP, when NH_4^+ is the incoming ion, probably occurs because of kinetic rather than thermodynamic reasons." [This is rationalized as follows:[10]]

"The incomplete conversion [of KMP] could be due to the random orientation of the incoming NH_4^+ ions. Figs. 6 and 7 show that neighboring NH_4^+ ions are oriented at right angles to one another. Unless there is a driving force favoring a certain orientation at a specific site, random exchange cannot be expected to result in any one regular arrangement for the entire crystal. Regions of mismatch between differently oriented ordered domains are more likely to occur, and unless annealing takes place, this will inhibit further exchange."

4. Lattice Disorder Caused by Ion Exchange

McDougall's single-crystal X-ray crystallography study[9,10] of the K/NH_4MP ion exchange system produced a further interesting and unexpected result: some excess electron density was observed in the difference maps, during refinement of the structures, beyond the positions of the four oxygen atoms associated with the phosphorus hetero atom, as well as around the phosphorus atom itself, but only for those crystals which had been subjected to ion exchange. This, together with the abnormally low temperature factors of the phosphorus atom in each of these cases, lead the authors to the conclusion that during ion exchange, another tetrahedral species had displaced part of the phosphate groups.

They concluded that part of the central PO_4^{3-} had been displaced by MoO_4^{2-}, to the extent of about 10%. This observation corresponds to similar effects reported by Fischer et al.,[39] who, for another molybdophosphate species, found some of the phosphate to occur in positions traditionally occupied by molybdate groups.

The implications of these observations for the ion-exchange properties of AMP and other LMC salts of MPAs are

1. AMP, for example, does not retain its structural integrity in the course of ion exchange, structural disorder in the form of a partial interchange of the phosphate hetero group, and some orthomolybdate accompanying the exchange.
2. While it would be thought that cation exchange takes place only through the channels in which the cations are situated, it now becomes evident that at least part of the exchange takes place by one of the exchanging cations moving right through the anion, probably entering the anion through one of its flat surfaces.
3. In so doing, the cation moving through the cage is probably responsible for the observed lattice disorder, by colliding with the PO_4 hetero-atom group and/or a MoO_6 cage group, rupturing bonds on a significant scale. In at least some of these instances, the bond-ruptured anion does not reform to its original state.

In fact, upon reflection, paying due regard to the relatively small dimensions of the channels in terms of atomic dimensions, it would appear doubtful if any two exchanging ions would be capable of passing one another in a channel without causing at least instantaneous deformation of one or more of the neighboring anions. It seems reasonable to surmise that, as an incoming ion penetrates into the interior of a crystal, for the various "jumps" it progresses from one cationic site to the next, a significant proportion involves the passing through the cage of one of the exchanging ions.

In the U-tube gel method for growing the starting crystals of McDougall's crystallographic studies,[9,10] the salts were formed in the absence of an excess of cationic salts. In at least some of the preparation methods for LMC salts of HPAs (e.g., the citromolybdate method), the crystals formed would inevitably be subjected to exchange reactions between the LMCs in the crystals and excess LMC cations present in solution throughout the preparation procedure. These ion-exchanged products would therefore be expected to be at hand already in the lattice-disordered state. The fact that different methods of preparation of the LMC HPA salt yield substances with not greatly differing ion-exchange properties (e.g., selectivities) suggests that lattice disorder of the kind discussed here may possibly have only a modest effect on ion-exchange properties. This is a consideration that may merit closer examination.

Tailing in elution peaks in column chromatography (Figure 8), particularly when involving the larger LMCs, almost certainly results from lattice disorder complications.

5. Basis for Selectivity in Ion Exchange on HPA Salts

Eisenman[40] (discussed at length by Sherry[33]) has shown that the selectivity sequence exhibited by an ion exchanger for ions in a particular group (e.g., the alkali metal ions), and the degree to which one ion is preferred over another (e.g., one alkali metal ion over that occurring above it in the Periodic Table), depends on the nature of the exchanger, i.e., whether it is a "weak-field" or a "strong-field" exchanger, and whether it is anhydrous or not. AMP has been shown by McDougall[9,10] to be an anhydrous exchanger, and although homologous exchangers such as AWP have not been proven to be anhydrous as well, they are at least nearly so.

With the small charge of -3 probably fairly uniformly spread over quite a large anion, the AMP lattice is not only a weak-field exchanger, but an extreme case of this type of exchanger, with a substantially weaker anionic field strength than any of the zeolite ex-

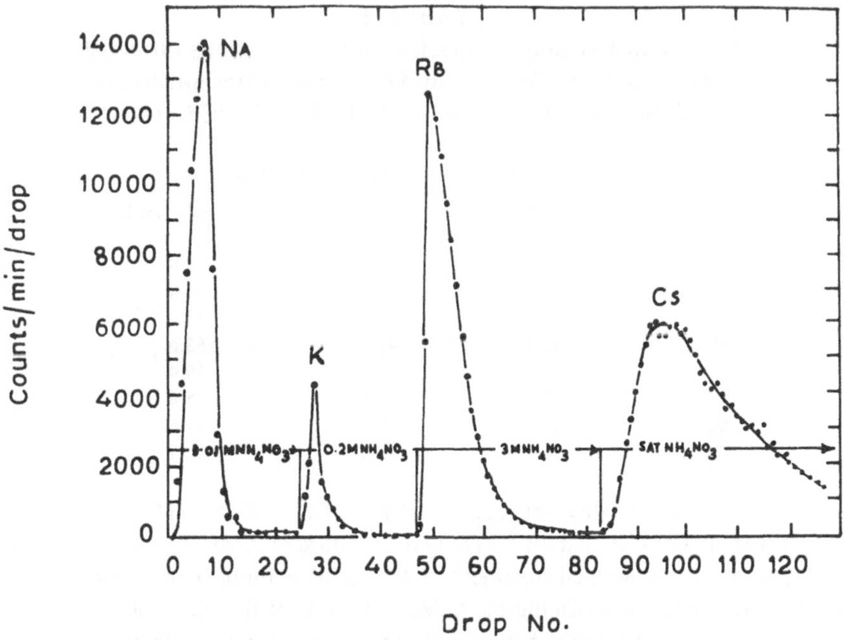

FIGURE 8. Separation of tracer quantities of the alkali metal ions on AMP by sequential elution with ammonium nitrate solutions of increasing concentration. Column dimensions: 1.6 mm × 5 mm in diameter.

changers known.[33,40] The selectivity sequence for the alkali metal ions exhibited by AMP and other HPA salts of the Keggin structure, namely

$$Cs^+ > Rb^+ > K^+ > Na^+ > Li^+$$

is therefore the same as that for the series I sequence (out of 11 observed sequences enumerated for various zeolite types) in the Eisenman classification.[40]

IV. ION-EXCHANGE PROPERTIES

A. EXTENT OF EXCHANGE AND SELECTIVITY SEQUENCE OF AMP WITH MONOVALENT CATIONS

A feature of the ion-exchange properties of the ammonium salts of the HPAs is that only part of the stoichiometric ammonium content is exchangeable. Earlier reports on the limited exchangeability of the NH_4^+ ions on AMP have been either of doubtful accuracy[2,3,41-47] or the authors' work was colored by their belief that exchange leads to a stoichiometric end product.[2-4,41,44] Table 1 reports values for the maximum percent of exchanges observed[9,11] with the various LMCs, as well as with H_3O^+, Li^+, and Na^+, when a large excess of a solution of the salt (or HNO_3) was passed through columns of AMP/asbestos. Coetzee's percentages exchanged[45] for potassium and rubidium, obtained under similar conditions, are interchanged, compared to those reported by McDougall,[9,11] for which the present author can vouch. (While the actual percentages of exchangeable NH_4^+ for a particular cation may possibly vary slightly with the method of preparation of the AMP, the chances of the sequence of exchangeabilities for the various cations varying between AMP samples obtained by different methods of preparation must be very slight indeed.)

The maximum exchange capacities realized (expressed as a percentage of the NH_4^+

TABLE 1
Maximum Exchange Capacities for M^+ Cations on AMP Obtained from Column Studies at Two Concentrations by McDougall,[9] Compared with the Results of Coetzee[43]

	Maximum exchange capacities (%)		
Cation	0.10 *M* MNO_3	0.25 *M* MNO_3	Coetzee's results
Li^+	19.2	19.4	20.9
Na^+	22.7	23.0	26.4
K^+	53.1	53.3	48.7
Rb^+	47.1	47.3	53.6
Cs^+	45.7	45.8	45.4
Tl^+	45.5	45.6	45.8
Ag^+	>80	>80	64.8
H_3O^+	21.7	21.9	30.7

replaced) vary according to the sequence $Ag^+>K^+>Rb^+>Cs^+\sim Tl^+>Na^+>H_3O^+>Li^+$. Silver is again anomalous, in that the McDougall structure of AMP is progressively converted into an AgMP phase. Prolonged treatment of the AMP column with silver nitrate solution therefore seems capable of replacing the bulk, if not all, of the NH_4^+ ions with Ag^+ ions.

There is no simple correlation between the maximum degree of exchange and either the ionic radius or any other thermodynamic parameter (such as hydration energy) of the incoming ion.

McDougall[9,46] determined ion-exchange isotherms for the exchange of the ions listed above with AMP at 25, 45, and 61°C (Figure 9). Isotherms for a given ion at the three temperatures are almost identical, except for the K^+/NH_4^+ and Ag^+/NH_4^+ systems (Figure 10). The smoothness and monotonical progress of the isotherms confirms that mixed crystals of nonstoichiometric composition are formed during the exchange.

From the isotherms, McDougall determined the rational selectivity coefficients, ${}^cK^M_{NH4}$ (following Helfferich[47]), defined for the ion exchange reaction

$$M^+ + \overline{NH_4^+} \rightleftharpoons \overline{M^+} + NH_4^+ \qquad (1)$$

(where the bar indicates the solid phase) as

$${}^cK^M_{NH4} = \frac{[\overline{M}]}{[\overline{NH_4}]} \cdot \frac{[NH_4]}{[M]} \cdot \frac{\gamma^2_{\pm\ NH4NO3}}{\gamma^2_{\pm\ MNO3}}$$

in which the ionic charges are left out for simplicity. The terms γ_{NH4NO3} and γ_{MNO3} are the mean molal activity coefficients. The values for ${}^cK^M_{NH4}$ are given in Table 2, and show the selectivity sequence to be $Cs^+\sim Tl^+>Rb^+(>NH_4^+)>K^+\sim Ag^+>Na^+\sim H_3O^+\sim Li^+$.

B. THERMODYNAMICS OF ION EXCHANGE OF MONOVALENT CATIONS ON AMP

The data of Table 2 have been used by McDougall[9,46] to calculate standard free energies, enthalpies, and entropies for (the ion-exchange) Reaction 1 (Table 3), using the methods of Sherry.[48] The ΔG° values are all relatively small. For unfavorable reactions, viz., the adsorption of Li^+, Na^+, H_3O^+, Ag^+, and K^+, ΔG° values are positive quantities, as expected, whereas for the reactions involving Rb^+, Cs^+, and Tl^+, the negative ΔG° values reflect strong preference for these ions over NH_4^+.

The relatively large positive ΔH° values for the systems Li^+/NH_4^+, Na^+/NH_4^+, and H_3O^+/NH_4^+ probably reflect the energy required to remove the hydration shells associated

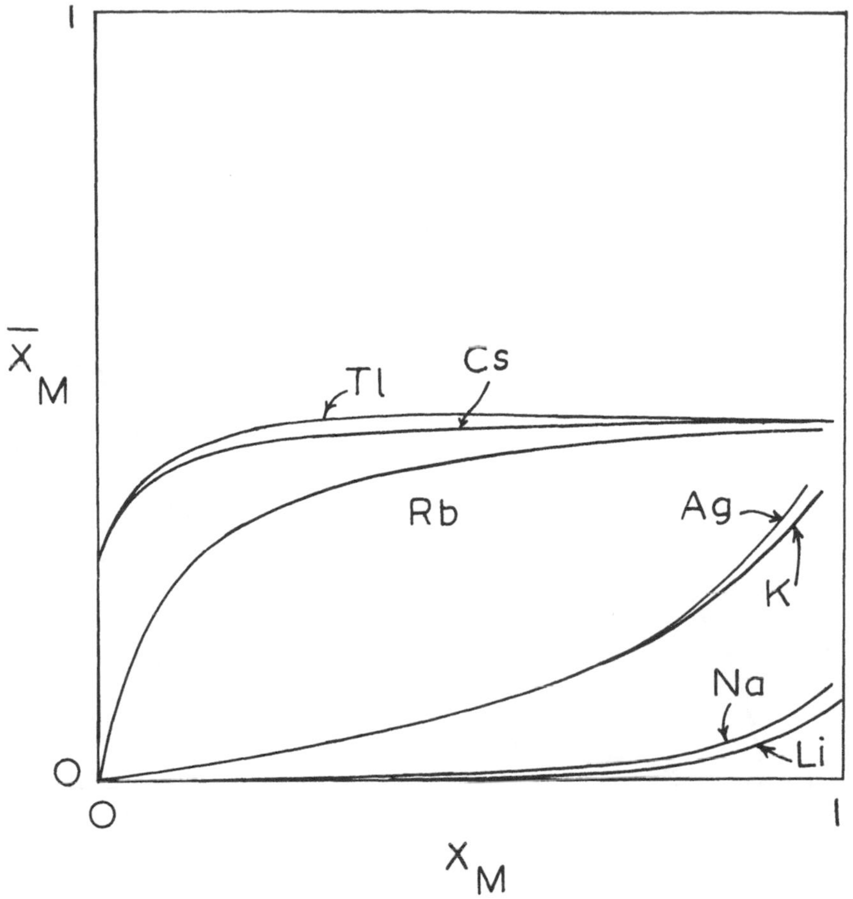

FIGURE 9. Ion-exchange isotherms for the exchange of various monovalent cations M^+ on AMP at 25°C. (From Williams, P. A. and Hudson, M. J., *Recent Developments in Ion Exchange*, Elsevier, London, 351. With permission.)

with these highly hydrated cations relative to the hydration energy of the NH_4^+ ion. For the other exchange reactions, ΔH° is small or zero, indicating[46] that electrostatic interaction between these cations and the lattice anions is weak.

The enthalpy term for the exchange involving Li^+, Na^+, and H_3O^+ predominates in determining the sign and magnitude of ΔG°. For the other exchange reactions, the selectivity depends almost entirely on the entropy term. Only for the exchange reactions involving Ag^+ and K^+ are the entropy terms negative, and for these ions exchange has been shown to be accompanied by a phase transition.

The entropy of a system may be considered[48] as consisting of two contributions — one for the aqueous phase and one for the exchanger phase. The entropy change for (the ion-exchange) Reaction 1 is thus given by

$$\Delta S^\circ = \Delta S^{hyd}_{M/NH4} + S^{ex}_{M/NH4} \quad (2)$$

and therefore

$$\Delta S^\circ = (S^{hyd}_{NH4} - S^{hyd}_{M}) + (S^{ex}_{M} - S^{ex}_{NH4}) \quad (3)$$

In Equation 3, S^{hyd} is the standard entropy of hydration, listed in the last column of Table 3, and S^{ex} is the standard entropy of the cations in the exchanger phase.

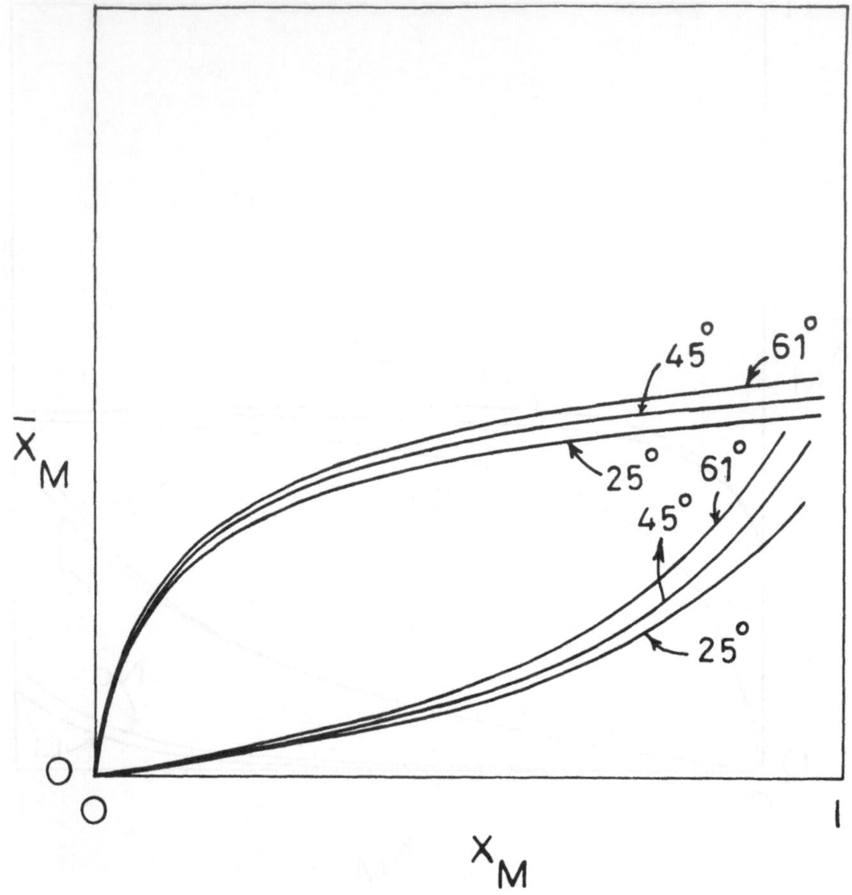

FIGURE 10. Ion-exchange isotherms for the exchange on AMP of Rb^+ (top set) and of K^+ (bottom set) at three different temperatures.

TABLE 2
Rational Selectivity Coefficients for the Cations M^+ at Different Temperatures

	Temperature			
Cation	**25°C**	**45°C**	**61°C**	**r+/Å[a]**
Li^+	0.072 ± 0.01	0.098 ± 0.01	0.134 ± 0.01	0.78
Na^+	0.077 ± 0.01	0.102 ± 0.01	0.143 ± 0.01	0.98
K^+	0.296 ± 0.04	0.270 ± 0.04	0.275 ± 0.04	1.33
Rb^+	10.5 ± 2.5	10.1 ± 2.5	9.4 ± 2.5	1.49
Cs^+	120 ± 30	120 ± 30	120 ± 30	1.65
Tl^+	110 ± 30	110 ± 30	110 ± 30	1.49
Ag^+	0.274 ± 0.05	0.206 ± 0.05	0.240 ± 0.05	1.13
H_3O^+	0.073 ± 0.01	0.098 ± 0.01	0.150 ± 0.01	—
NH_4^+	—	—	—	1.40

[a] Goldschmidt ionic radii according to Rosseinsky, D. R., *Chem. Rev.* 65, 467, 1965.

TABLE 3
Thermodynamic Quantities for the Exchange Reactions of Various Cations on AMP

Reaction	ΔG°_{av} J mol^{-1}	$\Delta G^{\circ}_{25^{\circ}}$ J mol^{-1}	$\Delta G^{\circ}_{45^{\circ}}$ J mol^{-1}	$\Delta G^{\circ}_{61^{\circ}}$ J mol^{-1}	ΔH° Jmol^{-1}	ΔS° eu	$S^{hyd}_{NH4} - S^{hyd}_{M}$	$S^{ex}_{M} - S^{ex}_{NH4}$ eu	S^{hyd}_{M} eu
AMP + Li^+ →		6,252 ± 400	6,150 ± 400	5,560 ± 400	14,800 ± 2,090	27.6	69.9	−42.2	−10.0
AMP + Na^+ →		6,360 ± 400	6,020 ± 400	5,400 ± 400	13,930 ± 2,090	25.5	38.5	−13.0	−24.5
+ K^+ →	3,140 ± 400			~0	−10.5	2.9	−13.4	21.3	
+ Rb^+ →	−5,690 − 630			~0	19.2	−9.2	28.4	69.0	
+ Cs^+ →	−11,880±1,250			~0	39.7	−12.1	51.9	71.9	
+ $T1^+$ →	−11,630±1,250				~0	38.9	−1.3	34.1	61.1
+ Ag^+ →	3,550 ± 400			~0	−12.1	44.3	−56.5	15.5	
+ H_30^+ →		6,480 ± 400	6,150 ± 400	5,270 ± 400	16,300 ± 2,090	33.0	59.8	−26.8	0

Note: Standard deviations have been omitted for the entropy values. Standard entropies of hydration, taken from Rosseinsky (see Table 2), are given in the last column. The value of S^{hyd}_{NH4} for ammonium (59.8 eu) was obtained by plotting ΔH^{hyd}_{M} values (according to Rosseinsky) for the monovalent cations against S^{hyd}_{M} for the same cations and interpolating $\Delta S^{hyd}_{NH_4}$ from the known value for $\Delta H^{hyd}_{NH_4}$.

The positive $\Delta S^{hyd}_{M/NH4}$ values for the reactions involving Li^+, Na^+, K^+, Ag^+, and H_3O^+ reflect the net freeing of water molecules in the aqueous phase as the above ions, during the exchange reaction, are replaced by the more weakly hydrated NH_4^+ ion. For Rb^+, Cs^+, and Tl^+, the $\Delta S^{hyd}_{M/NH4}$ values are negative, i.e., additional water molecules are bound in the aqueous phase when these ions are exchanged for the more highly hydrated NH_4^+ ion.

The $\Delta S^{ex}_{M/NH4}$ values are positive for Rb^+, Cs^+, and Tl^+, where the increase of entropy in the exchanger phase reflects the lattice distortion brought about by replacing NH_4^+ ions with the larger ions (see preceding section, as well as below). For the other ions, the negative $\Delta S^{ex}_{M/NH4}$ values result from a contraction of the lattice as NH_4^+ ions are displaced by smaller ions.

The values of $\Delta S^{ex}_{M/NH4}$ for the alkali ions (Table 3) increase with the increasing radius of the ingoing ion, as for the exchange reactions of the alkali ions on zirconium phosphate in the hydrogen form,[49] where this trend was interpreted as reflecting changes of hydration in the solid phase and differences in the degree of lattice distortion of the two forms of the exchanger. For AMP, only the latter cause can be applicable, as the lattice is anhydrous. Hence, $\Delta S^{ex}_{M/NH4}$ values for the alkali ions must arise entirely from lattice distortion effects.

1. Miscellaneous Experiments

McDougall,[9] apart from growing crystals of AMP and KMP by the gel plug technique, also attempted to prepare other single crystals by the same technique. All of these appeared to dehydrate rapidly in the atmosphere. Even after being sealed in Lindeman tubes, their quality was too poor for single-crystal work. Although the reflections were poor, it could be established that the compounds all had the McDougall/Keggin structure (except in the case of the silver salt).

Superimposed on the diffuse Keggin reflections for the RbMP crystal was a very sharp reciprocal lattice (in exact alignment with the Keggin reciprocal lattice), belonging to an hexagonal system. In view of the finding of Kraus,[50] that different compounds of the ferric tungstosilicate complexes intergrow in lamellar crystal edifices,[51] McDougall concluded that it would appear that the pure 12-molybdophosphates of the larger cations Rb^+, Cs^+, and Tl^+ possibly crystallized in a number of different hydrates, each hydrate belonging to a different crystal system.

The AgMP crystals formed large lumps of intergrown crystals which were difficult to separate. They also dehydrated rapidly. It was nevertheless confirmed, although the crystals were extensively twinned, that they belonged to a different, non-Keggin system.

It is interesting to note that AMP single crystals subjected to exchange with Cs^+ ions became fragmented,[9] probably because of the lattice expansion accompanying exchange of the NH_4^+ ions by the much larger Cs^+ ions. No obvious damage was, however, caused by exchange of AMP crystals with Rb^+ ions.

X-ray powder patterns of pure 12-molybdophosphates (obtained by adding solutions of MNO_3 to a solution of MPA) showed increasing diffuseness (and, hence, reducing crystallinity) in the sequence:

$$\text{AMP} \sim \text{KMP} > \text{RbMP} > \text{CsMP} \sim \text{TlMP}$$

McDougall[9] concludes from this behavior that, while the packing of NH_4^+ and K^+ ions in the McDougall/Keggin structure was close to ideal, the larger ions Cs^+ and Tl^+ were too large, resulting in the anions being pushed further apart to allow accommodation of these cations. This weakens ionic interactions in the lattice and promotes disorder (i.e., poor crystallinity).

McDougall[9] also conducted ion-exchange experiments at elevated temperatures with AMP and the ions listed in Table 4. Solutions containing an excess of the ingoing ion,

TABLE 4
Percentage of Ammonium Remaining in the AMP After Exchange in the Solid State with an Excess of Cation M

Exchange reaction	Percent residual ammonium			
	50°C	100°C	200°C	250°C
Li + AMP →	81	87	88	93
Na + AMP →	78	79	78	80
K + AMP →	36	27	3	2
Rb + AMP →	36	26	12	2
Cs + AMP →	58	43	18	10
Tl + AMP →	56	43	12	4
Ag + AMP →	47	81	15	a
H_3O^+ + AMP →	78	81	77	85

[a] AMP turned black.

together with a fixed amount of AMP, were placed in open crucibles, the contents mixed, and the crucible placed in an oven at a fixed temperature. After the water had evaporated, the mixture was kept in the oven for an additional 12 h. The precipitate was then collected on a Gooch crucible, the soluble salts washed out with absolute ethanol, and the ammonium content of the residue determined.

Table 4 records the results of the experiments. The ions involved in these solid-state reactions consist of two groups: in the case of the LMCs, increasing temperatures lead to an increasing degree of displacement of the NH_4^+ ions on AMP. The ions Li^+, Na^+, and H_3O^+ showed the opposite trend, or (in the case of Na^+) showed little evidence of temperature dependence. The results are entirely consistent with the expected effect of the entropy term on the various reactions with rising temperature.

For example, for the exchange reaction AMP + $Li^+ \rightarrow$, $\Delta S^{ex}_{M/NH4}$ (Equation 3) is a negative value, and since $\Delta S^{hyd}_{M\cdot iNH4}$ is zero, the reaction having taken place in the absence of water, a rise in temperature would, by the relationship

$$\Delta G = \Delta H - T\Delta S$$

make ΔG more positive.

The fact that ΔG for the reaction involving K^+ ions seems to decrease despite a small, positive $T\Delta S$ contribution with rising temperature is explained by McDougall[9] as being due to an overriding effect of the rising temperature on the breaking of H bonds which, at lower temperatures, provides an effective energy barrier to the exchange reaction going to completion.

C. KINETICS OF ION EXCHANGE ON AMP

Choudhuri and Mukherjee[52] studied the kinetics of ostensible ion-exchange adsorption of ammonium ions on to the pyridinium, picolinium, lutidinium, and collidinium salts of 12-tungstophosphoric acid. Equilibrium had not been established after 32 h. What these authors were observing was largely slow recrystallization rather than a true ion-exchange reaction.

Owing to the small dimensions of the channels in the lattice of HPA salts with the McDougall/Keggin structure, ion-exchange reactions involving these salts might have been expected to proceed rather slowly. The generally sharp breakthrough curves with AMP-asbestos columns[53] (Figure 11) and columns of AMP prepared by metathesis from MPA[23,54]

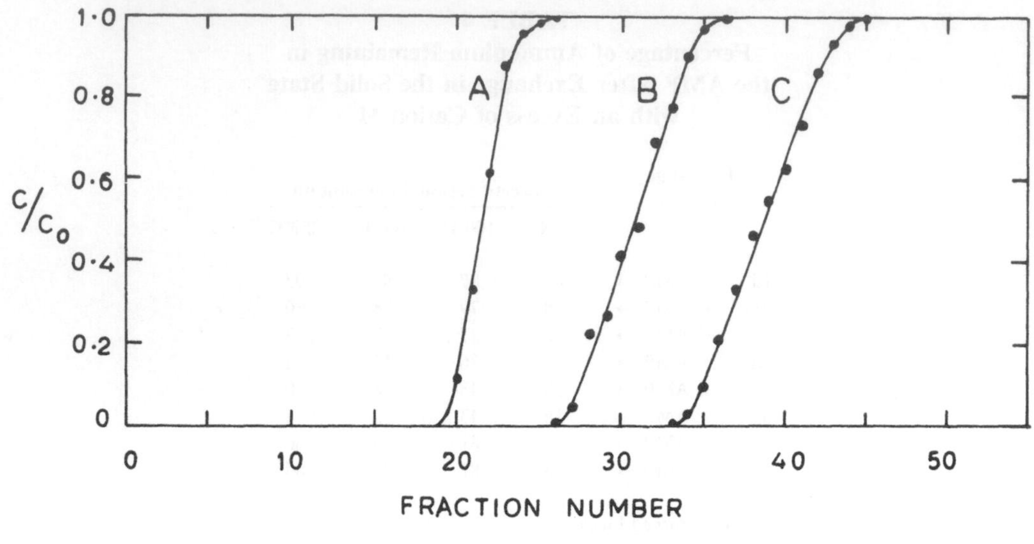

FIGURE 11. Breakthrough curves for Rb^+ and Cs^+ during ion exchange on separate columns (1 cm^2 × 7.4 cm) of AMP-asbestos (1 g of each). With (A) Cs^+ solution containing 496 mg Cs/100 ml; (B) Cs^+ solution containing 311 mg Cs/100 ml; (C)Rb^+ solution containing 160 mg Rb/100 ml. The flow rate was maintained at about 1 ml/min. (From *J. Inorganic Nuclear Chem.*, 12, 104, 1959. With permission.)

(Figure 12), as well as the excellent symmetry of most elution peaks in ion-exchange chromatography with AMP columns, nevertheless indicate relatively fast exchange kinetics, assisted, no doubt, by the small size of the particles.

Bisnath[12,13] studied the kinetics of exchange of K^+, Rb^+, Cs^+, and Tl^+ on AMP by a potentiometric method, i.e., by following the drop in pH caused by the instantaneous hydrolysis of ammonium ions, according to the reaction

$$NH_4^+ + H_2O \rightleftharpoons NH_3 + H_3O^+$$

released into the aqueous phase during the exchange. Dodecahedral crystals of AMP, of particle size up to 17.5 μm, were prepared by an adaptation of the citromolybdate procedure.[20] These were graded into particle size ranges by means of a Warman cyclosizer. At 25°C, particle diffusion was found to be the rate-controlling step at aqueous concentrations of the ingoing ion of about 0.05 *M*, whereas below about 0.005 *M*, film diffusion was the rate-controlling step.

The kinetic data are reported in Tables 5 and 6. Of particular interest is that the magnitude of the diffusion coefficient in AMP for Rb^+ is significantly larger than that for Cs^+, which is opposite to the situation applying for sulphonic acid ion-exchange resins and gel exchangers. This is obviously due to the anhydrous nature of the AMP lattice.

The kinetic behavior of K^+ ions in the solid phase was anomalous in that the plots of F vs. $t^{1/2}$, where F is the fractional attainment of equilibrium and t is the time, consists of two straight-line parts. The first part represents a fast exchange process up to about 11% of the fractional exchange, followed by a slower process. The latter is interpreted[12,13] as being due to the phase transition (see Section III.B.3) accompanying the K^+-NH_4^+ exchange reaction. (The diffusion coefficient for K^+ reported in Table 5 is that evaluated for the slow stage of the exchange.)

D. THE USE OF THE 12-HETEROPOLYACID SALTS IN SEPARATIONS

The most striking feature of the ion-exchange properties of the HPA salts is their extraordinary high selectivity, as already alluded to in the Introduction. Because of the small

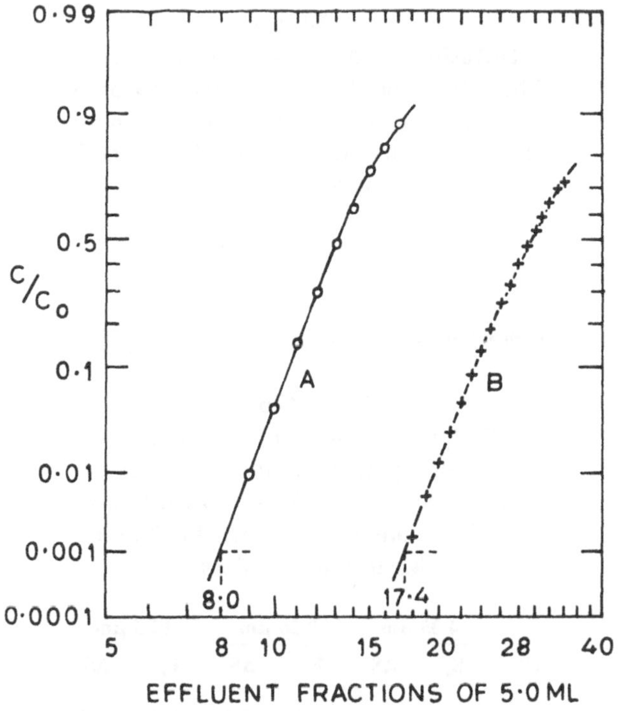

FIGURE 12. Log-probability plots of breakthrough curves for a trace cesium impurity in rubidium chloride solutions with 1 ml columns of "coarse" AMP (linear flow rate: 0.9 cm/min). For (A) 0.418 *M* RbC1; (B) 0.209 *M* RbC1. The values of 8.0 and 17.4 indicate the number of 5-ml fractions of effluent containing less than 0.1% of the original contaminant that may be collected. (From *J. South Afr. Chem. Inst.*, 17, 138, 1964. With permission.)

size of the pores in the McDougall/Keggin structure, only simple inorganic ions are exchangeable (also see Section IV.D.3). Amines, particularly large cations containing secondary or tertiary amine groups, such as 8-quinolinol and many alkaloids, form insoluble HPA salts, but these salts do not have the Keggin structure and their cations have no in-or-out mobility from the crystals. Mixing AMP, for example, with a solution containing such an amine salt can at best lead to surface exchange in the short term, and incorporation of the organic cation by slow crystallization in the longer term, particularly if the new HPA salt has a very low solubility.

Because of the small pore size and the absence (or near absence) of water in the interior of the crystals, Donnan exclusion of co-anions is probably close to 100% effective. Electrolyte sorption, typical of organic ion-exchange resins with concentrated electrolyte solutions, is therefore absent.

From the point of view of adsorbability, cations may be divided into the following three groups:

1. Those cations forming insoluble salts, viz., NH_4^+, K^+, Rb^+, Cs^+, Ag^+, and Tl^+ (i.e., the LMCs). As shown earlier, the relative affinity of an LMC salt of an HPA, e.g., the ammonium salt, for any one LMC, is generally determined by the insolubility of the HPA salt of the ingoing cation over that of the outgoing cation. The presence of even large concentrations of acids or salts of other (i.e., non-LMC) cations in solution has almost no influence on the adsorbability of an LMC, other than effects resulting from influences on activity coefficients in the external phase. Since solvents

TABLE 5
Diffusion Coefficients D^i in the AMP Phase for Monovalent Cations Measured with a Mean AMP Particle Size of 4.35 μm, a temperature of 24.60°C and C = 0.1 *M*

	Cation			
	K^+	Rb^+	Cs^+	Tl^+
$D^i/m^2s^{-1} \times 10^{14}$	0.83	1.98	1.64	2.50

TABLE 6
Enthalpies, E_a, and Entropies, ΔS, of Activation (given in kJ mol^{-1} and JK^{-1} mol^{-1}, respectively) for the Diffusion of K^+, Rb^+, and Cs^+ in AMP for Three Mean Particle Sizes

	4.35 μm		9.20 μm		14.5 μm	
C^+	E_a	ΔS	E_a	ΔS	E_a	ΔS
K^+	—	—	43.5	58.5	42.6	62.8
Rb^+	51.8	81.1	49.4	85.1	49.1	91.9
Cs^+	47.2	64.1	46.7	74.8	46.0	80.3

cannot penetrate the interior of the crystals, selectivity behavior in nonaqueous solvents or mixed solvents is again determined only by the influences of the solvent medium on the activity coefficients in the external phase.

The selectivity shown by the HPA salt exchangers toward the LMCs (i.e., their ability to discriminate between LMCs) is probably unsurpassed by any other known type of ion exchanger, as will be illustrated below.

2. Monovalent cations whose HPA salts are soluble, i.e., H_3O^+, Li^+, and Na^+, and bivalent cations. Compared to the cations of group 1, these cations are relatively weakly exchanged, and are usually capable of displacing only less than one of the three moles of ammonium or other LMC per mole of HPA anion. The extent to which this maximum exchange may take place increases with rising temperature, at least for AMP.[9] The selectivity of an HPA salt for any one of these cations over another in this group is quite modest,[9] probably of the same order as that for sulfonic acid-type ion-exchange resins.
3. Trivalent ions. These ions are adsorbed onto HPA salts very much as are the ions of group 2, but part of the ions taken up (or all of them if adsorption takes place from solutions buffered at a pH $\gtrsim$ 3.5) is adsorbed by an unknown mechanism almost certainly involving complex formation with HPA breakdown products.

1. Adsorption of Trace Quantities

An indication of the high selectivity of AMP and its ability to discriminate between adjacent members of, for example, the alkali metal ions is given by distribution coefficients,

TABLE 7
Distribution Coefficients, K_d, for the Alkali Metal Ions in 0.1 *M* NH_4NO_3 Solution with Dowex 50 (8% DVB) and AMP

	Cs	Rb	K	Na
Dowex 50	62	52	46	26
AMP	6000	230	3.4	~ 0

From *Nature,* 181, 1530, 1959. With permission.

K_d, for these ions between the solid and the aqueous phases determined in the same reference solution. Table 7 reports values for K_d, defined as

$$K_d = \frac{\text{Concentration of metal ion per gram of exchanger}}{\text{Concentration of metal ion per ml of solution}} \qquad (4)$$

determined in a 0.1 *M* NH_4NO_3 solution. Selectivity coefficients $\alpha^{Cs}_{Rb} = 26$ to 28 and $\alpha^{Rb}_{K} \sim 100$ are indicated, compared with values of 1.19 and 1.13, respectively, for the sulfonic acid cation exchanger Dowex 50 of 8% cross-linking. Corresponding separation factors[21] for AWP and ammonium molybdoarsenate (prepared via the free acid) were similar or somewhat smaller.

The value of log K_d on AMP for Rb^+ varied inversely[53] with the log of the ammonium nitrate concentration in static ion-exchange experiments, the plot, of slope −1, being a straight line up to $[NH_4NO_3] \sim 1.5$ *M*. In contrast, in solutions containing 0.1 mol/l of NH_4NO_3 as well as $LiNO_3$, the addition of the latter up to a concentration of about 1.0 *M* had no effect on K_d, showing that the NH_4^+/Rb^+ exchange reaction obeyed the mass action law, and that lithium ions did not materially compete with rubidium ions for exchange sites. Negative deviations above 1.0 *M* in both curves can be explained by activity coefficient effects. Krtil[56] demonstrated a similar relationship for the NH_4^+/Rb^+ exchange on AWP.

Krtil[56] studied the adsorption of carrier-free and trace quantities of ^{90}Sr and ^{90}Y onto AWP in static experiments. Figure 13 shows that maximum adsorption of strontium occurs at pH 2.1 to 3.3, and that of yttrium at pH ~ 1. Unfortunately, the solutions were very dilute and unbuffered, the AWP used had analyzed 0.72 mol of H_3O^+ per mole of PW, and equilibration was carried out for an unspecified period until "equilibrium" had been achieved. The left leg of curve A in Figure 14 is linear, with a slope of 2. In the pH range of 1.0 to 3.5, between 0.10 and 0.17 mol of Sr^{2+} adsorbed per mol of APW in static experiments.

Separation of trace amounts of strontium and yttrium (with appreciable cross-contamination of the Sr and Y fractions), and separation of trace amounts of cesium from trace amounts of strontium and yttrium, are reported by Krtil[56] on APW columns using asbestos as carrier (Figure 14).

For AMP, on the other hand, the ion-exchange behavior of bi- and trivalent ions is more clear-cut.[14] Figure 15 illustrates the sequential elution of the ^{137m}Ba daughter (of half-life 156 s) of ^{137}Cs from a column of AMP-asbestos onto which the ^{137}Cs had been adsorbed. Time was allowed for secular equilibrium to be re-established after each elution. Figure 15 shows that ammonium ions will elute barium, but ammonium ions plus acid is a better eluting agent. Acid alone (of $\geqslant 1$ *M*, not shown) is as good an eluting agent as 0.1 *M* NH_4NO_3 + 0.1 *M* HNO_3.

When 0.1 *M* NH_4NO_3 solutions containing trace amounts of yttrium as well as a sodium acetate/acetic acid buffer was buffered at various pH values and agitated overnight with a fixed quantity of AMP, the dependence of K_d on the pH measured is as shown in Figure

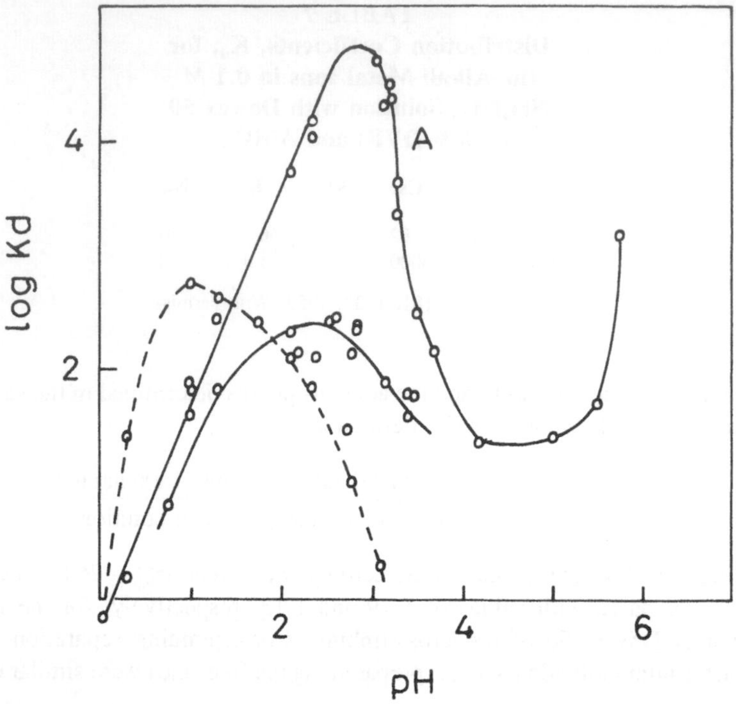

FIGURE 13. pH dependence of the distribution coefficient, K_d, for Sr(II) and Y(III) in ion exchange with AWP in static experiments. (A) Tracer amount of Sr: (B) 1.72 μmol Sr/ml: (C) Tracer amount of Y. (From *J. Inorganic Nuclear Chem.*, 22, 247, 1961. With permission.)

16. The left leg of the log K_d vs. pH curve has a slope of about 3, showing that the adsorption of yttrium up to a pH of about 3 is a true ion-exchange reaction between Y^{3+} and H_3O^+ ions. The curve reaches a maximum at a pH of about 3.3, and thereafter K_d rapidly decreases, reflecting no adsorption of yttrium ions beyond a pH of about 4.2.

The results of these static experiments are in conflict with those of dynamic experiments,[14] at least for pH values above 3. Yttrium adsorbed from a solution buffered above pH 4 is not only much more strongly adsorbed than indicated by Figure 17, but cannot be eluted with even concentrated ammonium nitrate solution. Instead, a moderately strong ($\gtrsim$ 0.5 *M*) solution of a mineral acid is required for the ready elution of the yttrium (Figure 17).

In fact, if the yttrium is adsorbed from an unbuffered solution, it behaves as if it were two species: part of it may be eluted with 1 *N* NH_4NO_3 (along with Sr^{2+}), and the remainder requires a strong acid eluting agent. Adsorption of a mixture of strontium and yttrium from a solution buffered at pH 4.5 permits an excellent separation of strontium and yttrium. A similar separation[46] of milligram quantities of Cd^{2+} and In^{3+}, after sorption from a solution buffered at pH 4.0, leads to the conclusion that this type of procedure is excellently suited to the group separation on AMP of LMCs, bivalent ions (plus Na^+ and Li^+), and trivalent ions. This is illustrated by the separation of fission products on AMP[57] (Figure 18).

Smit and Robb[14] explain the above phenomena as follows. When the solution was not buffered, that part of the yttrium ions that could be eluted with ammonium nitrate was adsorbed by conventional exchange with ammonium nitrate, along with strontium ions. The other part was taken up by another mechanism, which could be complex formation with heteropolyanion breakdown products, or some other unknown complex formation reaction. However, a comparison between elution curves A and D in Figure 17 suggests that buffering the solution at pH 4.5 also promotes the adsorption of strontium ions.

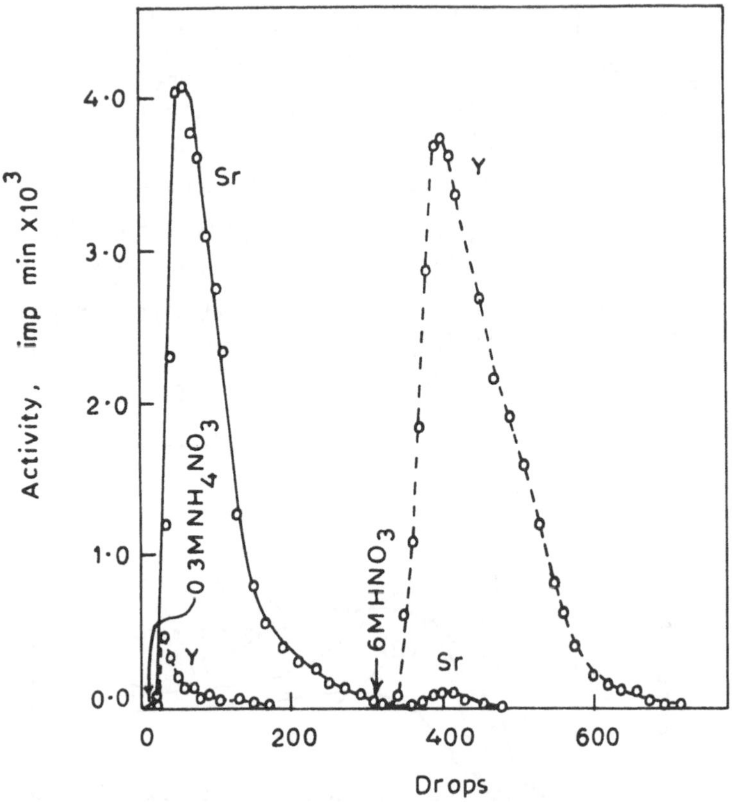

FIGURE 14. Column separation of trace amounts of ^{90}Sr and ^{90}Y on a 3-cm bed of AWP-asbestos (0.2 g of each). The drop interval was 17 s. (From *J. Inorganic Nuclear Chem.*, 22, 247, 1961. With permission.)

The anomalous decrease of K_d for yttrium ions above pH 3 in Figure 16 is also readily explained. At the higher pH values, the long equilibration time resulted in the appearance in solution of soluble degradation products, e.g., and to an 11- or 10-heteropoly anion,[58] as evidenced by a slight yellow hue in the aqueous phase. These breakdown products in solution form a more stable complex with the yttrium ion in solution than that formed in the solid phase, resulting in a gradual transfer of the yttrium to the aqueous phase with long equilibration times. If the mixtures of high pH value are equilibrated for short times (1 min, for example), the K_d values observed lie on an extension of the left leg of the curve for AMP in Figure 16.

The fact that a pH of $\gtrsim$ 4 is required for complete adsorption of yttrium ions onto AMP in the "complexed form", whereas an acid concentration of $\gtrsim 0.5\ M$ is required for its elution, means that the exchange/complex-formation equilibrium is not a dynamic one. Ion exchange on AMP therefore cannot be used to separate mixtures of the rare earths.

Figure 16 also shows the variation of adsorbability of yttrium on AWP as a function of pH. The curve shows a maximum at a pH of about 4.2, compared to a pH of about 1.0 reported by Krtil.[56]

2. Column Separations Involving Larger Quantities

The insoluble HPA salts, particularly the ammonium salts, are well suited for separating the alkali metals from one another and from non-LMC ions. Because of the unusual selectivity, small amounts of the more strongly adsorbed LMC may be picked up quantitatively from solutions containing large quantities of a more weakly adsorbed LMC. But the fact

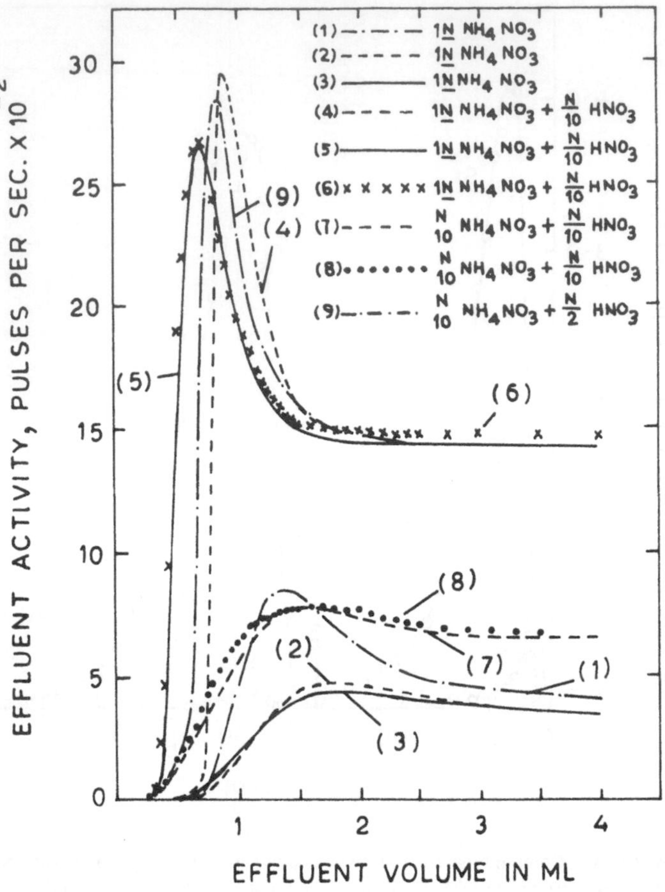

FIGURE 15. Elution of the ^{137m}Ba daughter of ^{137}Cs from a 2:1(w/w) column of AMP-asbestos with different eluents in the sequence shown. Successive elutions were done after allowing time (20 to 30 min) for secular equilibrium to be reestablished. Column dimensions were 0.21 cm^2 × 1.3 cm, and the linear flow rate was about 1.2 cm/min. (From *J. Inorganic Nuclear Chem.*, 26, 509, 1964. With permission.)

that non-LMC cations do not significantly compete for exchange sites on the insoluble HPA salt make these compounds quite unique in the field of ion-exchange separations.

AMP is by far the most useful exchanger in this group of compounds. Crystals of diameter 1 to 10 μm are readily prepared by a cyclic reaction. They are also available in the trade for example, from Bio-Rad. On their own, the material is too fine for column use. But in a 1:1 (w/w) mixture with Gooch-quality asbestos, AMP can be made into a column with adequate flow rates in laboratory-scale separations. The solubility of AMP is higher than that of AWP, but this advantage of the latter is largely offset by the latter's smaller particle size and the greater ease of preparation of AMP. As with the other HPA salts, column chromatographic development tends to produce tailing in peaks, particularly with Cs^+ and Tl^+, and to a lesser extent with Rb^+.

The incompleteness of separations due to tailing effects may, however, be overcome by dissolving the AMP in alkali (NaOH or NH_3 solution) before the most strongly held ion (e.g., Cs^+ or Tl^+) is eluted. The alkaline solution may then be partially neutralized to a pH of 2 to 5 by the addition of a weak acid, such as citric acid solution, and passing this solution through a fresh column of AMP. Incompletely eluted Rb^+ (for example) may then be eluted before recovering the strongly held Cs^+ or Tl^+, e.g., by column dissolution in alkali. Alternatively, if recovery of only the latter ions is of interest, the solution obtained

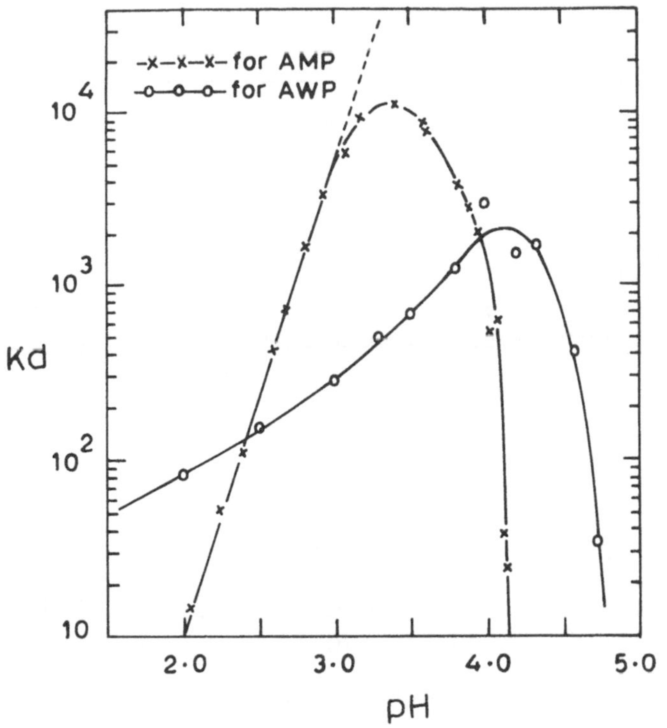

FIGURE 16. pH dependence of the distribution coefficient, K_d, for Y(III) with AMP and AWP in static ion-exchange experiments. The solutions were each 0.1 *M* in ammonium nitrate and were buffered with sodium acetate/acetic acid. "Equilibrium" was attained by agitation overnight. (From *J. Inorganic Nuclear Chem.*, 26, 509, 1964. With permission.)

by alkali dissolution of the first column may be charged with sufficient ammonium nitrate or chloride to promote passage of the smaller alkali metal ions through the column in the resorption step. Thallium may readily be separated from cesium by oxidizing Tl^+ to the trivalent state (e.g., with Br_2) and eluting adsorbed Tl^{3+} with acid.

In its use in the recovery and separation of K^+ (e.g., from Na^+), AMP probably has no peer. Figures 19 and 20 are typical examples.[53,59] Figures 21 and 22 show examples of K^+/Rb^+ and Rb^+/Cs^+ separations. Acceptably symmetrical elution peaks are obtained even with the relatively high linear flow rate of 0.5 to 1 cm/min. Some tailing is observed with the heavier alkali metals.

3. Other Separation Applications of AMP and Other HPA Salts

In the early 1960s, when the large-scale application of gamma radiation sources seemed just around the corner, Smit and co-workers[8,60] developed a process for the separation of macro quantities of ^{137}Cs from strongly acidic fission-product waste concentrates. On the basis of an assumed 24-h operation over 300 d, it was estimated that cyclical operation of two 32-l columns in series of AMP (prepared by the metathesis reaction[23]) would be adequate to produce 10 MCi of ^{137}Cs per year, at the time estimated at over ten times the annual world consumption. The cesium was to be eluted, after washing the column with 1 *M* HNO_3, with a 10 *M* NH_4NO_3 + 0.2 *M* HNO_3 eluent, followed by destruction of the ammonium nitrate by chloride-catalyzed decomposition in strong nitric acid solution.[61] Unfortunately, the envisioned large market for radiation sources never developed, and megacurie gamma sources were provided largely by by-product ^{60}Co, produced by using cobalt as a "flux flattener" in nuclear reactors.

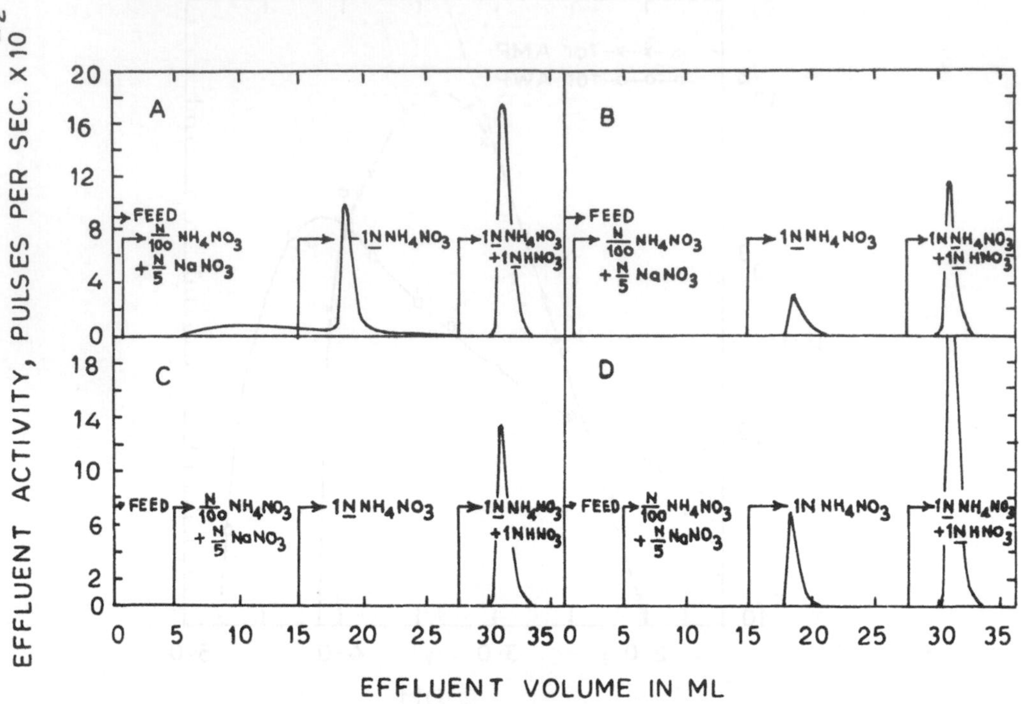

FIGURE 17. Behavior of macro quantities (1 mg each) of Sr and Y on a column of AMP-asbestos (0.72 g of each; column dimensions, 0.81 cm^2 × 3.6 cm). With (A) 1 ml feed solution tagged with ^{90}Sr-^{90}Y; (B) 1 ml feed solution and only Y tagged; (C) 5 ml of feed solution buffered with sodium acetate at pH 4.5 and only Y tagged; (D) same as C, and tagged with ^{90}Sr-^{90}Y. (From *J. Inorganic Nuclear Chem.*, 26, 509, 1964. With permission.)

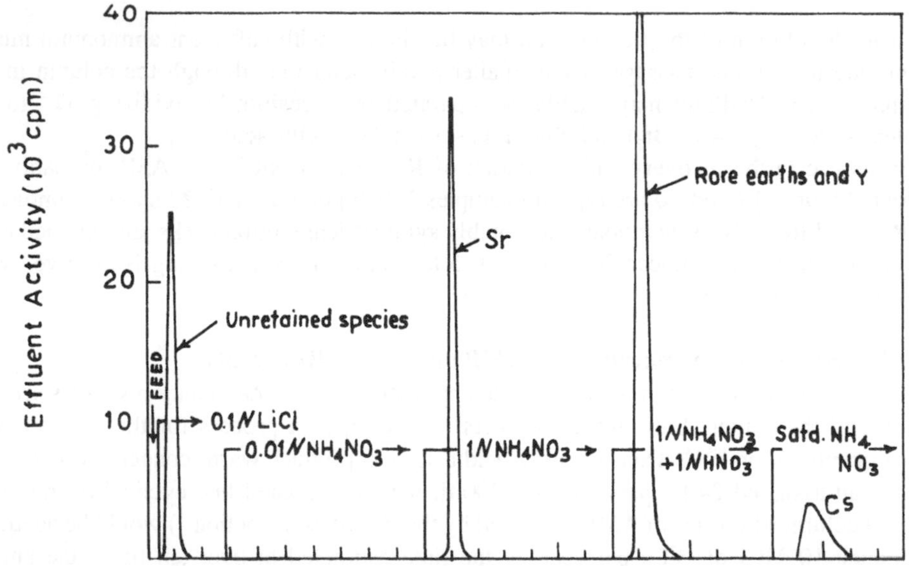

FIGURE 18. Separation of mixed fission products on an AMP-asbestos column, dimensions 0.61 cm$_2$ × 5.0 cm. The feed comprised one drop of fission-product solution in 1 ml of sodium acetate buffer of pH 4.0. Each fraction was five drops. The small peak at fraction number 55 represents a ruthenium activity. (From *Nucleonics*, 17(9), 116, 1959. With permission.)

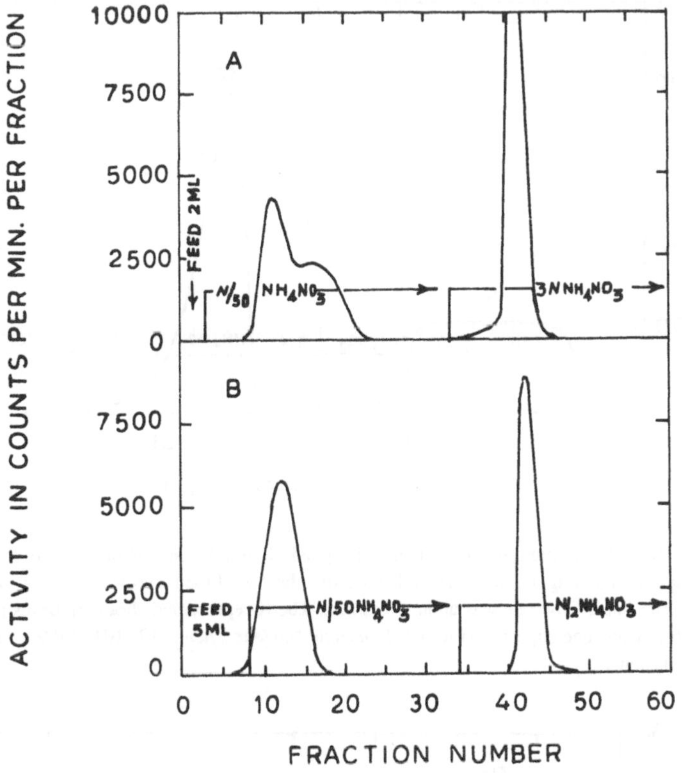

FIGURE 19. Separation of macro quantities of tagged Na and K on a column comprising 1.48 g of AMP and 0.74 g of asbestos, dimensions 0.81 cm^2 × 6.8 cm. Feed contained: (A) 1.3 mg of Na and 2.03 mg of K; (B) 104 mg of Na and 1.02 mg of K, the Na representing more than ten times the column ion-exchange capacity. (From *J. Inorganic Nuclear Chem.*, 12, 104, 1959. With permission.)

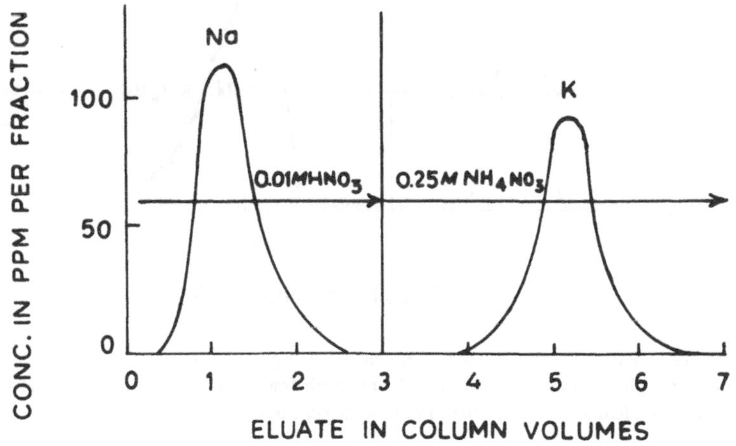

FIGURE 20. Separation of 0.038 meq of Na and K on a 5-ml column of AMP-asbestos (0.55 g of each) of 1 cm diameter. The flow rate was maintained at 1 ml/min. (From *Anal. Chim. Acta*, 44, 293, 1969. With permission.)

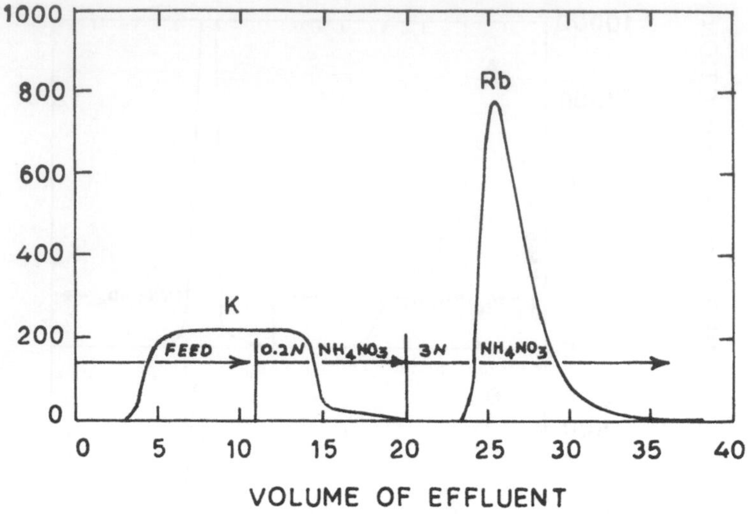

FIGURE 21. Separation of 600 mg of tagged K and 19 mg of tagged Rb on a column comprising 1 g of AMP and 0.5 g of asbestos, dimensions 0.81 cm^2 × 3.3 cm. The linear flow rate was about 1 cm/min. The K represented about ten times the column-exchange capacity. (From *J. Inorganic Nuclear Chem.*, 12, 104, 1959. With permission.)

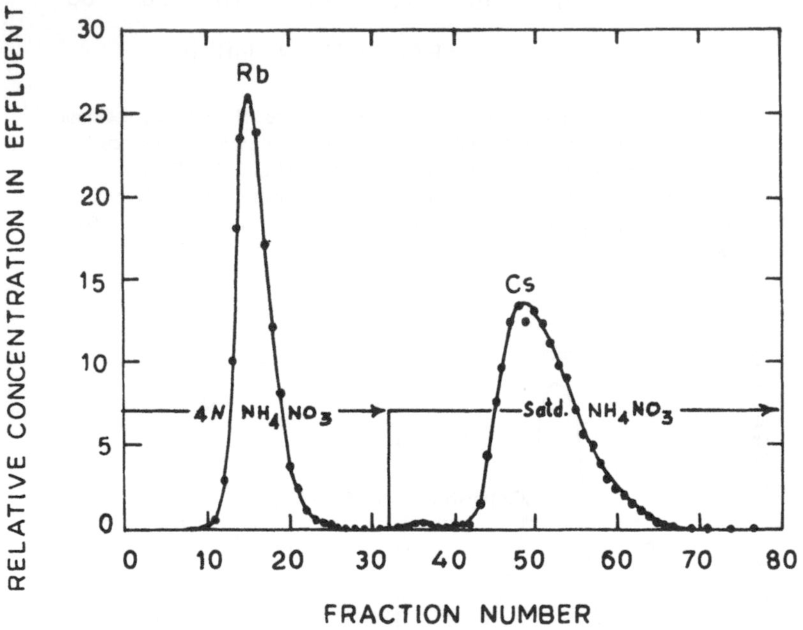

FIGURE 22. Separation of 12.3 mg of Rb and 13.5 mg of Cs on a column of (1:1 w/w) AMP-asbestos, dimensions 1.54 cm^2 × 8.9 cm. Each fraction represents ten drops. (From *J. Inorganic Nuclear Chem.*, 12, 104, 1959. With permission.)

AMP and other HPA salts are particularly well suited for the determination of the fission product ^{137}Cs (half-life, about 30 years) — or ^{137}Cs admixed with the 2.3-year ^{134}Cs — in a variety of liquids, including irradiated fuel,[62] milk,[63] water (together with some other fission products),[64] and sea water.[7] In the latter case, Bergstedt[7] determined the fallout ^{137}Cs in 350-l surface-water samples on board a ship, by acidifying the sample with nitric acid to pH 2, and then stirring into each sample 44 g of AMP (prepared by the citromolybdate method[20]). Tracer experiments showed that 96 ± 1% of the cesium was adsorbed after stirring for 15 min. The AMP was recovered by adding Gooch asbestos and removing the bulk of the liquid phase by filtration with suction using a "filter-stick" technique. The latter comprised a large buchner funnel with a muslin cloth strung across its wider end and the stem connected by a tube to a pump. Subsequent working up of the sample in the laboratory permitted the detection (by gamma spectrophotometry) of ^{137}Cs concentrations of between 100 and 400 pCi·m^{-3}, foreign activities (apart from some ^{134}Cs) being essentially absent. The same samples were used to estimate the cesium content in Table Bay waters as 0.35 μg·l^{-1}, in agreement with a similar value reported of sea water samples taken elsewhere.[65]

Of the methods for recovering and separating cesium from liquid samples, some used AMP with a carrier (e.g., asbestos), while others used thin layers of the exchanger on a filter support.[63,64] Broadbank and co-workers[63] studied the adsorption of cesium, strontium, and yttrium from neutral solution onto the potassium, rubidium, and cesium salts of MPA, but found no advantage over AMP. Other HPA salts studied for cesium separation include the thallous salts of MPA[66,67] and WPA.[68]

Several groups have performed separations of alkali metal ions and other cations using paper chromatography with paper impregnated with AMP and other HPA salts, or thin-layer chromatography with these salts applied to TLC plates.[69]

Insoluble HPA salts can be used conveniently for the preparation of a ^{137}Cs-^{137m}Ba "cow". AMP prepared by the metathesis reaction[23] is especially suited for preparing small-volume, highly active sources of the ^{137m}Ba daughter. Thus, Marais and co-workers[70] prepared a column bed of 100 to 200 mesh AMP, 5mm × 2mm in diameter, in a capillary column joined to a wider tube to which a separating funnel was fitted by means of a ground-glass fitting. By filling the column[70] plus separating funnel with a solution of about 1 *M* HNO_3 and 0.01 *M* NH_4NO_3 saturated with AMP, after adsorbing 10 MCi of ^{137}Cs on the column, 10-MCi quantities of ^{137m}Ba (half-life, 153.08 ± 0.04 s) could be collected repeatedly in one to two drops of effluent, by allowing about 30 min between "milkings" for secular equilibrium to be reestablished.

The "cow" was used for the routine determination of the resolving time of Geiger counters and scintillation counters, and the routine checking of the correct performance of counting assemblies.[71] The highly radioactive samples of ^{137m}Ba decayed essentially to background even after numerous milkings. The present author found this "cow" (with a few microcuries of ^{137}Cs) a very convenient source of a relatively short-lived nuclide for half-life determination by his undergraduate students. Deist[72] used the ^{137m}Ba milked from such a cow for the routine determination of the cation exchange capacity of soils.

V. CONCLUDING REMARKS: SCOPE FOR FURTHER RESEARCH

It has been shown in this chapter that the insoluble HPA salts represent a small, albeit in many respects unique, class of inorganic cation exchangers. Unlike other inorganic cation exchangers, the anions are discrete ions, and the cations interact with their anion neighbors very much as would the NH_4^+ ions with the PO_4^{3-} ions in the lattice of $(NH_4)_3PO_4$. The main difference lies in the size of the heteropolyanions, which results in voids and channels large enough to permit the diffusion of the cations in and out of the lattice.

The NH_4^+ and K^+ salts of MPA (and probably also WPA) are entirely anhydrous, but replacement of these two ions by the larger cations Cs^+ and Tl^+ (and possibly also Rb^+) leads to lattice expansion, and the expanded lattice is apparently capable of accommodating some water molecules.

The insoluble HPA salts of amines do not have the Keggin structure and, unless the contrary is proved, cannot act as ion exchangers, in the normally understood sense of the word. But Jain and Agrawal[73] recently reported the pyridinium salts of 12-tungstoarsenate and 12-molybdoarsenate to have a high affinity for cobalt complexes, but poor affinity for most of the other metal ions. Uptake is reported to be in the order $[Co(NH_3)_6]^{3+} > [Co(en)_3]^{3+} > Co^{2+}$, and the cobalt complexes are completely desorbed with 3 *M* NH_4NO_3. It is thus possible that a range of other non-Keggin HPA salts may prove to have useful cation-exchange properties.

The adsorption of trivalent ions on Keggin HPA salts has not received much attention. No data are available on the mechanism of the adsorption of these ions from solutions buffered at pH ~ 4, nor do we have any indication of the capacity of, say, AMP for yttrium ions when adsorbed by this mode of exchange. Information on the mechanism of this type of exchange is likely to become available only by single-crystal X-ray crystallography.

REFERENCES

1. **Baxter, G. P. and Griffin, R. C.,** *Am. Chem. J.*, 34, 204, 1905.
2. **Terlet, H. and Briau, A.,** Study of some methods for the determination of phosphoric acid by precipitation as ammonium phosphomolybdate, *Ann. Fals.*, 28, 546, 1935.
3. **Gisiger, L.,** Titrimetric determination of phosphoric acid upon the basis of the method of N Lorenz and with an immersion filter, *Z. Anal. Chem.*, 115, 15, 1938.
4. **Thistlethwaite, W. P.,** Composition and constitution of ammonium phosphomolybdate and the conditions affecting its precipitation, *Analyst,* 72, 531, 1947.
5. **Smit, J. van R.,** Ammonium salts of the heteropolyacids as cation exchangers, *Nature,* 181, 1530, 1958.
6. **Amphlett, C. B., McDonald, L. A., Burgess, J. S., and Maynard, J. C.,** Synthetic inorganic ion-exchange materials. III. Separation of rubidium and caesium on zirconium phosphate, *J. Inorg. Nucl. Chem.*, 10, 69, 1959.
7. **Bergstedt, R. P.,** Die gerbruik van ioonuitruiling op ammoniummolibdofosfaat vir die bepaling van sesium en tallium in seawater, M. Sc. thesis, University of South Africa, 1970.

7a. **Smit, J. van R. and Bergstedt, R. P.,** The determination of radioactive caesium in sea water by a batch ion-exchange method with ammonium 12-molybdophosphate: preliminary results, *Ann. Univ. Coll. West. Cape,* 7, 40, 1970.

8. **Smit, J. van R.,** The AMP process for caesium separation. I, Report R-3884, *United Kingdom Atomic Energy Authority,* 1961.
9. **McDougall, G. J.,** Ion-Exchange Properties and Crystal Structures of Some 12-Heteropolyacid Salts, Ph.D. thesis, University of the Witwatersrand, Johannesburg, 1975.
10. **Boeyens, J. C. A., McDougall, G. J., and Smit, J. van R.,** Crystallographic study of the ammonium/potassium 12-molybdophosphate ion-exchange system, *J. Solid State Chem.*, 18, 191, 1976.
11. **Boeyens, J. C. A., McDougall, G. J., and Smit, J. van R.,** The mechanism of ion exchange on ammonium 12-molybdophosphate (AMP), in *Recent Developments in Ion Exchange,* Williams, P. A. and Hudson, M. J., Eds., Elsevier, London, 1987, 291.
12. **Bisnath, B.,** Some Aspects of Ion Exchange on Ammonium 12-Molybdophosphate, M.Sc. thesis, University of Durban-Westville, Durban, 1983.
13. **Bisnath, B., Govinden, H. S., and Smit, J. van R.,** Kinetics of ion exchange on ammonium 12-molybdophosphate, in *Recent Developments in Ion Exchange,* Williams, P. A. and Hudson, M. J., Eds., Elsevier London, 1987, 357.
14. **Smit, J. van R. and Robb, W.,** Ion exchange on ammonium molybdophosphate. II. Bivalent and trivalent ions, *J. Inorg. Nucl. Chem.*, 26, 509, 1964.
15. **Amphlett, C. B.** *Inorganic Ion Exchangers,* Elsevier, London, 1964, chap. 4.
16. **Churms, S. C.,** Inorganic ion exchangers. V. Heteropolyacid salts and other inorganic ion exchangers, *S. Afr. Industr. Chem.*, 19, 145, 1965.
17. **Clearfield, A., Nancollas, G. H., and Blessing, R. H.,** New inorganic ion exchangers, in *Ion Exchange and Solvent Extraction,* Vol. 5, Marinsky, J. A. and Marcus, Y., Eds., Marcel Dekker, New York, 1973, 1.
18. **Wu, H.,** Chemistry of phosphomolybdic acids, phosphotungstic acids and allied substances, *J. Biol. Chem.*, 43, 189, 1920.
19. **Vogel, A. I.,** *A Textbook of Quantitative Inorganic Analysis,* 2nd ed., Longmans Green, London, 1951, 380.

20. **Kassner, J. L., Crammer, H. P., and Ozier, M. A.**, Determination of $P_2 0_5$ in phosphate rock, *Anal. Chem.*, 20, 1052, 1948.
21. **Smit, J. van R., Jacobs, J. J., and Robb, W.**, Cation exchange properties of the ammonium heteropolyacid salts, *J. Inorg. Nucl. Chem.*, 12, 95, 1959.
22. **Thistlethwaite, W. P.**, The "normal" 12-molybdophosphates of the alkali metals and ammonium, *J. Inorg. Nucl. Chem.*, 28, 2143, 1966.
23. **Smit, J. van R.**, Ion exchange on ammonium molybdophosphate. III. Preparation and properties of coarse ammonium heteropolyacid salts, *J. Inorg. Nucl. Chem.*, 27, 227, 1965.
24. **Stockdale, D.**, Determination of phosphorus, especially in basic slag and in steel, *Analyst*, 83, 24, 1958.
25. **Healy, T. V. and Ingham, G.**, Alkali Phosphotungstates. III. Sodium, Lithium and Hydrogen Phosphotungstates, Report AERE C/R 2594, United Kingdom Atomic Energy Research Establishment, 1958.
26. **Tourneux, C. and Devin, C.**, Sur les propriétés adsorbantes du phosphomolybdate d'ammonium, *C.R*, 232, 2430, 1951.
26a. **Tourneox, C. and Devin, C.**, Nouvelles observations sur les propriétés adsorbantes du phosphomolybdate d'ammonium, *C.R*, 233, 43, 1951.
27. **Gregg, S. J. and Stock, R.**, The adsorption of hydrocarbon vapours by ammonium phosphomolybdate, *Trans. Faraday Soc.*, 53, 1355, 1957.
28. **Keggin, J. F.**, The structure and formula of 12-phosphotungstic acid, *Proc. R. Soc.*, A144, 75, 1934.
29. **Bradley, A. J. and Illingworth, J. W.**, Crystal structure of $H_3PW_{12}O_{40} \cdot 29H_2O$, *Proc. Soc.*, A157, 113, 1936.
30. **Illingworth, J. W. and Keggin, J. F.**, Identification of the 12-heteropolyacids and their salts by means of x-ray powder photographs, *J. Chem. Soc.*, 575, 1935.
31. **Santos, J. A.**, X-ray study of the caesium salts of 12-heteropoly acids, *Proc. R. Soc.*, A150, 309, 1935.
32. **Ferrari, A. and Nanni, O.**, Researches on heteropoly acids. I. Structure of the phospho- and arsenododecamolybdates and of the phospho- and arsenododecatungstates of ammonium, potassium and thallium, *Gazz. Chim. Ital.*, 69, 301, 1939.
33. **Sherry, H. S.**, The ion-exchange properties of zeolites, in *Ion Exchange*, Vol. 2, Marinsky, J. A., Ed., Marcel Dekker, New York, 1969, 89.
34. **Helfferich, F.**, *Ion Exchange*, McGraw-Hill, New York, 1962, 187.
35. **Cotton, F. A. and Morehouse, S. M.**, Molecular structure of a diamagnetic, doubly o bridged, binuclear complex of Mo(V) containing a metal-metal bond, *Inorg. Chem.*, 4, 1377, 1965.
36. **Pauling, L.**, The molecular structure of the tungstosilicates and related compounds, *J. A. Med. Chem. Soc.*, 51, 2868, 1929.
37. International Union of Pure and Applied Chemistry, *Nomenclature of Inorganic Chemistry*, 2nd ed., Butterworth, London, 1970, 84.
38. **Schlemper, E. O. and Hamilton, W. C.** Neutron-diffraction study of the structures of ferroelectric and paraelectric ammonium sulfate, *J. Chem. Phys.*, 44, 4498, 1966.
39. **Fischer, J., Ricard, L., and Toledano, P.**, Novel phosphomolybdate structure. Crystal structure of the phosphomolybdate (NH_4) 5[$(MoO_3$ $_5(PO_4)$ $(HPO_4)] \cdot 3H_2 0$, *J. Chem. Soc. Dalton Trans.*, 941, 1974.
40. **Eisenman, G.**, Cation-selective glass electrodes and their mode of operation, *Biophys. J.*, 2, 259, 1962.
41. **Buchwald, H. and Thistlethwaite, W. P.**, Some cation exchange properties of ammonium 12-molybdophosphate, *J. Inorg. Nucl. Chem.*, 5, 341, 1958.
42. **Krtil, J. and Kourim, V.**, Exchange properties of ammonium salts of 12-heteropolyacids. Sorption of caesium on ammonium phosphotungstate and phosphomolybdate, *J. Inorg. Nucl. Chem.*, 12, 367, 1960.
43. **Meier, D. and Treadwell, W. D.**, Über den Austausch von Kalium-, Rubidium- und Cäsiumion an Ammonium = phosphormolybdat, *Helv. Chim. Acta*, 34, 155, 1951.
44. **Krtil, J. and Chanko, M.**, Ion exchange properties of ammonium salts of heteropoly acids. VII. Sorption of caesium-137 and rubidium-86 on acid and normal ammonium and thallous salts of phosphotungstic and phosphomolybdic acid, *J. Chromatogr.*, 27, 460, 1967.
45. **Coetzee, C. J. and Rohwer, E. F. C. H.**, Cation exchange studies on ammonium 12-molybdophosphate, *J. Inorg. Nucl. Chem.*, 32, 1711, 1970.
46. **McDougall, G. J. and Smit, J. van R.**, Thermodynamics of ion exchange on ammonium 12-molybdophosphate, in *Recent Developments in Ion Exchange*, Williams, P. A. and Hudson, M. J., Eds., Elsevier, 1987, 348.
47. **Helfferich, F.**, *Ion Exchange*, McGraw-Hill, New York, 1962, 154.
48. **Sherry, H. S. and Walton, H. F.**, The ion-exchange properties of zeolites. II. Ion exchange in synthetic zeolite Linde 4-A, *J. Phys. Chem.*, 71, 1457, 1967.
49. **Nancollas, G. H. and Tilak, B. V. K. S. R. A.**, Thermodynamics of cation exchange on semicrystalline zirconium phosphate, *J. Inorg. Nucl. Chem.*, 31, 3643, 1969.
50. **Kraus, O.**, Investigation of the crystal lattices of heteropoly acids and their salts. II. The constitution of the silicotungstates of trivalent metals, *Z. Krist. Allogr.*, 93, 379, 1936.

51. **Evans, H. T.,** in *Perspectives in Structural Chemistry,* Vol. 4, Dunitz, J. D. and Ibers, D. A., Eds., John Wiley & Sons, New York, 1971.
52. **Choudhuri, D. and Mukherjee, S. K.,** Mechanism of ion exchange in heteropoly salts, *J. Inorg. Nucl. Chem.,* 33, 1933, 1971.
53. **Smit, J. van R., Robb, W., and Jacobs, J. J.,** Cation exchange on ammonium molybdophosphate. I. The alkali metals, *J. Inorg. Nucl. Chem.,* 12, 104, 1959.
54. **Smit, J. van R.,** Purification of rubidium compounds from traces of caesium, *J. S. Afr. Chem. Inst.,* 17, 138, 1964.
55. **Krtil, J.,** Exchange properties of ammonium salts of 12-heteropolyacids. II. Separation of rubidium and caesium on ammonium phosphotungstate, *J. Inorg. Nucl. Chem.,* 19, 298, 1961.
56. **Krtil, J.,** Exchange properties of ammonium salts of 12-heteropolyacids. III. The sorption of strontium and yttrium on ammonium phosphotungstate and their separation from caesium, *J. Inorg. Nucl. Chem.,* 22, 247, 1961.
57. **Smit, J. van R., Robb, W., and Jacobs, J. J.,** AMP—effective ion exchanger for treating fission waste, *Nucleonics,* 17 (9), 116, 1959.
58. **Baker, M. C., Lyons, P. A., and Singer, S. J.,** Physical-chemical studies with heteropoly acids, *J. Phys. Chem.,* 59, 1074, 1955.
59. **Coetzee, C. J. and Rohwer, E. F. C. H.,** The determination of sodium and potassium after separation from each other and other ions by means of ammonium phosphomolybdate and other ion exchangers, *Anal. Chim. Acta,* 44, 293, 1969.
60. **Smit, J. van R.,** The AMP Process for Caesium Separation. II and III, Reports AERE-R 3700 and AERE-R 4006, United Kingdom Atomic Energy Authority 1961, 1962; Smit, J. van R., Jacobs, J. J., and Pummery, F. C. W., Report AERE-R 4039 United Kingdom Atomic Energy Authority, 1962; **Smit, J. van R. and Jacobs, J. J.,** Separation of cesium from fission product wastes by ion exchange on ammonium molybdophosphate, *Ind. Eng. Chem. Process Des. Dev.* 5, 117, 1966.
61. **Smit, J. van R.,** Chloride-catalysed destruction of ammonium nitrate by nitric acid, *Chem. Ind.,* 2018, 1964.
62. **Allison, G. M., Ferguson, R. A., and McLaughlin, D.,** The use of ammonium molybdophosphate for the determination of fission products. I. The determination of Cs-137 in solutions of irradiated fuel, Report CRRL—1069, Atomic Energy of Canada Limited, 1962.
63. **Broadbank, R. W. C., Hands, J. D., and Harding, R. D.,** A rapid assay of radioactive caesium in milk, *Analyst,* 88, 43, 1963.
64. **Broadbank, R. W. C., Dhabanandana, S., and Harding, R. D.,** A possible use of ammonium 12-molydophosphate for assaying certain radioactive fission products in water, *Analyst,* 85, 365, 1960.
65. **Folsom, T. R. and Sreekumaran, C.,** paper presented at a panel meeting on Reference Methods for Marine Radioactivity Studies, Vienna, November 18, 1968.
66. **Broadbank, R. W. C., Dhabanandana, S., and Harding, R. D.,** The ion-exchange and other properties of certain 12-molybdophosphates, *J. Inorg. Nucl. Chem.,* 23, 311, 1961.
67. **Hara, T.,** Capture of radioactive Cs on a phosphomolybdate precipitate layer and its determination, *Bull. Chem. Soc. Jpn.* 31, 635, 1958.
68. **Caron, H. L. and Sugihara, T. T.,** Highly specific method of separating Cs by ion exchange on thallous phosphotungstate, *Anal. Chem.,* 34, 1082, 1962.
69. See references 212 to 217 in Clearfield et al., Reference 17.
70. **Marais, P. G., Haasbroek, F. J., Lotz, E. van der S., and Smit, J. van R.,** Half-life of barium-137m, *J. S. Afr. Chem. Inst.,* 19, 1, 1966.
71. **Marais, P. G., Karsten, J. H. M., Lotz, E. van der S., and Smit, J. van R.,** Resolving time determination of Geiger-Mueller counters with barium-137m, *S. Afr. J. Agric. Sci.,* 9, 253, 1966.
72. **Deist, J.,** Use of barium-137m for determining the cation exchange capacity of soils, *Agrochemophysica,* 1, 27, 1969.
73. **Jain, A. K. and Agrawal, S.,** Sorption and elution of cobalt as hexamminecobalt (III) and tris(ethylenediamine) cobalt (III) ions on heteropoly acid salt exchangers, *Chem. Anal. (Warsaw),* 26, 341, 1981; *Chem. Abstr.,* 96, 78988e.

Chapter 4

INORGANIC ION EXCHANGERS IN RADIOCHEMICAL ANALYSIS

Laszlo Szirtes

TABLE OF CONTENTS

I. INTRODUCTION

A great number of inorganic materials, both synthetic and natural, are known which are capable of ion exchange.[1-17] In connection with this, a wide variety of applications have been found for synthetic inorganic ion exchangers, and new uses are constantly being suggested. Numerous applications can be found not only in the field of water purification—the original major application—but also in analytical chemistry, organic chemistry and biochemistry, food technology, the isolation of elements in hydrometallurgy and, of course, in many specialized fields related to the utilization of atomic energy.[18]

The application of organic resins in radiochemical practice is limited by virtue of their limited stability under various conditions. For example, the breakdown of their polymer chain at higher temperatures gives rise to monomers in the effluent. In addition, they are very sensitive to high radiation doses, which cause significant changes in capacity and selectivity, prescission, and the degree of cross-linking.[19]

In the field of radiochemistry, therefore, there is a need for ion-exchange materials possessing a high degree of stability with regard to temperature (especially temperatures higher than 150°C), with regard to the effect of ionizing radiation, and which are able to maintain an appreciable ion-exchange capacity over a wide range of acidity.

In this chapter, the applications of synthetic, inorganic ion exchangers for various (radiochemical) purposes will be discussed with the intention of showing the wide range of their potential uses.

II. APPLICATIONS OF METAL OXIDE AND HYDROUS OXIDE ION EXCHANGERS

A. OXIDES AND HYDROUS OXIDES OF MULTIVALENT METALS

A very large number of scales of insoluble metal oxides and hydrous oxides possess ion-exchange and other important properties which make them very useful for the various separation processes or analytical determinations:

1. The treatment of high-temperature water for the removal of corrosion products and ionic impurities. This application is analogous to boiler-water treatment as practiced at present.
2. Purification and conditioning of moderator and cooling water in pressurized water nuclear reactors
3. Chemical separation and purification in intense radiation fields
4. Catalysis reactions taking place at temperatures at which organic resins are decomposed
5. Separation of ions from mixtures, metal recovery, and industrial waste-treatment processes

In recent years, a number of review papers were written in which the ion-exchange properties of hydrous oxides such as SiO_2, TiO_2, ThO_2, and ZrO_2 were described.[20-30] Among these materials, hydrous zirconium oxide was used in high-temperature water treatment (HTWT).

In the case of the coolant water of a closed-cycle nuclear reactor system, the temperature of the water is about 300°C and the pressure is approximately 100 atm. After considering data obtained by various experts[31-34] on numerous synthetic inorganic ion exchangers, zirconium phosphate and hydrous zirconium oxide were selected for this purpose, and are used with success.

Hydrous zirconium oxide in alkaline media (pH $>$ 7) works as a cation exchanger. It is capable of the sorption of alkali and alkaline metal ions. In addition, it has very good sorption properties toward the polyvalent hydrolyzable metal ions, such as Fe^{3+}, Cr^{3+}, and Ni^{2+}.

Amphlett[35] gave a method for separating micro amounts of radioactive cesium (^{137}Cs) and radiostrontium (^{90}Sr) nuclides from fission products. Kraus and Phillips,[36] using the same ion exchanger, eluted the Cs^+ ions from the column with 0.5 *M* NH_4NO_3 solution. Other authors[37-39] have described the separation of transuranium elements (in anion form), such as Np, Pu, Am, and Bk, using a hydrous zirconium oxide ion exchanger.

In a series of radioactive "mother-daughter" element separations, the hydrous zirconium oxide ion exchanger was used[40-42] with success as a sorbent, e.g., in the ^{113}Sn-^{113}In system. During the separation process for producing ^{188}Re radionuclide, the latter was eluted with water; the efficiency of separation was found to be 99.99%.

Using hydrous zirconium oxide as an ion exchanger, the alkaline earth metal ions have been eluted with 1 *M* NH_4Cl solution and the rare earth with a 1 *M* HCl solution.[43] Uranium has been extracted from sea water samples using hydrous titanium oxide as the sorbent.[44]

Other authors[45-47] have reported data on the other hydrous oxides, but the results showed that among the investigated materials, the best properties were exhibited by hydrous titanium oxide. Using the above-mentioned separation system, Bettinale[48] described the production of 1000 t/year. From another work[49] it is known that the I^-, Br^-, and Cl^- anions may be separated quantitatively, using a zirconium oxide ion exchanger.

Earlier, it was not clear whether the ion exchange taking place in the hydrous metal oxides was a surface or a volumetric process. Now, by comparing the surface area and the total ion-exchange capacity values of the given material, it can easily be determined. The hydrous metal oxides are generally amphoteric in character, relating to the pH value, i.e., they are able to work as an anion or a cation exchanger. This phenomenon is described by the classical formula $Al\,(OH^+)_2 + OH^- \leftrightarrows Al\,(OH_3) \leftrightarrows AlO(OH)^-{}_2 + H^+$, or by the general reaction mechanism:

$$M\text{–}OH \rightleftharpoons M^+ + OH^- \qquad \text{acidic solution}$$

$$M\text{–}OH \rightleftharpoons M\text{–}O^- + H^+ \qquad \text{basic solution}$$

Since most hydrous metal oxides strongly adsorb the polyvalent anions, as described by Kraus and Phillips,[36] their elution may be carried out only with strong (concentrated) reagents, such as NaOH. Tucher and co-workers[50] have demonstrated that alumina is eminently suitable for the separation, in anion form, of fluorine (^{18}F), molybdenum(^{99}Mo), technetium (^{99m}Tc), iodine(^{132}I), and tellurium(^{132}Te) radionuclides.

In order to produce the radioactive nuclide ^{99m}Tc, which is in widespread use in medical practice, the alumina is generally applied as the sorbent in the so-called radionuclidic generator system.[51] Alumina was also used by Shishkov[52] to separate molybdate and tungstate anions, eluting the former with 4 *M*, and the latter with 10 *M*, HCl solutions. From the works of other authors,[53-55] it is well known that freshly synthesized hydrous manganese oxide can successfully be employed for separating various elements from fission products. Gerevini and Soigliana[56] reported that the ion-exchange capacity of such a material is 0.73 mval/g and that the selectivity order for alkali metals is $Cs^+ > NH_4^+ > K^+ > Na^+ > Li^+$. This order of selectivity followed the same sequence predicted by Eisenman's theory for dehydrated ions. Puskariev and co-workers[57] studied the sorption of ^{144}Ce, ^{91}Y, ^{65}Zn, ^{106}Ru, ^{134}Cs, ^{95}Nb, and $^{95}Nb + ^{95}Zr$ radionuclides on hydrous manganese dioxide. The authors determined that the Ce and Y nuclides are sorbed as chlorides. Subsequently, Ru (pH = 4 to 5), Zn (pH = 6 to 7), and Nb + Zr (pH = 2 to 9) radionuclides were sorbed in the form of radiocolloids; the Cs^+ ions were not sorbed under the conditions of the investigations.

Benton and Horsfall[58] presented the selectivity order of synthetic magnetite for various anions:

$$Cl^- << SO_4^{2-} < HCO_3^- < CrO_4^{2-} < HPO_4^{2-} < Fe(CN)_6^{4-}$$

For determining the amount of transition elements in hydro-organic media, a hydrous tin oxide ion exchanger was applied by other authors.[59,60] The same hydrous oxide was used in a fully automated process for determining impurities in high-purity iron metal.[61]

Ahrland and co-workers[62] observed the cation-exchange properties of silica gel. They pointed out that the distribution coefficient (k_d) changes with pH. Based on these data, the authors presented[63] the results of the separation of rare earths from U and Pu in acidic media.

The same ions were separated by Cvjeticanin[64] by means of hydrous manganese oxide. The author mentioned that after the separation, 99.2% of the initial amount of fission products was retained on the column. Kraus and co-workers[43] showed that alkali metal ions are eluted with a 1 *M* NH_4Cl solution and the rare earths, with a 1 *M* HC1 solution. Silica was also used by Milone and co-workers[65] to separate micro amounts of Cs^+ from Sr^+. Details regarding the sorption order in a sulfate solution for Cu^{2+} on various inorganic ion exchangers were described by Laskorin and co-workers;[66] they stated that silica is less efficient than the other oxides, e.g., tin, titanium, and zirconium oxides.

B. HYDROUS OXIDES OF SIX-VALENT METALS

The hydrous oxides of multivalent metals, such as Nb, Ta, Sb, Mo, and W, have cation-exchange properties. Among them, hydrous antimony (V) oxide (HAO), antimonic acid (AA) and polyantimonic acid (PAA) have been among the most intensively studied.

Jander and Simon,[67] as long ago as 1923, reported that the pentavalent AA adsorbs the Li^+ ion. Jander[68] indicated that this sorption behavior might be attributed to the chemical action of the oxide gel. The coagulation values of antimony pentoxide gel obey the Shulz-Hardy rule (see Ghosh and Dhar[69]). It has been reported by other authors[70-73] that the adsorption and desorption of K^+ ions on AA is due to as exchange between the H^+ ions of the acid and the K^+ ions in the external solution. Lefebvre and Gaymard[74] have also reported that the adsorption of alkali and alkaline earth metal ions on AA is regulated by an ion-exchange mechanism with a selectivity sequence of the order:

$$Na^+ < K^+ < NH_4^+ < Rb^+ < Cs^+ \text{ and } Ba^{2+} < Sr^{2+} < Ca^{2+} < Mg^{2+}$$

The adsorption properties of AA samples strongly depend on the preparation conditions, i.e., aging, drying, etc.[75]

A great number of studies are associated with determining the distribution coefficients (K_d) on the mentioned materials. These results are useful from the viewpoint of separations of trace and/or micro amounts of various elements in analytical chemistry, radiochemistry, and bioorganic chemistry. Most of these determinations were carried out using HNO_3 media in order to obtain fundamental ion-exchange data in a noncomplexing medium, which may also be useful for applications involving the reprocessing of nuclear fuel.

A systematic study of K_d for various metal ions on crystalline AA(AAc) has been reported by Abe.[76] Girardi and co-workers[77] have also reported that about 3000 adsorption-desorption cycles were carried out in a standardized way over 11 adsorbents, including 9 inorganic ion exchangers from various media, for preliminary screening of possible useful materials for radiochemical separations. At the same time, it must be mentioned that Krishnan and Crapper[78] have also studied the retention of 20 trace elements on a column, with a diameter of 1 cm and a length of 3 cm, containing hydrated antimony pentoxide (HAP), using HC1, HNO_3, and H_2SO_4 of various concentrations as eluants. During the work, the extended plate theory was utilized for evaluating the K_d values of different metal ions on HAP from their retention results. It was found that the evaluated K_d values on HAP are essentially compatible with those of AAc for about 20 metals ions.

Baetsle and co-workers[79,80] earlier reported that AAc is a promising adsorbent for the recovery of ^{90}Sr and ^{137}Cs radionuclides from acidic affluent in reprocessing the spent fuel elements. Other authors have demonstrated that AAc shows extremely high values of K_d for sodium,[81,82] strontium,[79,83-85] cadmium,[83,86,87] and mercury[88] ions. Moreover, hydrated antimony pentoxide (HAP) also exhibits extremely high adsorptive ability for Na^+, Ag^+, and Tl^+ ions, even in concentrated mineral acid solutions.[89] These properties have wide applications for the selective removal of Na^+ ions from biological materials or high-purity substances for determining trace elements by neutron-activation analysis.

1. Data Concerning the Separation of Alkali Metal Ions

Caletka and co-workers[90] found the following sequence of sorption for various AA:

AAc

$Li^+ << K^+ < Cs^+ < Rb^+ < Na^+$ in HNO_3 medium

$Li^+ << K^+ < Rb^+ < Cs^+ < Na^+$ in NH_4NO_3 medium

These are similar to sorption orders for organic resins of the strong acidic type.

HAP

$Cs^+ < K^+ < Rb^+ < Na^+$ in HNO_3 medium

$K^+ < Cs^+ < Rb^+ < Na^+$ in HCl medium

Caletka and co-workers[90] and Abe and co-workers[81,82,91,92] observed an extremely large value of K_d for Na^+ ions on both ion exchangers. The authors reported that Na^+ ions cannot be removed from the AAc column, even with concentrated HCl, HNO_3, H_2SO_4, and $HClO_4$. It is eluted only with 1 *M* NH_4NO_3 solution.

Furthermore Abe and Tsuji[91] and Higuchi and Takehira[92] found that when the AAc is heated, the K_d values increase, first slightly up to 200°C and then markedly to about 240°C, reaching the maximum value at 330°C. In this case, the $K^+ < Rb^+ < Cs^+ < Na^+$ selectivity series exists, as described by the author.

The mutual separation of microamounts of alkali metal ions was performed successfully on a small column (diameter = 0.8 cm, length = 6 cm) of AAc by Abe[82] and Abe and Ito.[81] The efficiency of the separation might be concluded from the elution curves (see Figures 1 and 2).

The Na^+ ions are strongly adsorbed on AAc, and may be easily eluted only with NH_4NO_3 solution. The elution order is slightly different from that of HNO_3 that is: $Li^+ < K^+ < Rb^+ < Cs^+ < Na^+$. Incomplete separation is obtained for Rb^+-Cs^+ ions in these conditions. On the amorphous material, the same elution order was found as observed on the strong acid-type organic resins.

The separation of alkali metal ions can be performed on a column containing amorphous antimony acid (AAa), using HNO_3 solution as the eluant.[79] However, the Na^+ ions became difficult to remove from the column with a few cycles of regeneration, or with increasing time of immersion in HNO_3 because of the gradual transformation of AAa to AAc.[80,93]

2. Data Concerning the Separation of Alkaline Earth Metal Ions

Baetsle and Huys[94] have reported the following affinity sequence for polyantimony acid (PAA) in HNO_3 media: $Ra^{2+} < Ba^{2+} < Ca^{2+} < Sr^{2+}$. The same selectivity order was found for AAc by Abe and Uno.[84] They reported that the H^+-Ba^{2+} exchange is relatively

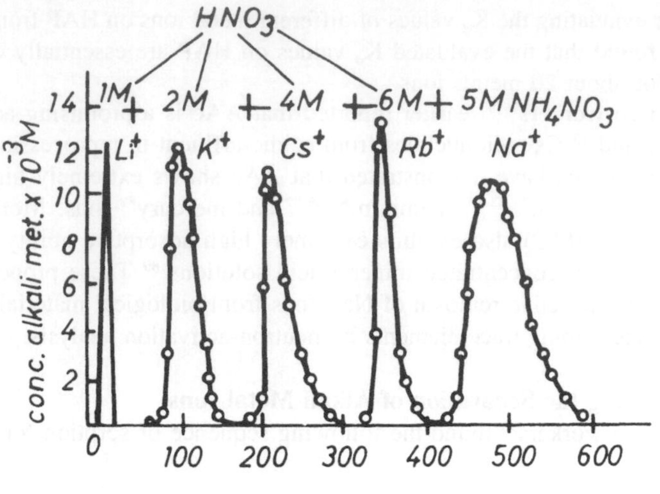

FIGURE 1. Separation of alkali metal ions on AAc using HNO_3 eluant with various concentrations and NH_4NO_3 solution. (From Abe, M., *Bull. Chem. Soc. Jpn.*, 42, 2683, 1969. With permission.)

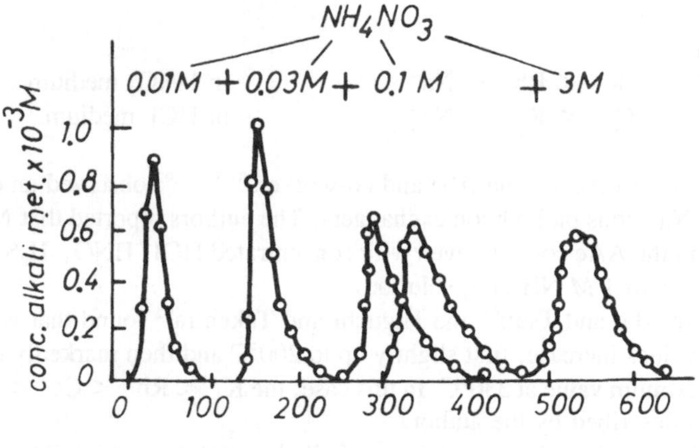

FIGURE 2. Separation of alkali metal ions on AAc using NH_4NO_3 solution with various concentrations. (From Abe, M., *Bull. Chem. Soc. Jpn.*, 42, 2683, 1969. With permission.)

fast: the equilibrium is reached in 24 h, whereas the other H^+-Me^{2+} exchanges are very slow. As reported by other authors,[85,94] much higher K_d values are obtained for the Ca^{2+} and Sr^{2+} ions on AAc than for those on PAA. Concerning the selectivity data, the separation of ^{90}Sr radionuclide from fission products has been reported by Baetsle and Huys,[94] the ^{90}Sr radionuclide recovery from acidic solution of fuel reprocessing has also been reported by Aubertin and co-workers.[95] Abe and Uno[84] described the separation of $Ca^{2+}Mg^{2+}$-Cs^+ ions on a small-size column (diameter = 0.4 cm, length = 2 cm) containing AAc. As mentioned by the authors, the separation of Mg^{2+} and Cs^+ from Ca^{2+} and Sr^{2+} is performed quantitatively, but the adsorbed Ca^{2+} and Sr^{2+} ions cannot be removed, even with concentrated HNO_3, 10 *M* NH_4NO_3, and 0.2 EDTA solutions—the last adjusted to pH = 10.

3. Data Concerning the Separation of Transition Metal Ions

The affinity of AAc toward the transition metal ions was examined by Abe and Kasai,[87] Abe and Sudoh[93] and Girardi and co-workers.[77] In HNO_3 media, the various authors found the following selectivity series: $Ni^{2+} < Mn^{2+} < Zn^{2+} < Co^{2+} < Cu^{2+} < Cd^{2+}$. The time for the equilibrium change of Cu^{2+}, Zn^{2+}, and Mn^{2+} is 4 d, whereas more than 1 month is necessary for Ni^{2+} and Co^{2+}.

The above authors also reported the following affinity sequences for tervalent metals in HNO_3 medium: $Al^{3+} < Ga^{3+} < Yb^{3+} < Fe^{3+} < La^{3+} < In^{3+}$.

Abe and Kasai[87] achieved satisfactory separation with 99% recovery of pure Ni^{2+} and Cu^{2+} ions by using a column (diameter = 0.4 cm, length = 2 cm) containing AAc and then eluting with 0.1 *M* and 6 *M* HNO_3 solutions, respectively.

According to Abe,[86] the Cd^{2+} ions adsorbed on an AAc column cannot be removed completely because of a tailing effect; total elimination is possible only by using 0.5 *M* KCl in 1 *M* HCl-acetone solution as an eluant. The rapid Cd^{2+} elution in the presence of KCl may be due to the lattice expansion, because the lattice distance for the H^+ form of AAc is 1.026 mm and that for the K^+ form is 1.047 mm. Belinskaiya and co-workers[96] have reported the use of PAA for the removal of Ag^+ and Cd^{2+} ions from concentrated radioactive solutions containing various other ions.

4. Application of Hydrous Metal Oxides to Real Samples

Most real samples such as biological and mineral substances, sea-water, and glassy-type materials, contain a large amount of Na^+ ions, compared to the other ions.[97]

a. Neutron Activation Analysis

Because of its high sensitivity to many elements, neutron activation analysis should be a suitable method for the simultaneous determination of trace elements or materials in micro amounts. However, these determinations in the low-activity level are often made difficult or impossible by the presence of a radiosodium background. It is difficult to eliminate sodium without affecting the concentration of other cations.[98,99] In many cases, the successful elimination of radiosodium permits a simultaneous determination of other cations and many trace elements by instrumental means. Girardi and co-workers[77,100,101] have developed a method using HAP for the removal of radiosodium from mixed radiotracer solutions obtained by wet digesting the irradiated samples. The adsorbent was used in the form of a chromatographic column or in batch experiments. A retention capacity of the order of 30 mg Na^+/g HAP has been obtained. The decontamination factor (DF) in this experiment was found to be 10^{10}.

The authors reported that in each chromatographic experiment 2 ml of the radiotracer solution (6 and 12 *M* HCl) was passed through the column at room temperature at a flow rate of 1.5 to 2 ml/min. Then the column was washed with two 15-ml portions of the acid solution of the same concentration at the same flow rate.

Of the 60 radionuclides tested, only Ta^{4+} and Na^+ ions were quantitatively retained at both HCl concentrations. Fluoride ions were retained only partially. Prolonged washing (60 ml) with 12 *M* HCl resulted in a 92% recovery, while the remaining 8% seemed to be irreversibly retained by the column. At the lower HCl concentration (6 *M*), a small percentage of Co^+, Sr^{2+}, Ba^{2+}, and Ra^{2+} was retained. Prolonged washing (50 ml) resulted in the complete recovery of all these elements. The elution of retained Na^+ ions was attempted by various eluting agents, such as concentrated NH_4NO_3, $NaNO_3$, $Na_2H_2Sb_2O_7$, H_2O, and MetOH in quantities of approximately 30 column volume. Washing with 200 column volume of mineral acids (HF, HCl, H_2SO_4, and HNO_3) and organic acids, e.g., tartaric and oxalic, was also ineffective. The total retention capacity of HAP was determined in 1 *M* HNO_3 (Figures 3 and 4).

At a trace Na^+ level, a column (diameter = 6mm, length = 20 mm; 0.6 g HAP) was successfully operated at a flow rate of 3 ml/min. The total separation process (adsorption and washing) was completed in about 5 min. The retention values of different ions are shown in Table 1.

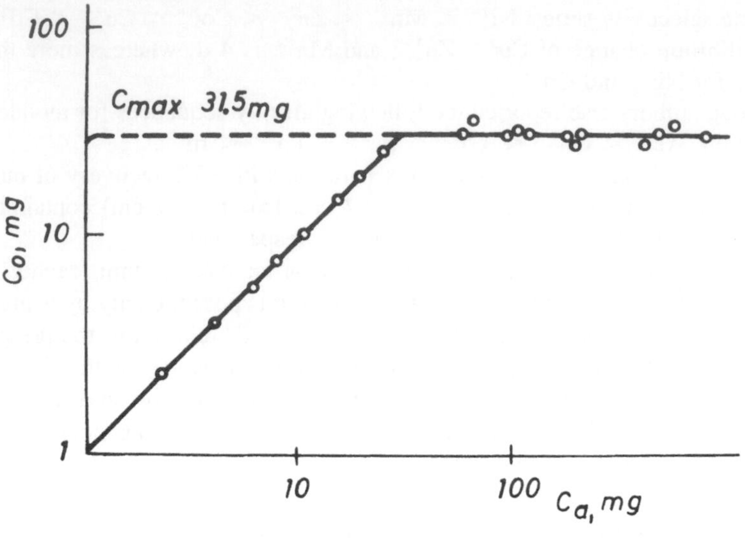

FIGURE 3. Retention capacity on HAP in batch equilibration. C_o — total amount of Na^+ in the solution (20 ml) before equilibration. C_a — total amount of Na^+ on HAP (1 g) after equilibration (From Girardi, F. and Sabbioni, E., *J. Radioanal. Chem.*, 1, 169, 1968. With permission.)

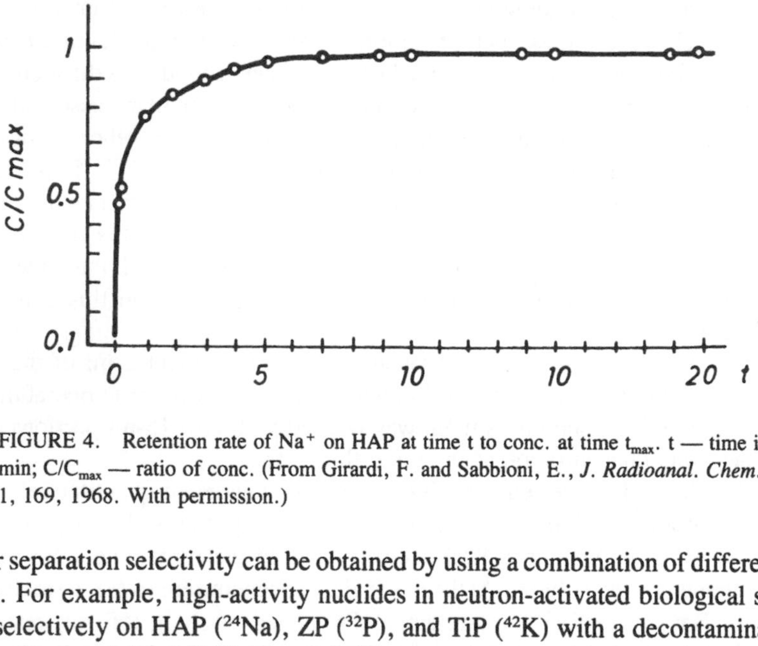

FIGURE 4. Retention rate of Na^+ on HAP at time t to conc. at time t_{max}. t — time in min; C/C_{max} — ratio of conc. (From Girardi, F. and Sabbioni, E., *J. Radioanal. Chem.*, 1, 169, 1968. With permission.)

Higher separation selectivity can be obtained by using a combination of different selective adsorbents. For example, high-activity nuclides in neutron-activated biological samples are separated selectively on HAP (^{24}Na), ZP (^{32}P), and TiP (^{42}K) with a decontamination factor higher than 10^{-4} and 10^{-2}.[78,102] Clearfield[103] reported the applications of HAP and AAc to some samples (Table 2).

The simultaneous determination of multielements is not very effective without the separation of radiosodium and/or radio-bromine. As all radiochemists know, the development of a separation procedure is often a long and painful experience.

Cuypers and co-workers[100] have established schemes for the systematic analysis of trace impurities in high-purity Al and Zn metals using synthetic inorganic ion exchangers. A typical separation scheme (Scheme 1) is shown below.

Tjioe and co-workers[104] have also developed an automated chemical separation in routine

TABLE 1
Retention of Different Ions on HAP (7 mm × 30 mm column)

Ion	Radiotracer	6 *M* HCl		12 *M* HCl	
		15 ml	30 ml	15 ml	30 ml
Ag^{+}	^{110m}Ag	97	99	98	100
Al^{3+}	^{28}Al	99	100	96	98
As^{a}	^{76}As	96	99	97	98
Au^{3+}	^{198}Au	95	98	95	99
Ba^{2+}	^{131}Ba, ^{133}Ba, and ^{140}Ba	86	95	95	97
Br^{-}	^{82}Br	92	99	97	100
Ca^{2+}	^{49}Ca	92	97	96	99
Cd^{2+}	^{115}Cd	96	99	96	99
Ce^{3+}	^{141}Ce and ^{144}Ce	95	99	98	99
Cl^{-}	^{38}Cl	97	99	98	99
Co^{2+}	^{60}Co	96	99	97	99
Cr^{3+}	^{51}Cr	98	100	96	98
Cs^{+}	^{134}Cs	88	95	88	97
Cu^{2+}	^{64}Cu	97	100	97	99
Eu^{3+}	^{154}Eu	97	100	97	100
F^{-}	^{18}F	82	90	60	70
Fe^{3+}	^{59}Fe	98	100	98	100
Ga^{3+}	^{68}Ga	96	99	96	99
Ge^{4+}	^{77}Ge	96	98	98	100
Hf^{4+}	^{175}Hf	98	99	98	99
Hg^{2+}	^{203}Hg	99	100	96	99
I^{-}	^{128}I and ^{131}I	95	99	91	99
In^{3+}	^{114m}In and ^{113m}In	98	99	97	99
Ir^{3+}	^{192}Ir	96	99	96	98
K^{+}	^{42}K	90	97	93	99
La^{3+}	^{140}La	97	98	98	99
Lu^{3+}	^{177}Lu	—	—	98	100
Mg^{2+}	^{27}Mg	98	99	94	99
Mn^{2+}	^{56}Mn and ^{54}Mn	99	100	97	99
Mo^{6+}	^{99}Mo and ^{101}Mo	97	98	95	99
Nb^{5+}	^{95}Nb (^{95}Zr)	96	99	96	99
Np^{4+}	^{239}Np	98	99	97	99
Na^{+}	^{24}Na and ^{22}Na	0	0	0	0
Ni^{2+}	^{65}Ni	98	100	98	99
Os^{4+}	^{191}Os	98	99	94	99
Pa^{5+}	^{233}Pa	70	93	96	99
Pd^{2+}	^{109}Pd	97	99	97	98
Pt^{4+}	^{199}Pt	98	99	97	99
Ra^{2+}	^{226}Ra	88	96	94	100
Rb^{+}	^{86}Rb	25	27	85	97
Re^{7+}	^{186}Re	99	100	96	98
Ru^{4+}	^{103}Ru and ^{106}Ru	96	99	96	99
S^{6+}	^{35}S and ^{37}S	98	99	99	100
Sb^{a}	^{124}Sb	97	99	97	99
Sc^{3+}	^{47}Sc and ^{46}Sc	98	99	97	99
Se^{4+}	^{75}Se	95	98	99	100
Sn^{a}	^{123}Sn and ^{125m}Sn	96	98	98	100
Sr^{2+}	^{85}Sr and ^{87m}Sr	92	96	92	97
Ta^{5+}	^{182}Ta	0	0	0	0
Tc^{7+}	^{99m}Tc (^{99}Mo)	100	100	100	100
Tb^{3+}	^{160}Tb	—	—	98	99
Te^{4+}	^{131}Te	95	99	96	98
Ti^{4+}	^{51}Ti	98	99	98	100

TABLE 1 (continued)
Retention of Different Ions on HAP (7 mm × 30 mm column)

		6 *M* HCl		12 *M* HCl	
Ion	**Radiotracer**	**15 ml**	**30 ml**	**15 ml**	**30 ml**
U^{6+}	^{239}U	98	100	99	100
V^{a}	^{52}V	97	98	98	99
Y^{3+}	^{90}Y	—	—	98	99
Yb^{3+}	^{160}Yb	96	98	96	98
W^{6+}	^{187}W	95	97	97	100
Zn^{2+}	^{65}Zn	99	100	98	99
Zr^{4+}	^{95}Zr and ^{97}Zr	97	99	96	99

Note: Results are reported as the percent eluted in the first + second fractions.

[a] Uncertain oxidation state.

From Girardi, F. and Sabbioni, E., *J. Radioanal. Chem.*, 1, 169, 1968. With permission.

neutron activation analysis for biological samples based on a combination of distillation and the use of HAP and Dowex 2 × 8 ion exchangers (Scheme 2). Numerous works were presented by various authors[105-121] using the ion exchangers for solving separations of Na from biological samples.

Schumacher and co-workers[122] have recently reported a semi-automated and non-time-consuming radiochemical separation scheme for determining 25 trace elements in biological specimens by means of 16 different organic and inorganic ion exchangers. The total time needed for one separation is 3 h and the simultaneous processing of four samples takes only 4.5 h. It is, of course, known that the chromatographic separation of Li^+ from Na^+ on sulfonated cation-exchange resin is not favorable because of the low values of the separation factor (between 1.5 and 2.0), whereas AAc exhibits an extremely large separation factor ($\alpha Li^+/Na^+ = 2.6 \times 10^5$ and $\alpha Li^+/K^+ = 1.5 \times 10^3$). This fact indicates that selective separation is feasible for the Li^+ ion from these metal ions, the technique of which was developed by Abe and co-workers.[123,124]

It was mentioned earlier that a promising potential application of inorganic ion exchangers is their use in radioisotope generator systems. An alumina column is generally employed for these purposes. Some years ago, Arino and co-workers[125] developed a new $^{68}Ge/^{68}Ga$ generator system using an HAP column. They postulated that HAP could selectively adsorb the ^{68}Ge radionuclides and then ^{68}Ga may be eluted with 2% Na oxalate solution. This miniaturized generator can provide a high recovery of ^{68}Ga (80 ± 10%) with 99.9% radionuclidic purity.

b. Water Purification

A few years ago, attention was directed toward the application of synthetic inorganic ion exchangers for water electrolysis and the production of Na^+ and Cl^- gas. One of the main problems in the development of advanced alkaline water electrolysis, which operates at about 150°C, is to find a replacement for the chrysotile asbestos which is currently used as the gas separator material. Vandenborre and Leysen[126] attempted to develop an electrolysis cell based on PAA. They obtained thin (25 mm) sheets of PAA with a surface area of 500 cm^2. An evaluation of the PAA membrane under electrolysis conditions has been performed on an electrolysis unit cell (EUC) having a commercial 5% Pt catalyst support as the cathode

TABLE 2
Application to Real Samples by Means of HAP and AAc

Element	Separation from	Eluant	Elution order	Exchanger	Sample	Note	Ref.
Na	Other elements	6—10 *M* HCl or HNO_3	Na retain	HAP	Biol.	For NAA	76, 89, 101, 104—107
Na, K	Other elements	10 *M* HCl	Na retain	HAP/resin	Biol.	For NAA	108—110
Na	Transition metals	7—10 *M* HCl	Na retain	HAP/TiP/ZrP	Biol.	For NAA	78, 102, 107
Na	43 elements	7 *M* HCl	Na retain	HAP/Al_2O_3	Biol.	For NAA	111
Na	25 elements	Various	Element isolate	HAP/various exchangers	Biol.	For NAA	105
Na	Transition metals	6 *M* HCl	Na retain	HAP	Rock	For NAA	105, 112, 113
Li	Na, K, Ca, etc.	0.4 *M* HCl	Li isolate	AAc	Rock	For determination Li	114
Na	Transition metal	12 *M* HCl	Na retain	HAP	Opt. glass	For NAA	115, 116
Na	Other elements	12 *M* HCl	Na retain	HAP	Pure metals	For NAA	100—117
Na	Cu, Mn, Co, Fe, Cr	12 *M* HCl	Na retain	HAP	Pure metal carbonate	For NAA	118
Na	Alkali metals	HCl	Na isolate	HAP	Pure water	For detn. Na, Cl	119
^{43}K	Other elements	HNO_3	^{43}K isolate	HAP	V target	For isolation	120
Na, K	Li	0.1 *M* HNO_3	Na, K, Ca retain	AAc	LiCl	For purification	106
Na, K	Other elements	9 *M* HCl	Na, K retain	HAP/KPW	K salt	Removal radio K	121
^{68}Ga	^{68}Ge	0.2% Na oxalate	^{68}Ga elute	HAP	Generator	$^{68}Ga/^{68}Ge$	108

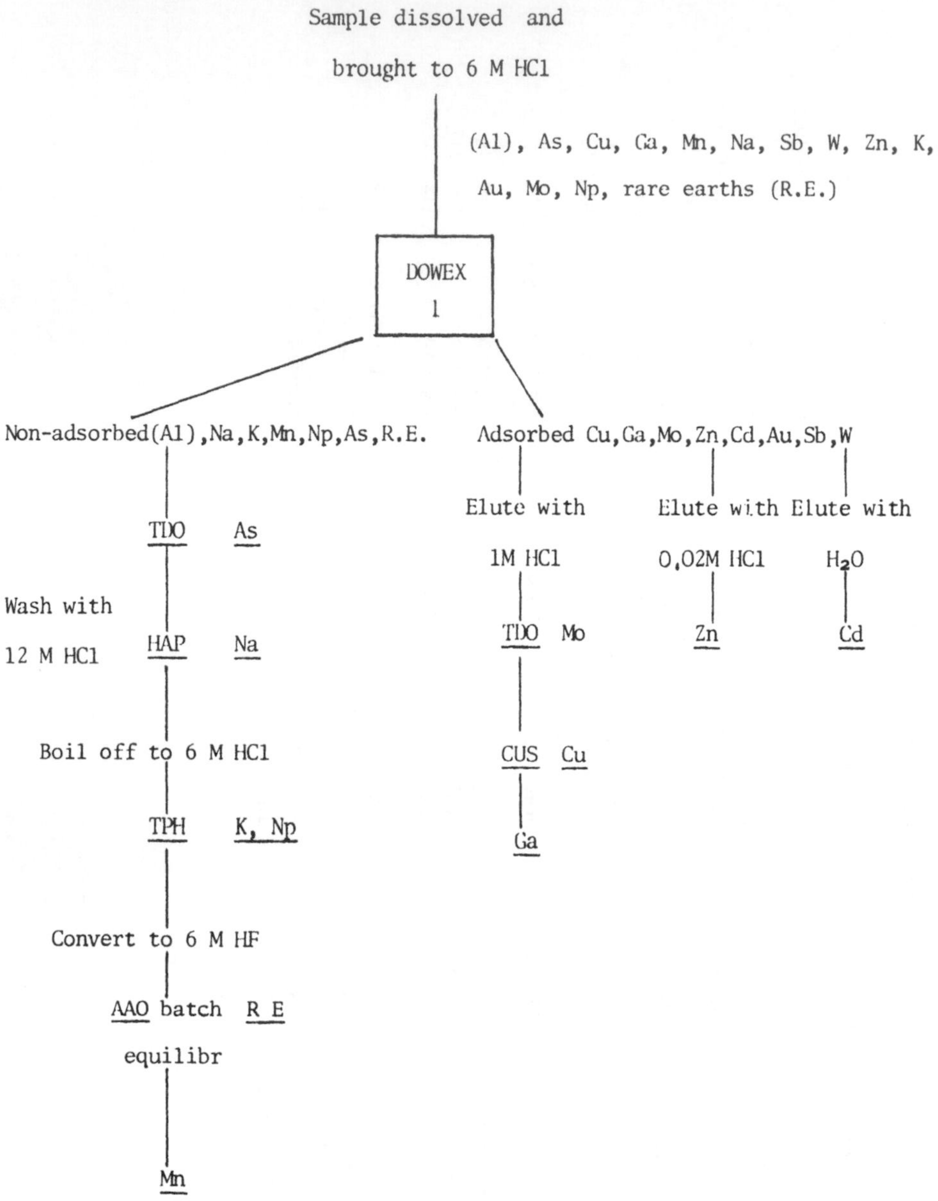

SCHEME 1.

and a 50-mesh Ni grid as the anode. The result indicates that the PAA membrane resists chemical degradation in a hot alkaline solution up to 150°C and, in contrast to nafion, no special problem will have to be faced for a period of 300 h continuous operation at room temperature. The ohmic drop in the cell was low as a result of using a thin membrane which possesses excellent conducting properties.

c. Other Separations

Amphlett et al.[127] and Abe and Ito[70] indicated that tungstic acid (TA) exhibits a pH titration curve with a break-point, indicating both an ion-exchange reaction at pH values below 5 and the dissolution of the exchanger at higher pH values. For hydrous tungsten (VI) oxide (HTO), De and Chowdhury[128] found no breakpoint in the pH titration curve.

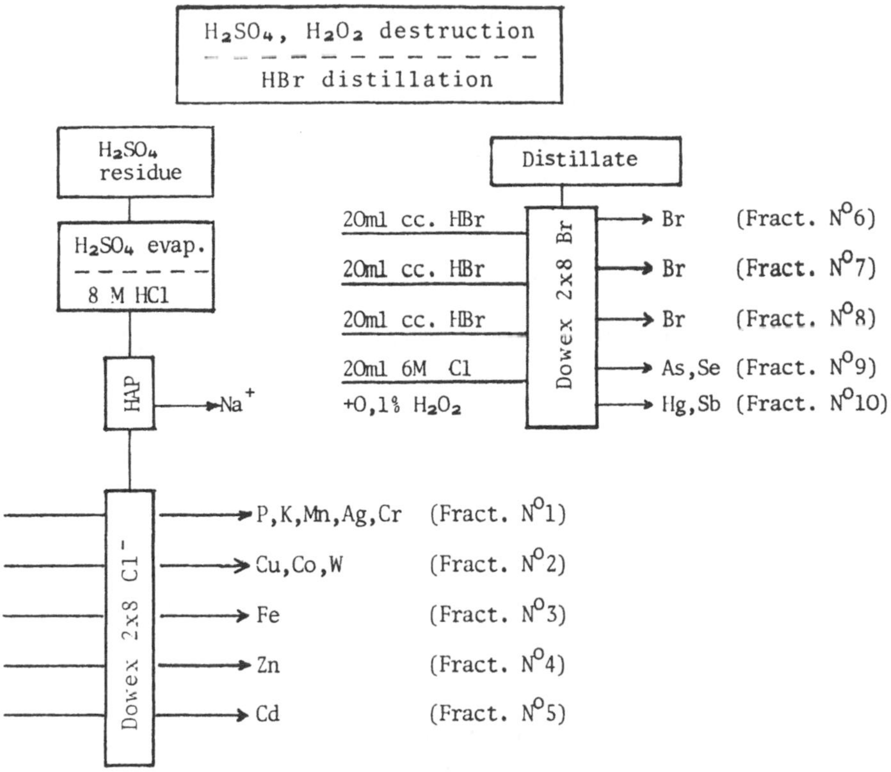

SCHEME 2. (From Tjioe, P. J., et al., *J. Radioanal. Chem.*, 16, 153, 1973. With permission.)

Because of the solubility of TA and HTO in alkaline solutions, the exchange capacity of these ion exchangers must be determined at a pH less than 5. Abe and Ito[70] reported that the maximum uptakes for K^+ was about 0.4 and 0.6 meq/g (at$\mu = 0.1$) on white and yellow tungstic acid, respectively. The amount of H^+ ions liberated by the addition of a neutral salt solution is very small because the oxides are weakly acidic exchangers. The exchange capacities of two different oxides for Na^+ at different pH values are given in Table 3.[129]

When a 2 M NaCl solution in buffer media at pH = 6 was passed through the HTO column, the elution curve of H^+ ions showed a long tailing effect because of the presence of acid sites with different acidities. The anion exchange capacities for batch no. 1 were found to be 0.13, 0.08, and 0.05 meq Cl^-/g at pH = 0, 1, and 3, respectively.

Szalay and Bartha[130] studied the equilibrium in aqueous media for the exchange of K^+, Tl^+, and Tl^{3+} ions on the yellow and blue acids ($WO_{2.87}$ H_2O). The equilibrium could be characterized by a Langmuier-Hueckel isotherm. At 25°C, the adsorption is dependent on the HCl concentration. The authors reported that Tl^{3+} did not adsorb on the yellow acid, but did on the blue one, as a Tl^+ ion, by virtue of a redox process. From the isotherm, the number of active sites on the surface was calculated, and from this and the lattice constant of the oxide, the active specific surface area for K^+ ions was found to be 9.4 and 0.4 m^2/g in 0.05 and 2.87 M HCl solutions, respectively, for the blue acid; the corresponding values for the yellow one were 1.8 and 0.04.

The K_d values of various metal ions were determined and possible separations on HTA were studied by De and Chowdhury.[128] The ion exchanger (0.5 g) was immersed in 50 ml of a solution containing 2×10^{-4} M metal ions in 1 M NaCl solution adjusted to pH = 5 with a buffer solution. They found that, among the metal ions, Ni^{2+} and Mn^{2+} are completely adsorbed not only in this concentration range, but also when larger concentrations were used. The results obtained are shown in Table 4.

TABLE 3
Cation-Exchange Capacity of TO at Different pH[a]

Batch No.	pH of solution		Exchange capacity (meq H^+/g)
	Initial	Final	
1[b]	2	1.8	0.2
1	3	2.7	0.3
1	4	3.65	0.35
1	5	4.55	0.92
1	6	5.2	2.12
3[c]	2	1.8	0.2
3	3	2.7	0.3
3	4	3.6	0.4
3	5	4.3	2.76
3	6	4.9	2.98

[a] With NaCl solutions.
[b] Batch 1: 0.1 *M* Na_2WO_3 (100 ml) + 3 *M* HCl (100 ml), aged for 2 h.
[c] Batch 3: 0.1 *M* Na_2WO_3 (100 ml) + conc. HCl (50 ml), aged for 4 h.

TABLE 4
Distribution Coefficients of Some Metals

Metal ion	K_d values	
	Batch No. 1	Batch No. 3
Zn^{2+}	51.37	21.6
Cu^{2+}	52.23	46.97
Co^{2+}	35.0	12.0
Ca^{2+}	134.0	104.2
Sr^{2+}	116.0	91.5
Ba^{2+}	28.57	12.51
Mg^{2+}	8.7	4.132
Ni^{2+}	ta[a]	ta
Mn^{2+}	ta[a]	ta

Note: The batches are the same as in Table 3.

[a] ta = total adsorption.

The authors[128] mentioned that some useful separations can be achieved for Mg^{2+} from a mixture of Zn^{2+}, Cu^{2+}, Ca^{2+}, Sr^{2+}, Mn^{2+}, and Ni^{2+}. Co^{2+} can be separated from Ni^{2+} and Mn^{2+}, by using a glass column of 1.5 cm diameter containing about 5 g of the ion exchanger. A very effective separation of Ca^{2+} has been carried out from a large amount of Mg^{2+} (Figure 5).

The adsorption capacities of Np^{4+} on TA (prepared as a bluish-green glassy gel) were dependent upon the concentration of both HNO_3 and NP^{4+}. The adsorption of vanadium on TA was considerably less than that of Np^{4+}. The breakthrough points of Np^{4+} were 2000 or 1000 column volumes at 1 or 2 *M* HNO_3 with 1×10^{-2} mol/1 of vanadium concentration, respectively. Thus, TA as an ion exchanger was suitable for separating Np^{4+} from $VO_2(NO_3)_2$ solution.[130,131]

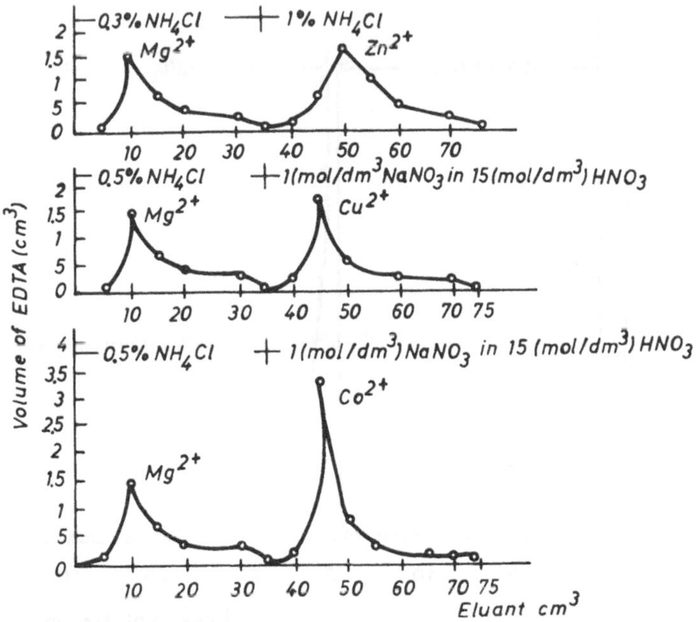

FIGURE 5. Separation of Zn^{2+}, Cu^{2+}, and Ca^{2+} from Mg^{2+} on TO column (Batch No. 1). (From De, A. K. and Chowdhury, K., *Chromatographia*, 11, 586, 1978. With permission.)

III. APPLICATIONS OF ZIRCONIUM PHOSPHATE AND RELATED ION EXCHANGERS

A. ZIRCONIUM PHOSPHATE

The selectivity of amorphous zirconium phosphate (ZPa) was found to be the same as that of the organic resins with sulfonic groups,[132,133] i. e. $Li^+ < Na^+ < K^+$, $NH_4^+ < Rb^+ < Cs^+$. Among them, the Rb^+ and Cs^+ metal ions are better adsorbed and the Tl^+ and Ag^+ metal ions are less well adsorbed than the other alkali metal ions. From acidic solutions, the alkaline earth and rare earth metal ions are much less adsorbed than the alkali metal ions and could easily be separated on ZPa.

Amphlett and McDonald[134] have shown that ZPa can also work as an ionic sieve. The sorption data of more than 60 ions were presented by Maeck and co-workers.[135] They found that on the H^+ form of the exchanger, alkali metal ions are adsorbed much more strongly than the alkaline earth ions. For that reason, the alkali metal ions can easily be separated. On the NH_4^+ exchanger, the opposite was true.

Amphlett[136] reported a method for separating macro amounts of Rb^+ and Cs^+ ions using ZPa as ion exchanger. The separation was carried out using a column with diameter = 6 mm and length = 25 mm. The ions were eluted with 0.1 *M* NH_4NO_3 (Cs^+) and NH_4NO_3 + HNO_3 (Rb^+) solutions. One of the fractions contained 97% of the initial amount of Rb^+ (without Cs^+) and the other contained 99% of the initial amount of Cs^+ (without Rb^+). Kraus and co-workers[137,138] used the same ion exchanger for separating micro amounts of Rb^+ and Cs^+ ions. By this process, Rb^+ was eluted with 1 *M* and the Cs^+ with concentrated NH_4Cl solutions, with a flow-rate of 8 to 9 ml/h.

Amphlett and co-workers[139] also described the separation of Rb^+ from Sr^{2+}, eluting them from the column with 1 *M* NH_4NO_3 (Rb^+) and 0,.1 *M* NH_4NO_3(Sr^{2+}) solutions. Also, on a similar column, ^{137}Ba and ^{137}Cs radionuclides may have been separated, the latter having been eluted with 1 *M* HCl solution.

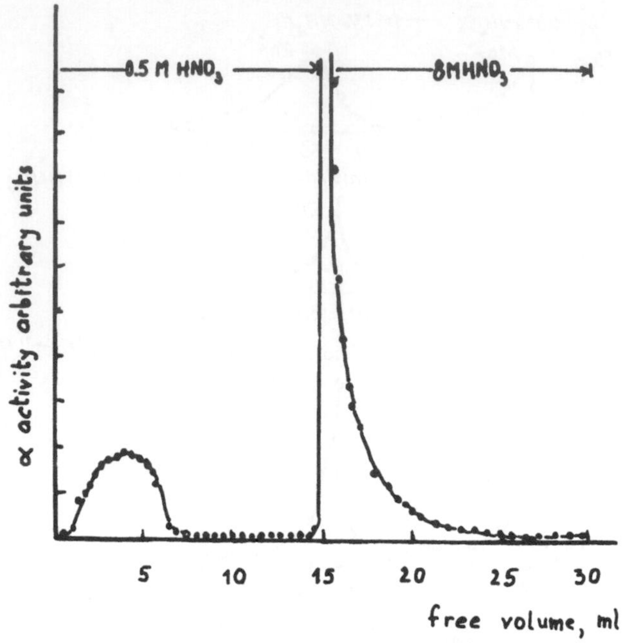

FIGURE 6. Separation of U, Pu, and fission products using ZPa column. (From Gal, I. and Ruvarac, A., *J. Chromatogr.*, 13, 549, 1964. With permission.)

Gal and Ruvarac[140] reported the separation of Pu from U and from the fission products, using a ZPa column. A mixture of 0.5 *M* HNO_3 and 0.02 *M* $NaNO_3$ solutions containing U, Pu, and fission products was percolated through a column (diameter = 0.4 cm, length = 15 cm) containing ZPa (flow rate = 0.73 ml/min). With these conditions, the author had good separation results (Figure 6).

The following decontamination factors (DF) were found for Pu products:

	UO_2^{2+}	^{90}Sr	^{144}Ce	^{95}Nb	^{95}Zr	^{137}Cs
$DF_{20}°$	~10^3	~10^3	178	24	>10^3	1.5
$DF_{50}°$	~10^3	~10^3	172	15	>10^3	1.1

Using ZPa in both the H^+ and NH_4^+ forms, Zsinka and the author[141,142] reported the separation of Sr^{2+} from Rb^+. The dynamic experiments were carried out as follows: column length = 300 mm, diameter = 8 mm; flow rate = 0.05 ml/min, ion-exchanger weight = 3 g; particle size = 0.4 mm. The concentrations of the ions were $C_{RB}{}^+ = 3.3 \times 10^{-5}$ *M* and $c_{Sr}{}^{2+} = 3.0 \times 10^{-1}$ *M*.

Based on the elution data (Figure 7), it can be stated that at the ion concentration mentioned, a good separation of Ba^{2+} and Cs^+ can be achieved on a ZPa column. The separation of the same ions and that of Sr^{2+} from Rb^+ were carried out using crystalline zirconium phosphate (ZPc). The column operations were absolved under the following conditions: column length = 150 mm, diameter = 5 mm; flow rate = 0.1 ml/min; concentration = 0.2 mval/g per ion. The observed distribution coefficients for the ions are shown in Tables 5 and 6.

The ions were eluted from the column with both HC1 and $HC1O_4$ solutions of various concentrations (Figures 8 and 9). In both cases, the elution curves showed well-separated peaks containing the given ions in a relatively small volume of the eluant. Making the usual

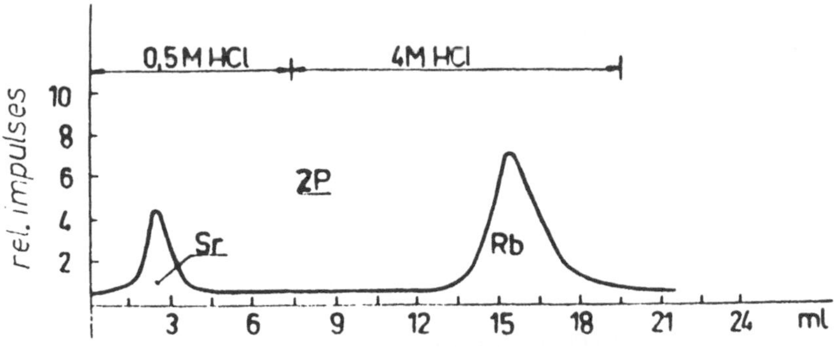

FIGURE 7. Separation of Sr^{2+} — Rb^+ on ZPa column. (From Zsinka, L. and Szirtes, L., *Radiokhimiya*, 12, 774, 1970. With permission.)

TABLE 5
Distribution Coefficients for Ba^{2+} and Cs^+

	$HClO_4$		HCl		HNO_3		NaCl	
Concentration (*M*)	Ba	Cs	Ba	Cs	Ba	Cs	Ba	Cs
0.001	0.22	3.85	0.03	3.91	2.20	3.71	1.30	3.96
0.01	0.05	3.67	0.01	3.44	0.92	3.28	1.09	3.82
0.1	>0.01	1.20	0.01	3.16	0.20	3.24	0.81	3.67
1.0	>0.01	0.62	>0.01	2.40	>0.01	2.37	0.35	3.11
5.0	>0.01	0.10	>0.01	0.85	>0.01	0.53	—	—
10.0	>0.01	>0.01	>0.01	0.64	>0.01	0.21	—	—
15.0	>0.01	>0.01	—	—	—	—	—	—
20.0	>0.01	>0.01	—	—	—	—	—	—

TABLE 6
Distribution Coefficients for Sr^{2+} and Rb^+

	$HClO_4$		HCl		HNO_3		NaCl	
Concentration (*M*)	Sr	Rb	Sr	Rb	Sr	Rb	Sr	Rb
0.001	1.61	2.52	0.03	2.35	0.95	2.42	0.72	2.62
0.01	0.83	2.38	0.01	2.13	0.50	2.22	0.65	2.25
0.1	>0.01	1.78	>0.01	1.58	0.03	1.68	0.42	2.03
1.0	>0.01	1.04	>0.01	1.04	>0.01	1.06	>0.01	1.67
5.0	>0.01	0.33	>0.01	0.41	>0.01	0.06	—	—
10.0	>0.01	0.11	>0.01	0.09	>0.01	0.01	—	—
15.0	>0.01	>0.01	—	—	—	—	—	—
20.0	>0.01	>0.01	—	—	—	—	—	—

evaluation of the gamma spectra, 99% of the initial radioactivity was adapted to the elution peaks. Also from the peaks of the gamma spectra, the radionuclidic purity was determined by conventional methods, to be 99.0 to 99.6% for various solutions. In such a way, the efficiency of the separations was determined indirectly, and a relatively good value was found. In connection with the separation of products of parallel nuclear reactions /(n,γ) and (n,p)/, the separation of Co^{2+} from Fe^{3+} was carried out using ZPa.[143] The dynamic studies were carried out under the following conditions: column length = 170 mm, diameter = nm; weight of the ion exchanger = 5 g, particle size = 0.4 mm; flow rate = 0.2 ml/min. The metal ions were eluted with 1 *M* HC1 (Co^{2+}) and concentrated HC1 (Fe^{3+}) solutions as shown in Figure 10.

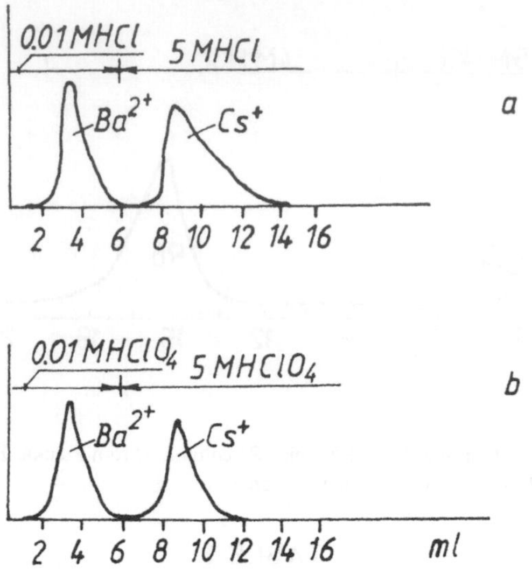

FIGURE 8. Elution curves of Ba^{2+} and Cs^+ in various media. (a) In HCl medium; (b) in $HClO_4$ medium. (From Szirtes, L. and Zsinka, L., *J. Radioanal. Chem.*, 30, 131, 1976. With permission.)

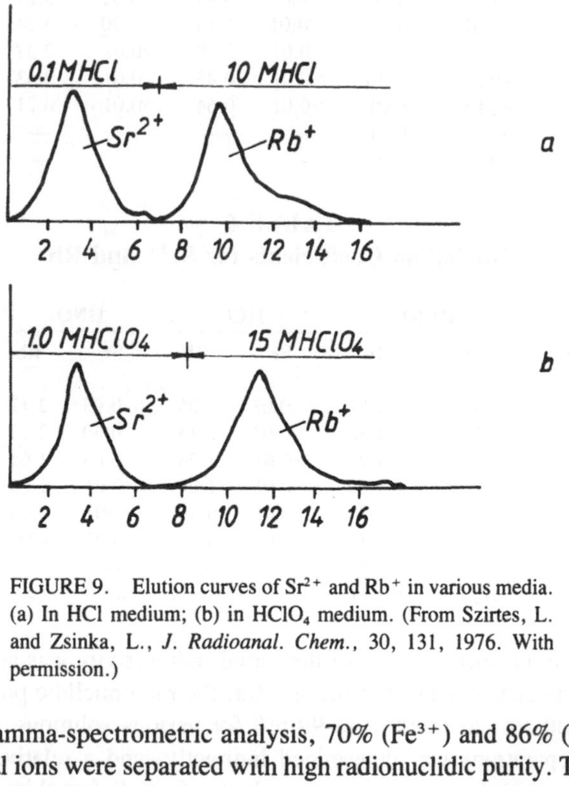

FIGURE 9. Elution curves of Sr^{2+} and Rb^+ in various media. (a) In HCl medium; (b) in $HClO_4$ medium. (From Szirtes, L. and Zsinka, L., *J. Radioanal. Chem.*, 30, 131, 1976. With permission.)

Based on the gamma-spectrometric analysis, 70% (Fe^{3+}) and 86% (Co^{2+}) of the initial amounts of the metal ions were separated with high radionuclidic purity. The same separation was repeated for irradiated target material. The irradiated 0.1 g cobalt metal (target material) was dissolved in concentrated HCl. After evaporation, the residue was taken up in 5×10^{-3} *M* HCl, then separated on the column described above. The same elution results were found as in the case of the model separation. The elution curves for other metal-ion pairs,

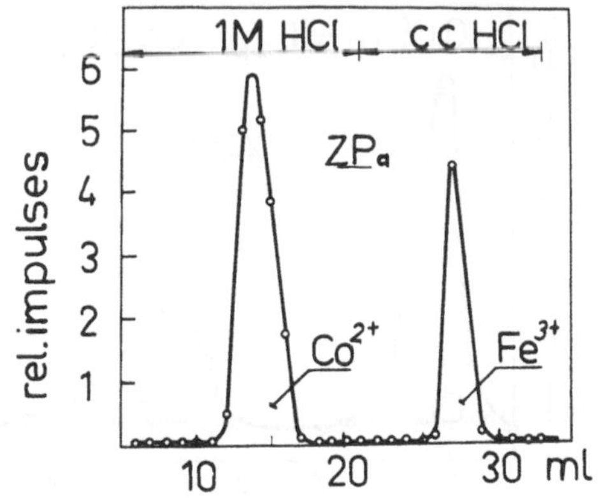

A

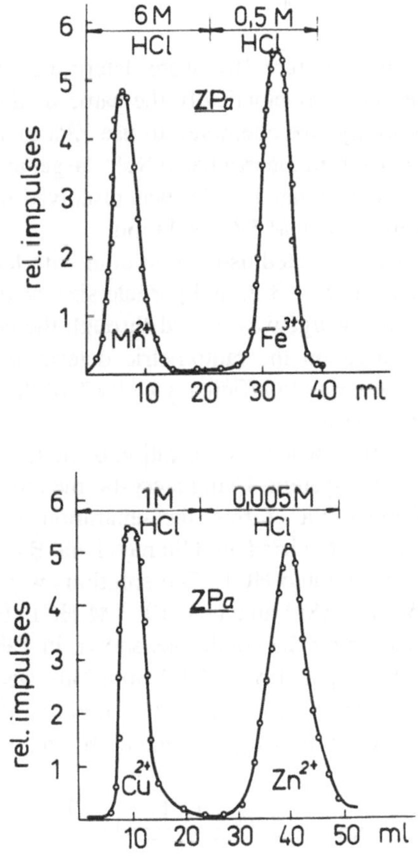

B

FIGURE 10. (A) Separation of Co^{2+} and Fe^{3+} on ZPa column. (From Iofa, B. Z., Szirtes, L., and Zsinka, L., *Radiokhimiya*, 10, 491, 1968. With permission.) (B) Separation of transition metal ions on ZPa column.

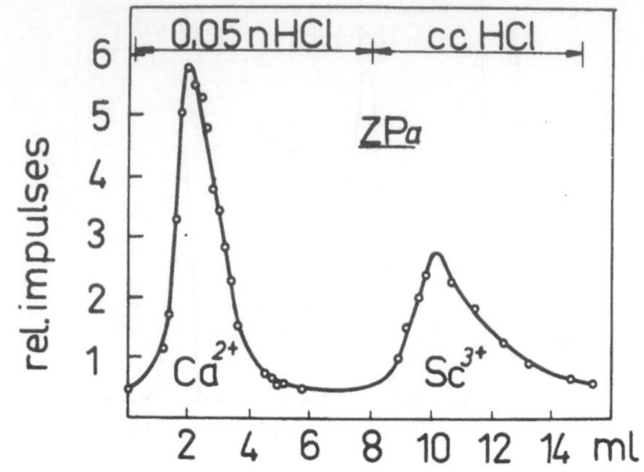

FIGURE 11. Elution curves of Ca^{2+} and Sc^{3+} on ZPa column. (From Zsinka, L. and Szirtes, L., *Mag. Kem. Lapja,* 21, 536, 1966. With permission.)

i.e., Cu^{2+}-Zn^{2+} and Mn^{2+}-Fe^{3+}(Figure 10), were determined under the same conditions. The efficiency of the separation was practically the same as that found in the Co^{2+}-Fe^{3+} separation. As another possibility, we attempted to use ZPa to separate a micro amount of Ca^{2+} (carrier-free ^{45}Ca nuclide) from the irradiated Sc^{3+} target material.[144] The investigation was carried out as follows: the irradiated Sc_2O_3 was dissolved in concentrated HC1, evaporated, and the residue taken up in 1 *M* HC1 solution.

The column operation was perfected using a column with length = 170 mm, diameter = 5 mm, weight of the exchanger = 3 g, and particle size = 0.4 mm. The solution, with a Ca^{2+} concentration of 0.08 mval/g, was passed through the column, with a flow rate of 0.18 ml/min (Figure 11). Based on the radiometric determinations of the solutions, the elution peaks showed a 99% separation efficiency (~99% of the initial amount of ^{45}Ca was found with high radionuclidic purity).

As mentioned earlier, another possible application of inorganic ion exchangers is in the so-called radionuclidic generator system. Using a crystalline zirconium phosphate (ZPc) ion exchanger in both H^+ and mono-Na^+ forms, the separation of ^{113}Sn-^{113}In and ^{137}Ba-^{137}Cs was presented. Before separation, the irradiated Sn metal and $BaCO_3$ targets were chemically treated, i.e., dissolved in concentrated HC1. The solutions were then evaporated, and the residues taken up in 0.01 *M* HC1 (Sn) and 1 × 10^{-3} *M* HC1 (Ba) solutions, respectively. The generators were operated under the conditions shown in Table 7.

For the elutions, 1 *M* $HC1O_4$ and 5 *M* HC1 solutions were used, respectively. The elutions were repeated three times, in each case after attaining the radioactive equilibrium. The time necessary for this equilibrium was calculated by the well-known formula:

$$t_{max} = 3.32 \frac{T_1 T_2}{T_1 - T_2} \log \frac{T_1}{T_2} \quad (1)$$

where T^1 and T^2 are the half-lives of the mother and daughter nuclides, respectively. With regard to the results from the radiometric investigation of the eluants, about 90% of the initial amount of the radionuclides (^{113}In and ^{137}Cs) was eluted with a radionuclidic purity higher than 98%. On repeating the same work with the solutions having initially higher ion concentrations, a worthwhile separation efficiency was found. For this reason, these generator systems can be used only where low radioactivity is necessary, (e.g., for educational purposes). Also, ZPc in monosodium form was used to determine the amount of radiostrontium

TABLE 7
Conditions of the Separation

	^{113}Sn—^{113m}In	^{137}Ba—^{137}Cs
Column length	100 mm	100 mm
diameter	10 mm	10 mm
Ion exchanger	ZPc in H^+ form	ZPc in HNa form
Interlayer dist.	0.76 nm	1.105 nm
weight	2 g	2 g
Flow rate		
Adsorption	0.1 ml/min	0.5 ml/min
Elution	1.0 ml/min	0.5 ml/min
Concentration of initial solution	0.06 mval/g	0.01 mval/g

(^{90}Sr) in wastewater.[145] For the examination, 0.25 g of the ion exchanger with a particle size of 0.4 mm was used. Then 5-ml $SrCl_2$ solutions of various concentration (0.0225, 0.045, 0.06, and 0.25 mol/dm^3) were added and shaken for 3.5 h to attain equilibrium. (According to Alberti et al.,[146] this takes about 20 min.) The conversion was then calculated by means of the formula:

$$w = \frac{C\ V}{b\ g}\left(1 - \frac{I_e}{I_0}\right) 100\% \tag{2}$$

where C is the concentration of the solution in millimoles per cubic centimeter, where V is the volume of solution in cubic centimeters, where b is the weight of ion exchanger in grams, where g is the amount of Na^+ ions in millimoles per gram, I_0 and I_e are the radioactivity (impulses) of the samples before and after equilibrium (2 ml), respectively (Figure 13).

Additionally, the measurements were repeated at various pH values in the range of 2.8 to 11 with a constant Sr^{2+} concentration (0.06 mol/dm^3) of the solution, as shown in Figure 14.

It can be seen that the Sr^{2+} conversion is independent of the pH, and its value is higher than shown by the 50%, which means that, besides Na^+-Sr^{2+}, some H^+-Sr^{2+} exchange also takes place. During the ion-exchange process, the pH of the solution decreases from 4 to 2.

With these results, a wastewater sample with a saline concentration of 0.9 mol/dm^3 and a 0.4 Bq/dm^3 radiostrontium (^{90}Sr + ^{90}Y) content was investigated by the same method. Initially, the sample was filtered (to free it from mechanical impurities); then, the solution concentration was adjusted with $SrC1_2$ solution to 1.06 mol/cm^3. A 53% Sr^{2+} conversion was found for the sample and the decontamination factor (DF) as a function of the Sr^{2+} concentration was determined (Figure 15).

The DF was 41, i.e., 2.5% of the initial amount of radiostrontium remained in the sample solution after its treatment. Under the given conditions, the total amount of radiostrontium may quantitatively be retained from the wastewater by using a two-step separation.

Gal and Gal[147] described a method for separating U and Pu using both ZPa and titanium phosphate (TPa). The dynamic studies were executed using a column with a length of 8 cm and a diameter of 0.26 cm; adsorption and elution were carried out at a flow rate of 0.2 ml/min. The ^{90}Sr, ^{90}Y, ^{144}Ce, UO_2^{2+}, and ^{137}Cs radionuclides were separated from the NH_4C1 solution with various concentrations (0.1, 1, 6, and 5 *M* to the nuclides, respectively). Figure 16 shows that the authors obtained a good separation for U. A good thermal stability in both the dynamic and static conditions (e.g., when the ion exchange is proceeding at about 300°C) of the inorganic ion exchangers, first zirconium phosphate, means that they are

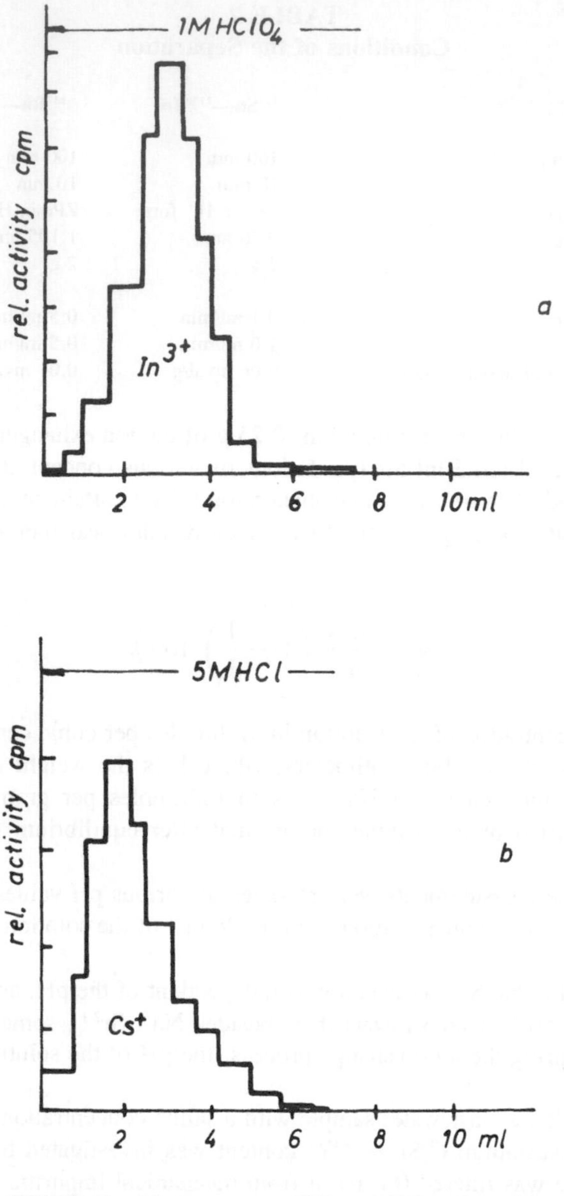

FIGURE 12. Elution curves of ^{113}In and ^{137}Cs radionuclides on ZPa column. (From Szirtes, L. and Zsinka, L., 1st Danube Symp. Chromatography, Szeged, Hungary, 1978. With permission.)

capable of decontaminating cooling water circulating in a closed cycle at a high temperature.[148,149] While these water samples are generally of a neutral or basic character, for the sorption of hydrolyzed phosphates (under these conditions), it is necessary to have another column filled with hydrous ZrO_2.

Grover and Chidley[150] reported that this system enabled ^{60}Co, ^{90}Sr, ^{137}Cs, and ^{131}I radionuclides to be separated from the cooling water of a nuclear reactor. As a matter of fact, the authors did not repeat the work at about 300°C, so it cannot be known whether the separation system works at such a temperature with the same efficiency. Here, this work can be postulated as a theoretical rather than a practical one.

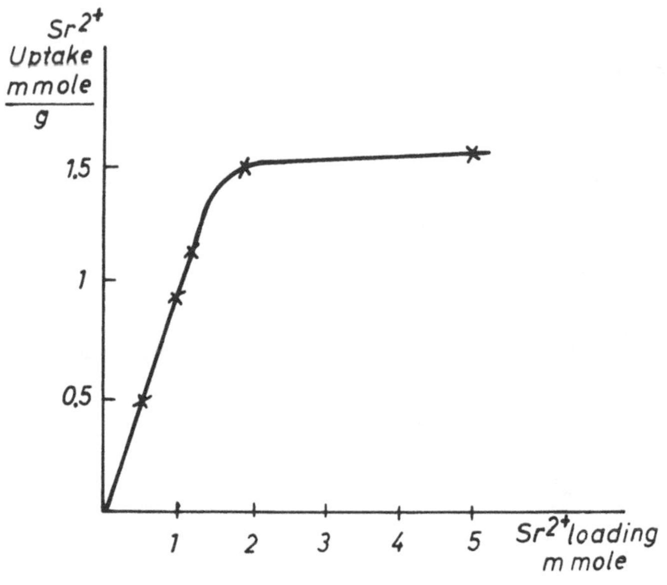

FIGURE 13. Sr^{2+} uptake versus Sr^{2+} concentration. (From Kornyei, J. and Szirtes, L., *Izotoptechnika* 23, 243, 1980. With permission.)

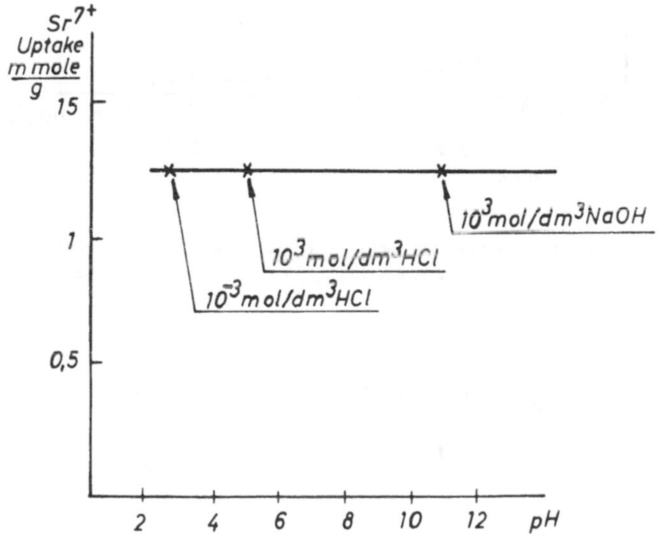

FIGURE 14. Sr^{2+} uptake at various pH at Sr^{2+} concentration of 0.06 mol/dm^3. (From Kornyei, J. and Szirtes, L., *Izotoptechnika*, 23, 243, 1980. With permission.)

B. OTHER ION EXCHANGERS CONTAINING ZIRCONIUM

The alkali metal and alkaline earth metal ions have been successfully separated using zirconium molybdate (ZM), titanium molybdate (TM), and titanium tungstate (TW) ion exchangers.[151] During these experiments, the column operations were carried out under the following conditions: column length = 300 mm, diameter = 8 mm, weight of the ion exchanger = 3 g, particle size 0.4 mm, and flow rate = 0.15 ml/min. The total ion concentration of the initial solution was 0.25 mval/g. The elution curves are shown in Figures 17 and 18.

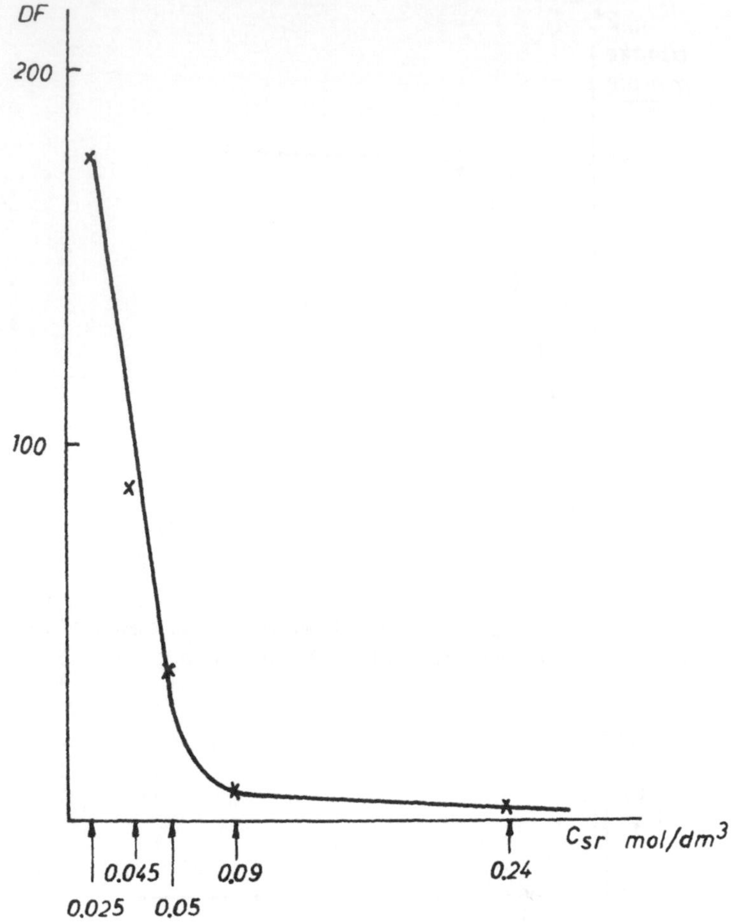

FIGURE 15. Decontamination factor as a function of Sr^{2+} concentration. (From Kornyei, J. and Szirtes, L., *Izotoptechnika*, 23, 243, 1980. With permission.)

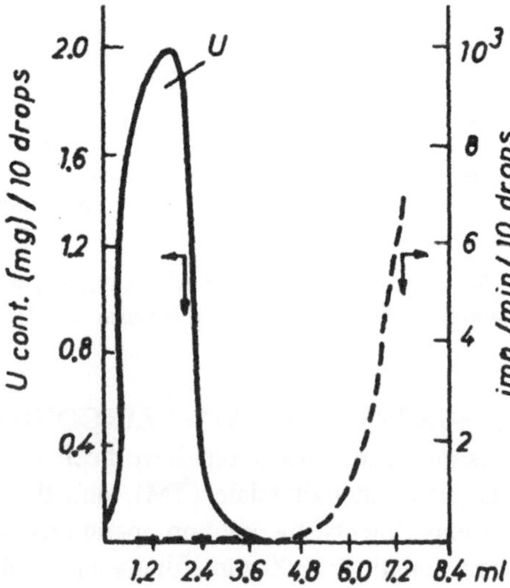

FIGURE 16. Separation of U on ZPa column. (From Gal, I. and Gal, O., in *Proc. 2nd Int. Conf. PUAE*, 28, 24, 1958. With permission.)

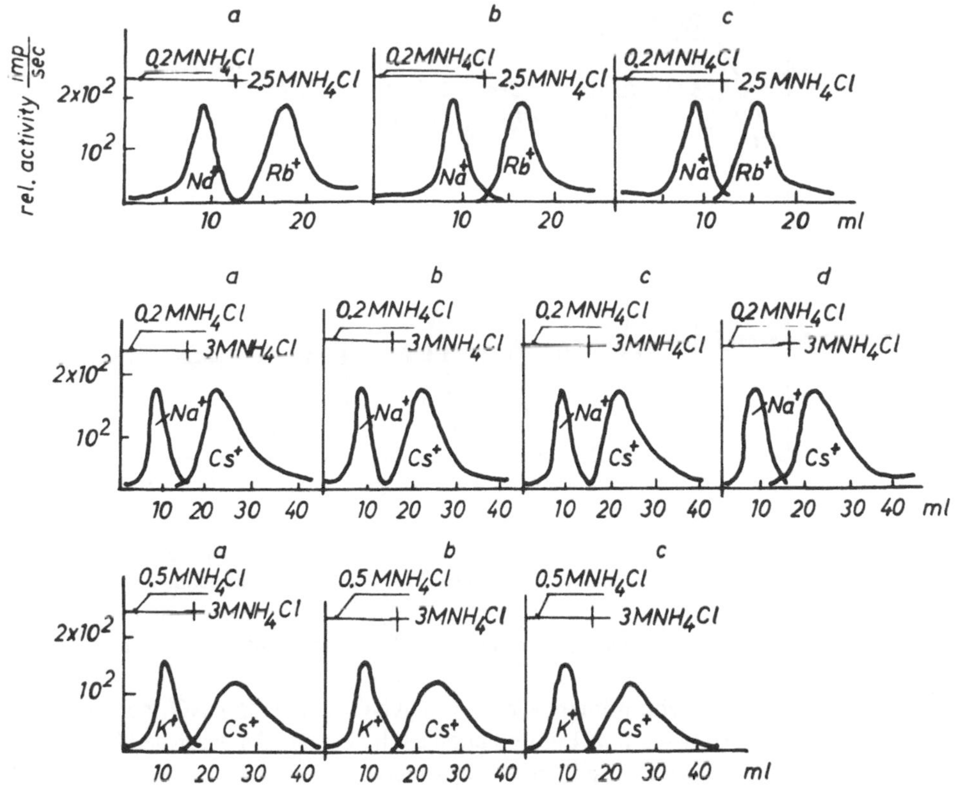

FIGURE 17. Separation of alkali metal ions on columns containing various inorganic ion exchangers. (a) ZMa; (b) TMa; (c) TWa. (From Cziboly, Cs., Zsinka, L., and Szirtes, L., *Mag. Kem. Lapja*, 24, 470, 1969. With permission.)

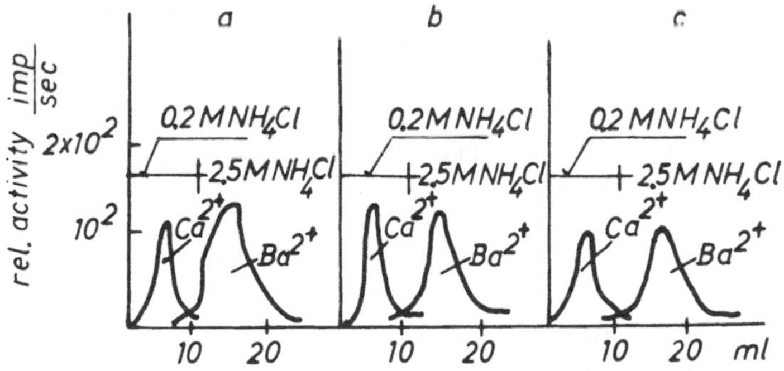

FIGURE 18. Separation of alkaline earth metal ions on columns containing various inorganic ion exchangers. (a) ZMa; (b) TMa; (c) TWa. (From Cziboly, Cs., Zsinka, L., and Szirtes, L., *Mag. Kem. Lapja*, 24, 470, 1969. With permission.)

Taking the elution data into consideration, the efficiencies of the separation for alkali metals and alkaline earth metals were found to be 95 and 89%, respectively.

It seems that in the low ion-concentration range of the initial solution, a relatively good separation of the ions may be carried out.

Kraus and co-workers[152] reported the separation of transition and heavy-metal ions using zirconium tungstate (ZW) as the ion exchanger. The dynamic determinations were carried

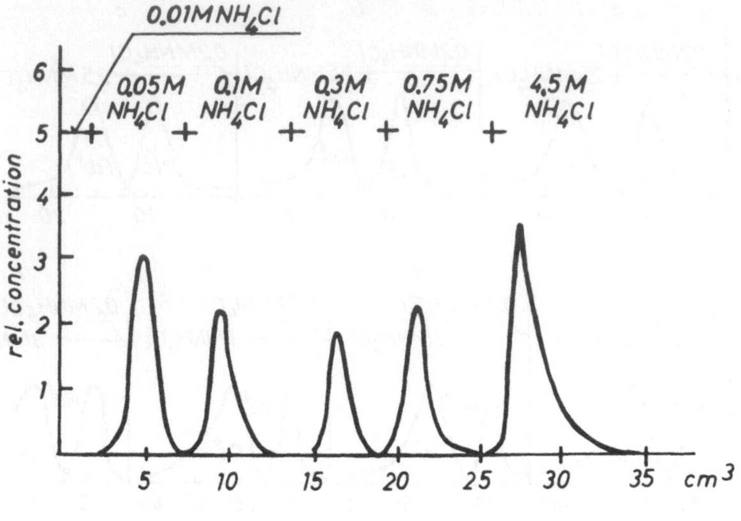

FIGURE 19. Separation of traces of alkali metal ions on ZWa. (From Kraus, K. A., Carlson, T. A., and Johnson, J. S., *Nature*, 177, 1128, 1956. With permission.)

out under the following conditions: column length = 550 mm, diameter = 9 mm, and flow rate = 0.5 ml/min.

The Co^{2+} was eluted with a mixture of 0.5 *M* KNO_3 and 1 *M* HNO_3 solutions while other ions were desorbed using 8 *M* LiC1 solution as the eluant. As described by Nunes da Costa and Jeronimo,[153] similar conditions were used to elute Cu^{2+} – Au^{3+} in an HC1 medium, and Ag^{+} and Cu^{2+} in an H_2SO_4 medium, with complex solutions as eluants. In order to separate tracer amounts of alkali metal ions, ZWa was used as an ion exchanger by Kraus and co-workers[152] and by Crouch and co-workers[154] (Figure 19).

The sorption properties of amorphous zirconium tellurate (ZTa) have been investigated[155] in our laboratory. The K_d were determined by equilibrating 0.1g of the ion exchanger with the various radioactive tracers taken in 10 ml HC1 and NH_4C1 solutions of various concentrations. The ion concentration of the initial solution in each case was adjusted to 0.03 mval/10 ml, with the exception of Hg, Au, Cd, Ag, Ga, As, Sb, W, Sc, Tc, In, and the rare earth metals, which are present in tracer amounts. The results are presented in Table 8 and shown in Figure 20. These studies indicate the following sorption order for the alkali metal and alkaline earth metal ions:

$$Na^{+} << K^{+} < Rb^{+} < Cs^{+} \text{ and } Sr^{2+} < Ca^{2+} < Ba^{2+}$$

Bresler[156] stated that the use of zirconium chromate (ZCr) is strongly limited, although Sr^{2+} and Ca^{2+} have been successfully separated on it from other elements. For determining the radiostrontium content of various sea-water samples, Sinotskin and co-workers[157] used amorphous zirconium oxalate (ZOa).

The procedure for the separation of radiostrontium was reported as follows: a column with a diameter of 0.4 cm was used, containing the ion exchanger (ZOa) in Ca^{2+} form, the flow rate of the solution was chosen to be 0.15 ml/min, and the Sr^{2+} content of the water sample was adjusted with 0.02 *M* $SrC1_2$ solution. The Ca^{2+}, Mg^{2+}, and Sr^{2+} ions were precipitated together. The carbonates were then dissolved in the minimum possible amount of HC1 solution. After its pH was adjusted to 7, this solution was passed through the column three times. During this process, the authors found that 75% of the initial amount of ^{90}Sr and 95% of that of the ^{90}Y radionuclides were sorbed on the ZOa. Following these experiments and using this method, the radiostrontium contents of various samples originating from the Black Sea and the Atlantic Ocean were determined.

TABLE 8
Distribution Ratios for Some Alkali Metal Ions

Concentration (M)	D_{Na}		D_K		D_{Rb}		D_{Cs}	
	NH_4Cl	HCl	NH_4Cl	HCl	NH_4Cl	HCl	NH_4Cl	HCl
0.0001	13.0	11.0	400	420	350	450	570	995
0.001	11.0	11.0	260	265	260	210	310	660
0.01	8.0	10.0	95.0	53.0	65.0	97.0	98.0	96.0
0.1	5.0	7.0	11.0	9.0	10.0	20.0	17.0	2.5
0.5	4.0	5.0	7.0	6.0	6.5	10.0	1.0	0.2
1.0	3.0	4.0	3.0	1.0	2.0	7.0	0	0
2.0	2.9	3.8	1.2	0.8	1.4	2.1	0	0
3.0	3.0	4.0	1.3	0.7	1.5	2.4	0	0
4.0	—	3.9	—	0.7	—	2.3	—	0

IV. APPLICATIONS OF OTHER ION EXCHANGERS

A. CHROMIUM PHOSPHATE ION EXCHANGER

The possibility of preparing and the behavior of amorphous chromium phosphate (CPa) were investigated in detail by Zsinka and the author.[158] Following this work, we studied[159] the selectivity behavior of CPa toward ions with various valencies. The distribution coefficients were determined in the same way as that used for ZWa, and the K_d values are presented in Figure 21. On the basis of K_d values for various metal ions, CPa was used to separate Co^{2+} from Fe^{3+} (Figure 22). During the separation, a column, made from 3 g of CPa, was used having a particle size of 0.4 mm. For the initial solution, with the concentrations $Co^{2+} = 1.4 \times 10^{-2}$ M and $Fe^{3+} = 2 \times 10^{6}$ M, a 0.2 ml/min flow rate was chosen.

It was found that the given ions can be separated in a relatively small volume of effluent with a good selectivity (as shown by radiometric determinations). Together with the various kinds of zirconium phosphate ion exchangers; the use of CPa for a generator system was also investigated,[160] namely, the use of CPa for ^{113}Sn-^{113}In and ^{99}Mo-^{99}Tc generator systems.

The target materials (containing the mother elements) were chemically treated, as in the case of zirconium phosphate, i.e., the irradiated metallic tin was dissolved in concentrated HC1 solution. After evaporation, the residue was taken up in 0.01 M HC1 solution, while the radioactive MoO_3 dissolved in 0.1 M NH_4OH was then evaporated and the Mo taken up in 1×10^{-3} M NH_4OH solution.

The generators operated as follows: columns with length = 300 mm and diameter 12 mm were used; the weight of CPa, with a particle size of 0.4 mm, was 2 g. Flow rates of 0.6 ml/min (^{113}In) and 0.3 ml/min (^{99m}Tc) were chosen. The primary elution cuves are shown in Figure 23. In each case, the elutions were repeated three times, after reaching radioactive equilibrium between the mother/daughter element activity (Table 9).

Using CPa, about 93% of the initial amount of radioindium (^{113}In) with a radionuclidic purity of approximately 98%, was eluted. The same radionuclidic purity for radiotechnetium (^{99m}Tc) was achieved with only five fractions (10 ml) containing only approximately 28% of the calculated initial amount of radiotechnetium. Based on these data, CPa was regarded as a less efficient ion exchanger for separating technetium radionuclides. However, in the case of Sn-In, the separation results are very promising.

B. HETEROPOLYACIDS AND OTHER RELATED ION EXCHANGERS

As a result of an extremely careful investigation of heteropolyacids, Smit and coworkers[161] made the following observations:

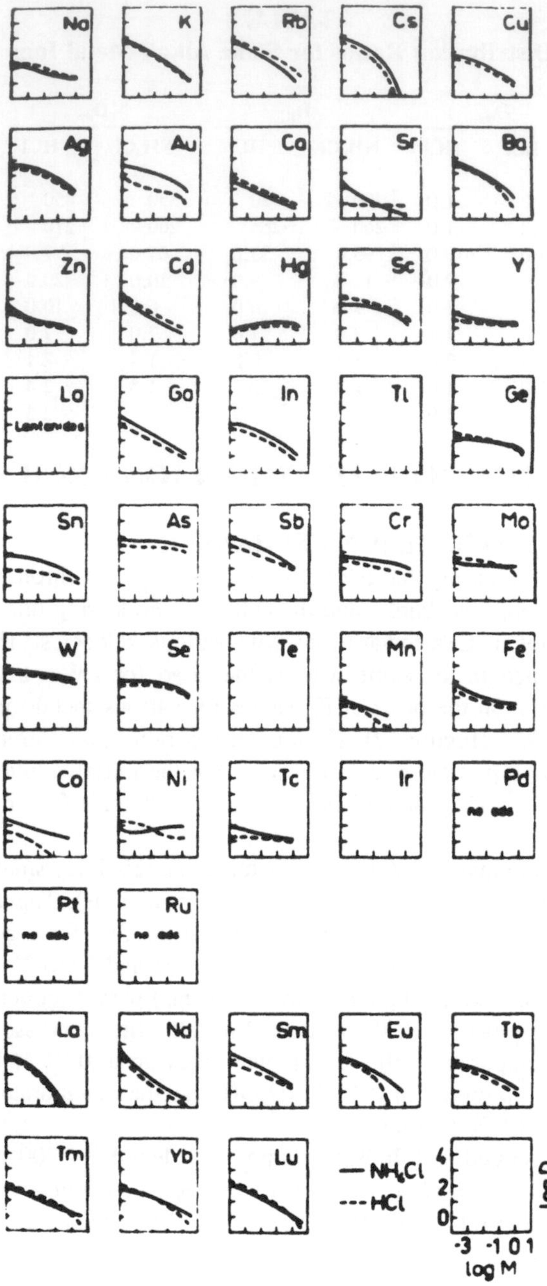

FIGURE 20. Distribution coefficients (k_d) of various metal ions on ZTa. (From Szirtes, L. and Zsinka, L., *Radiochem. Radioanal. Lett.*, 7, 61, 1971. With permission.)

1. The retention increases with the increase in the non-hydrated ion diameter.
2. The Tl^+ and Ag^+ ions are more strongly adsorbed than other alkali metal ions having the same diameter; this is related to the existence of covalent bonding.
3. The distribution factors of the neighboring ions are tenfold higher than those for Dowex-50 cation exchange resin.
4. The K_d values are practically independent of pH in the 1.1 to 4.5 range.

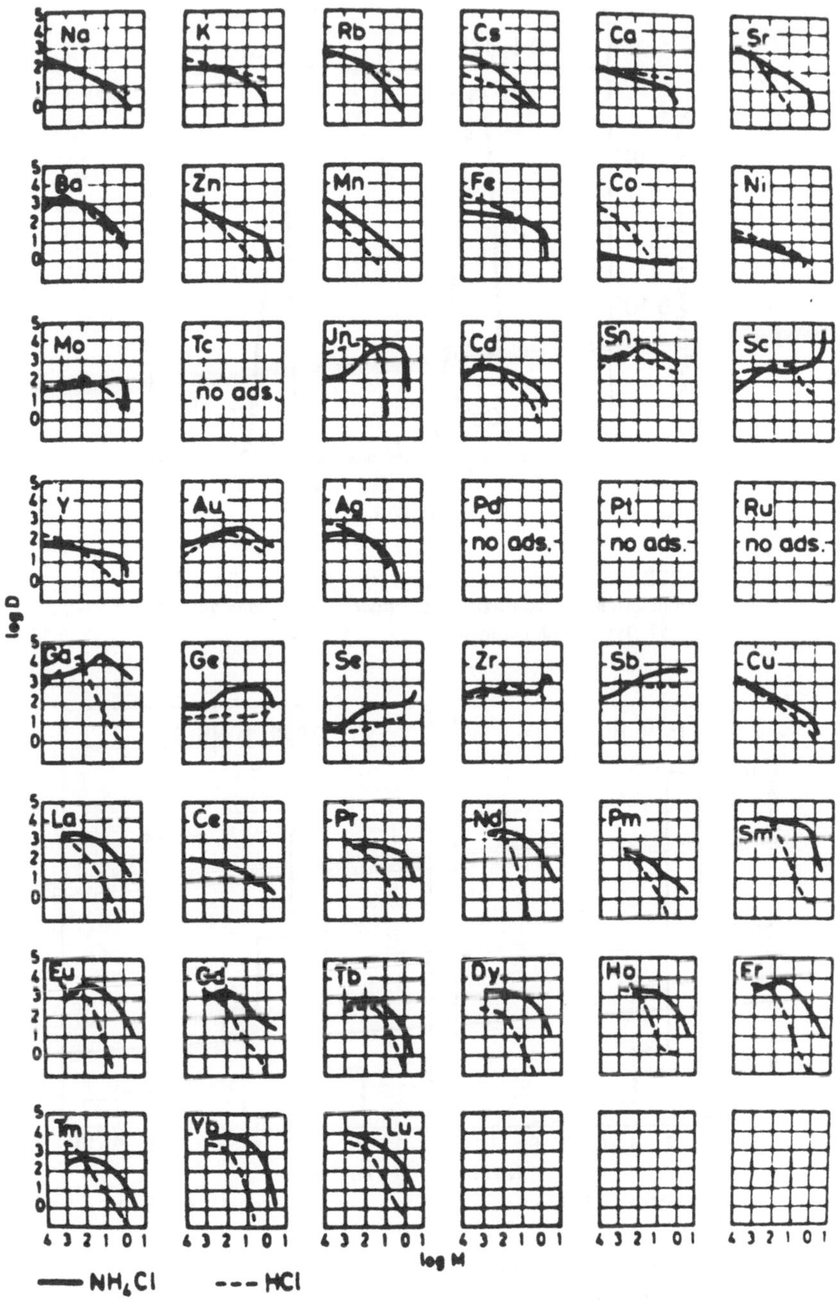

FIGURE 21. Distribution coefficients of various metal ions on CPa. (From Zsinka, L. and Szirtes, L., *Radiochem. Radioanal. Lett.*, 2, 257, 1969. With permission.)

Smit[162] carried out the separation of alkali metal ions under dynamic conditions. During these experiments, he separated micro amounts of these metals using a column containing ammonium molybdophosphate (AMP) having a particle size of approximately 200 mesh. Macro amounts of the same metals were also separated by Smit using a mixture of AMP and asbestos (1:1) as an ion exchanger. In this case, the elution was carried out using a NH_4NO_3 solution with various concentrations (0.01 *M* for Na^+, 0.2 *M* for K^+, 3 *M* for Rb^+, and a concentrated solution for Cs^+ ions), as shown in Figure 24.

An extremely high selectivity of AMP toward the Cs^+ ion was found by Smit and co-

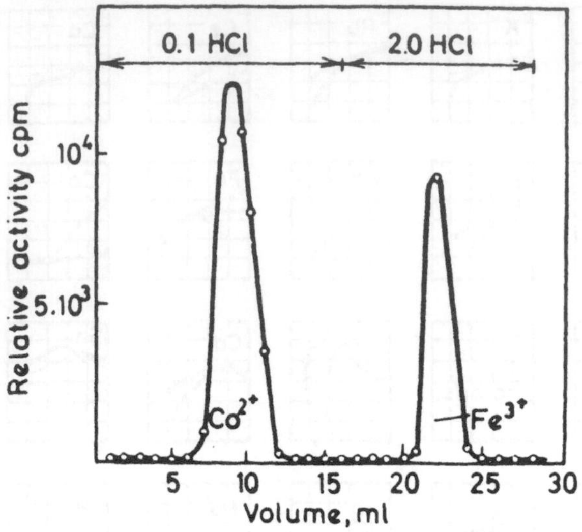

FIGURE 22. Separation of Co^{2+} from Fe^{3+} on CPa column. (From Zsinka, L. and Szirtes, L., *Radiochem. Radioanal. Lett.*, 2, 257, 1969. With permission.)

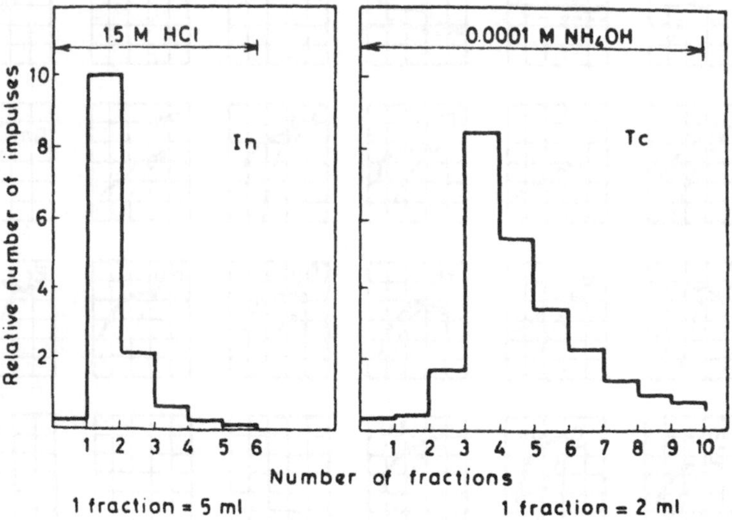

FIGURE 23. Elution curves of ^{113}In and ^{99m}Tc radionuclides on CPa column. (From Füssy, É. and Szirtes, L., *Radiochem. Radioanal. Lett.*, 8, 19, 1971. With permission.)

workers.[163] In connection with this fact, a column with 2 g of AMP was used for the sorption of the Cs^+ content of 20 l of sea water. In addition, the authors reported that this column was capable of determining the Cs^+ content of water samples from rain and rivers.

Smit reported that after sorption on AMP from a neutral solution, the ^{90}Sr and ^{90}Y radionuclides may be separately eluted with mixed solutions of 0.1 *M* NH_4NO_3 + 0.1 *M* HNO_3 (^{90}Sr) and 1 *M* NH_4NO_3 + 1 *M* HNO_3 (^{90}Y). He also showed that if a buffer solution is used, the same separation may be repeated with a macro amount of the same radionuclides. The author mentioned that the Cd^{2+}-In^{3+} metal ions may also be separated from each other on the same column after their sorption from a solution of pH = 4. Their elution may be

TABLE 9
Data of Elution Experiments

Nuclides	Nuclide activity (μCi) Mother	Daughter	Eluted %	Radionuclide purity (%)	Impurity (based on T 1/2)
^{113}Sn—^{113m}In	11.0	10.3	93.6	98.2	^{113}Sn
		10.4	94.5	98.1	
		10.1	92.0	98.6	
^{99}Mo—^{99m}Tc	360	104	28.8	92.8	^{99}Mo
	105	29	27.7	92.1	
	83	23	27.8	93.2	

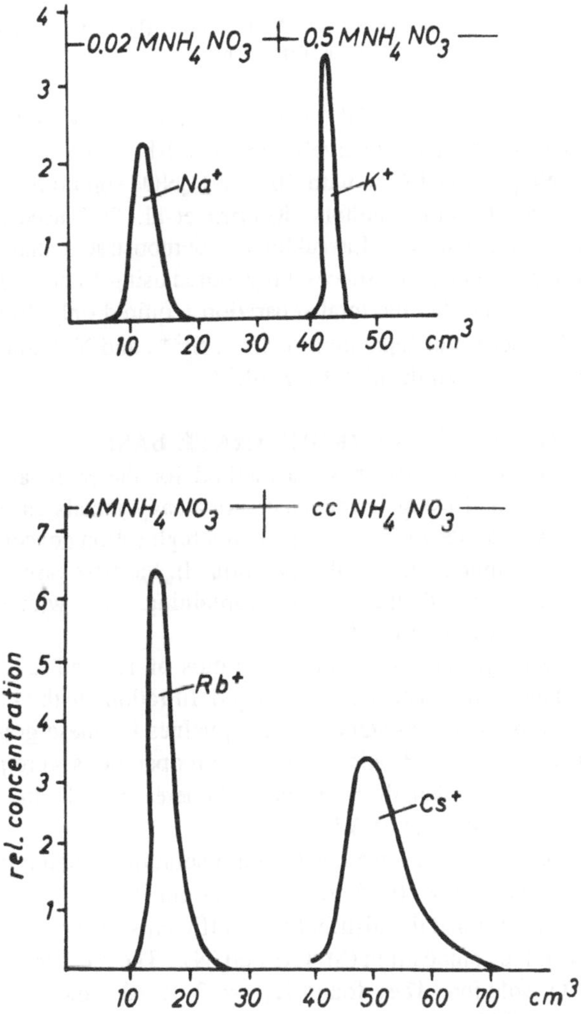

FIGURE 24. Separation of alkali metal ions on mixed AMP/asbestos ion exchanger. (From Smit, J. van R., *Nature*, 181, 1530, 1958. With permission.)

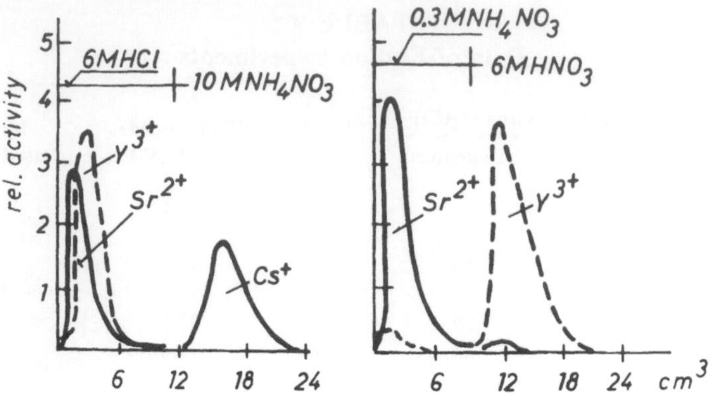

FIGURE 25. Separation of Sr^{2+} on AWP column. (From Krtil, J., *J. Inorg. Nucl. Chem.*, 22, 247, 1961. With permission.)

carried out using 0.125 *M* NH_4NO_3 (Cd^{2+}) and a mixture of 1 *M* NH_4NO_3 and 1 *M* HNO_3 solutions. Krtil and Krivy[164] have effectively separated Rb^+ and Cs^+ ions on AMP, eluting the Rb^+ with 1 *M* NH_4NO_3 and Cs^+ with 10 *M* NH_4NO_3 solutions. The same separation has also been presented by other authors. Kourim et al.[165,166] used ammonium tungstophosphate (AWP) for this purpose. Lavrukhina[167] demonstrated that Rb^+ and Cs^+ ions adsorbed on an AWP column may be successfully eluted using 1 and 6 *M* NH_4NO_3 solutions, respectively. It was reported that the total separation required only 40 to 50 min.

A method for the successful separation of Cs^+, Sr^{2+}, and Y^{3+} ions using an AWP ion exchanger (Figure 25) has been detailed by Krtil.[168]

C. ION EXCHANGERS ON A FERROCYANATE BASE

Rodinskii and co-workers[169] described a method for the preparation of a concentrate containing radiocesium, and using ferrocyanates of heavy metals as ion exchangers. The separation system was developed for solutions with a high cation concentration. The authors used both batch and dynamic methods of separation. In the first case, they mixed 1.5 g of ferrocyanate ion exchanger with the solution containing Cs^+; with this quantity of ion exchanger, 0.2 g of Cs^+ was extracted.

On mixing the ferrocyanate with cement in ratios of 1:4 and 2:3, they obtained very durable granules which could withstand prolonged filtration in thick layers without any noticeable increase in hydraulic resistance. The K_d values for these granules were found to be ~51 (1:4 ratio) and ~6100 (2:3 ratio). The column operations were performed under the following conditions: column length = 45 cm, diameter = 0.1 cm, and volume = 4.5 cm^3. The initial solution had a pH of 6.8.

A good separation of Cs^+ was achieved from a solution containing 20 to 25% Na^+ + K^+ salts and approximately 5×10^{-4} g/ml radiocesium.[170]

Other authors[170] determined the distribution coefficients for Sr^{2+}, Cs^+, and Eu^{3+}, using an ion exchanger with the composition $(MoO_2)_2Fe(CN)_6$. The measurements were performed using a 0.5 *M* HNO_3 solution. They found $K_d^{Cs} = 2 \times 10^4$, and $K_d^{Eu} = 8 \times 10^2$. These values suggest the possibility of separating Cs^+ and Sr^{2+} from wastewater after extraction of the rare earth metals.

A method for separating radiocesium arising from mother barium nuclides has been developed.[171] The irradiated target ($BaCO_3$) is first chemically treated by dissolving it in 2 *M* HCl and diluting the solution with distilled water. Ba^{2+} was precipitated with 1 *M* H_2SO_4 below the boiling point of the solution; following this, the solution was evaporated and the residue taken up in 0.25 *M* Li-EDTA solution. The solution was then flown up under a Co-ferrocyanate column. Here, the Cs^+ was completely adsorbed, whereas the other cations remained in solution. Elution of Cs^+ can be carried out with a 2 *M* HC1 solution at a flow rate of 1 ml/min. The very pure Cs^+ fraction can be utilized for preparing injection solutions ($^{131}CsCl$) for human medical purposes.

D. ZEOLITE TYPE ION EXCHANGERS

About 15 to 20 years ago, the application of synthetic inorganic ion exchangers was virtually restricted to the use of synthetic and natural zeolites. Ames and co-workers[172] used apatite to separate radiostrontium (^{90}Sr) from wastewater. In column operation, they used 50 g apatite with a particle size of 0.25 to 1 mm. The Sr^{2+} concentration was adjusted with Sr $(NO_3)_2$ solution to 2 mg/ml. The flow rate chosen for the eluant was 7 cm/h. Under these conditions, 99% of the initial amount of the radiostrontium was eliminated with a 3 *M* $NaNO_3$ solution.

Herr and Riedel[173] presented a separation system using synthetic zeolites to distinguish daughter ^{223}Fr nuclides from the mother ^{227}Ac element and from other radioactive nuclides. The nuclides ^{227}Ac, ^{227}Th, ^{215}Po, and ^{211}Pb were sorbed on zeolite, whereas the nuclides ^{223}Fr, ^{207}Tl, and ^{223}Ra immediately passed through the column where they were separated from each other (in the second step). In addition, the ^{223}Fr nuclide which formed due to alpha decay after attaining radionuclidic equilibrium, was eluted with a 0.5 *M* NH_4 Cl solution.

As described by Amphlett,[174] vermiculite (South Carolina) has successfully been used to separate Na^+ and Cs^+ ions from various solutions. It was shown that from a 0.1 *M* NaCl solution 8.3% of the initial amount of Na^+ was sorbed, while from a 0.01 *M* CsC1 solution, 96% of the Cs^+ was sorbed. Based on these data, vermiculite was used for separating Cs^+ from various solutions.

Rodinskii and co-workers[175] presented a method for preparing concentrates of Cs^+ ions with glauconite. The adsorption of Cs^+ ions was studied under dynamic conditions. The conditions for the column operations were: column length = 70 cm, volume = 100 cm^3, and pH of initial solution = 9.1. Based on the results, the authors obtained the following sequence of cations: $Mg^{2+} < Ca^{2+} < Sr^{2+} < Na^+ < La^{3+} < NH_4^+ < K^+ < Cs^+$.

In addition, they reported that within the range of a micro amount of Cs^+ and a macro amount of other cations, Cs^+ sorption is independent of pH in the range of 2.7 to 13. Further, a high specific-activity radiocesium concentrate could be obtained by eluting the adsorbed radiocaesium with 0.1 *M* $(NH_4)_2CO_3$ solution.

E. SYNTHETIC INORGANIC ION EXCHANGERS ON A SILICA MATRIX

Various inorganic exchangers were prepared on a silica gel matrix in such a way that both the exchanger and the gel were synthesized in parallel, and the wet products were mixed at about pH = 5, followed by drying at 60°C.

These types of mixed ion exchangers were made in the form of spheres using a column (length = 4 m) filled with paraffin oil. The spheres were collected on the bottom of the column and washed in toluene. After that, the wet ion exchanger was dried at 60°C for 10 h.[176] The given quantities of Si and Me for the preparation were determined by the ratio:

$$S_{Me} = \left(\frac{Me}{Me + Si}\right) \cdot 100\% \qquad (3)$$

By varying this ratio, ion exchangers with various properties have been produced.[177,178] The silica gel was prepared by mixing the buffer (1.25 *M* NaOH + 12 *M* CH_2COOH) and sodium silicate solutions at room temperature. Crystalline zirconium phosphate (ZPc), ammonium molybdophosphate (AMP), and hydrous manganese dioxide (MDOH) were synthesized by the well-known "classical" methods. In such a way, crystalline silica-ZP (Si-

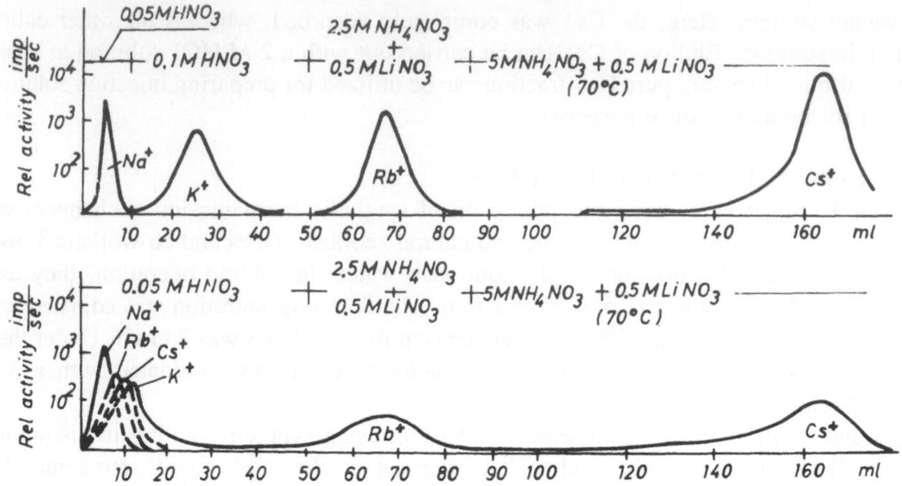

FIGURE 26. Elution curves of alkali metal ions. Form of ion exchanger: (a) Li^+; (b) H^+. (From Le Van So and Szirtes L., *J. Radioanal. Nucl. Chem.*, 99, 55, 1986. With permission.)

ZPc), silica-AMP (Si-AMP), and silica-MDOH (Si-MDOH) were prepared in the form of spheres.

Some separations were investigated using these ion exchangers. The separation of the alkali metal ions was carried out on both the Li^+ and H^+ forms of Si-ZPc.[179] The column operations were performed under the following conditions: column length = 170 mm, diameter = 7 mm, and weight of the ion exchanger = 1.5 g, with a particle size of ~ 100 mesh. Flow rates of 0.1 ml/min for adsorption and 0.5 ml/min for elution were chosen (Figure 26).

In all cases, the solution containing the metal ions was adjusted with a buffer solution to pH = 6. The Li^+-form ion exchanger was prepared by titration of the H^+ form of Si-ZPc with a 1 *M* $LiNO_3$ solution. It was found that the alkali metal ions could be separated on the Li^+-form ion exchanger, but not on the H^+ form. The failure of the separation on the H^+ form of Si-ZPc can be attributed to the steric hindrance of the ion exchanger, as a result of which Rb^+ and Cs^+ ions are eluted together with the Na^+ ions. On the other hand, a significant amount of Rb^+ and Cs^+ ions is retained on the column, due to the catalytic effect of Na^+ and K^+ ions.

Under the same experimental conditions, the separation of transition metal ions was investigated using the Na^+-form Si-ZPc ion exchanger (Figure 27).

During the separation of transition metal ions, the above-mentioned "blocking" effect was not observed, possibly because these ions have a smaller unhydrated diameter and a higher heat of hydration. Because of the high affinity of Si-ZPc toward transition metal ions, they can easily be separated from each other using eluants with complexing agents.

Le Van So and Do Minh Vung[180] presented data concerning the separation of Be^{2+} from other metal ions using an H^{+-} form Si-ZPc ion exchanger. The column operations were effected under the following conditions: column length = 60 mm, diameter = 8 mm, weight of exchanger = 1.5 g (with a particle size of 50 to 100 mesh), and flow rate = 0.5 ml/min. One hundred micrograms of metal ions were adsorbed on the column from 30 ml of 0.1 *M* EDTA solution, with a pH previously adjusted with NH_4OH or HNO_3 solutions. After sorption, the ions were eluted with 20 ml of the same solution. The effect of pH is shown in Figure 28.

The separation is based on the fact that the Be-$(EDTA)_2$ complex is very strongly bound on the Si-ZPc, while the other transition metal ions, in the form of anion complexes, remained in the solution (Figure 29).

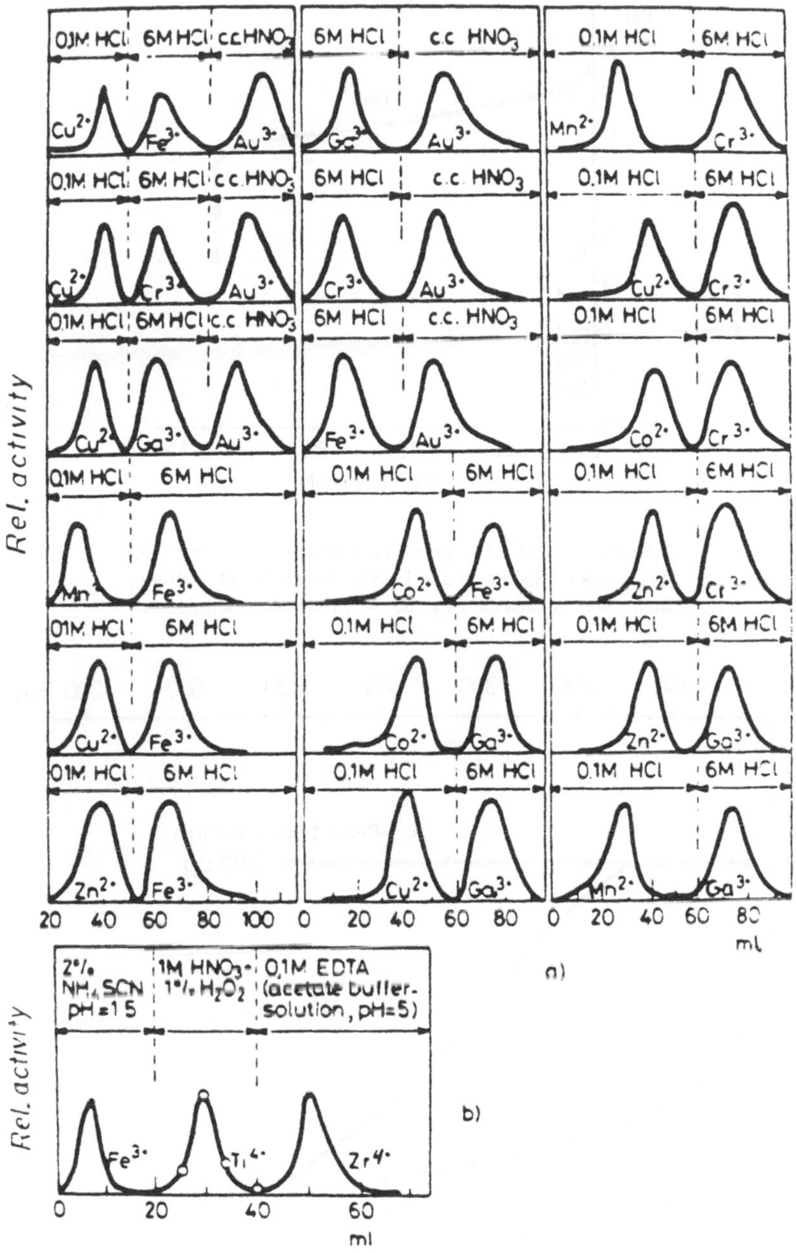

FIGURE 27. Separation of transition metal ions on Na^+ from Si-ZPc column. (From Le Van So and Szirtes, L., *J. Radioanal. Nucl. Chem.*, 99, 55, 1986. With permission.)

After that, and under the same experimental conditions, the separation of Be^{2+} from Mg^{2+}, Sr^{2+}, and Ba^{2+} was effected, taking into consideration the fact that these metal ions form much more stable complexes with EDTA than do those of Ca^{2+} (Figure 30).

The results were satisfactory for the quantitative separation of Be^{2+} from Mg^{2+}, Ca^{2+}, Sr^{2+}, Ba^{2+}, Y^{3+}, La^{3+}, Ce^{3+}, TiO^{2+}, Zr^{4+}, Th^{4+}, Cr^{3+}, Mn^{2+}, Fe^{3+}, Fe^{2+}, Co^{2+}, Ni^{2+}, Cu^{2+}, Ag^+, Zn^{2+}, Cd^{2+}, Hg^{2+}, Al^{3+}, Ga^{3+}, In^{2+}, Th^{3+}, Sn^{2+}, Pb^{2+}, and Bi^{3+}. Using this method, the Be^{2+} content of a natural rock sample was determined (Figure 31 and 32).

The intersection of the "a" curve with the ordinate in Figure 31 gives the Be^{2+} content

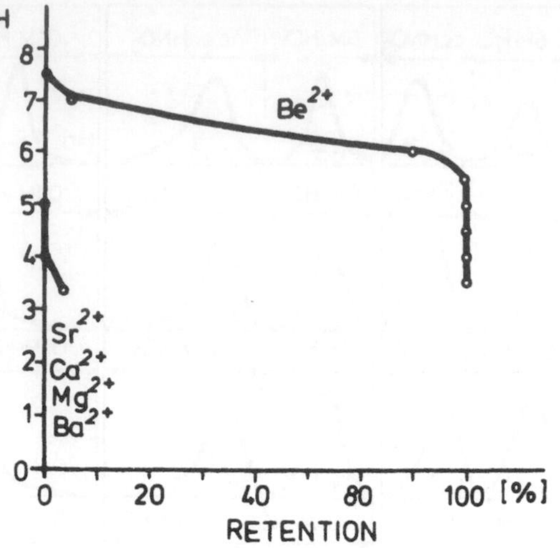

FIGURE 28. Effect of pH on the retention of alkaline earth metal ions on Si-ZPc. (From Le Van So and Do Minh Vung, *Radiochem. Radioanal. Lett.*, 56, 343, 1982. With permission.)

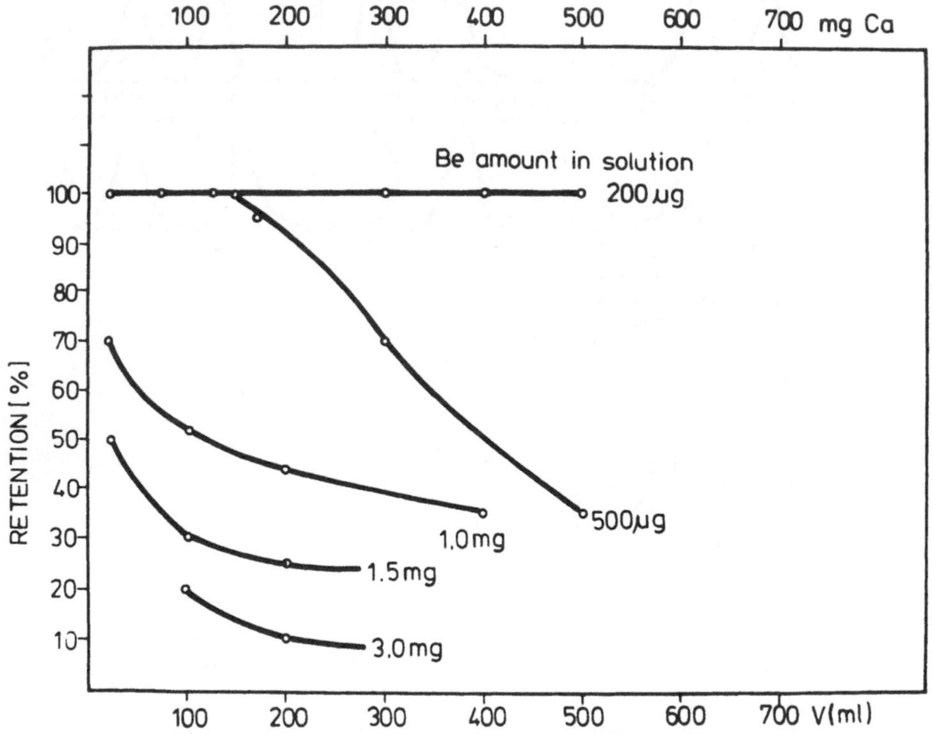

FIGURE 29. Effect of Be^{2+} and Ca^{2+} concentrations and the volume of eluant on the retention of Be^{2+} ions. (From Le Van So and Do Minh Vung, *Radiochem. Radioanal. Lett.*, 56, 343, 1982. With permission.)

of the investigated rock sample. Le Van So presented[181] data concerning the separation of Sr^{2+} from Ba^{2+}, Y^{3+}, La^{3+}, Ce^{3+}, Bi^{3+}, and Pb^{2+} using Si-MDOH as the ion exchanger.

Separation was realized under the following conditions: a column with a length of 120 mm and a diameter of 8 mm was used that contained 3 g of ion exchanger, with a particle size of 100 mesh, and a flow rate of 0.5 ml/min was chosen for both adsorption and elution.

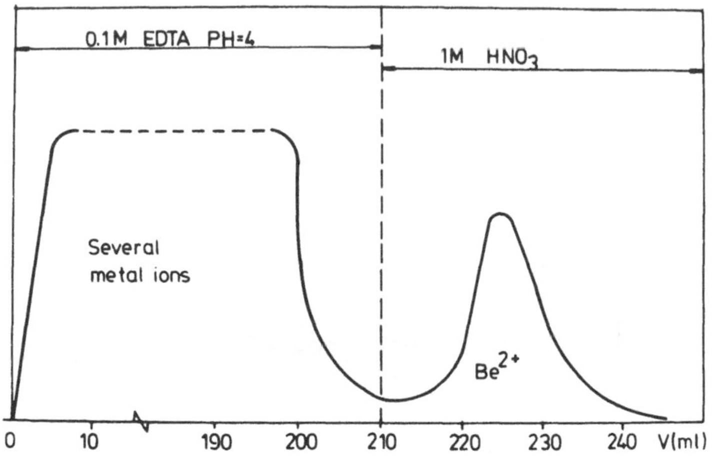

FIGURE 30. Elution curve of Be^{2+}. (From Le Van So and Do Minh Vung, *Radiochem. Radioanal. Lett.*, 56, 343, 1982. With permission.)

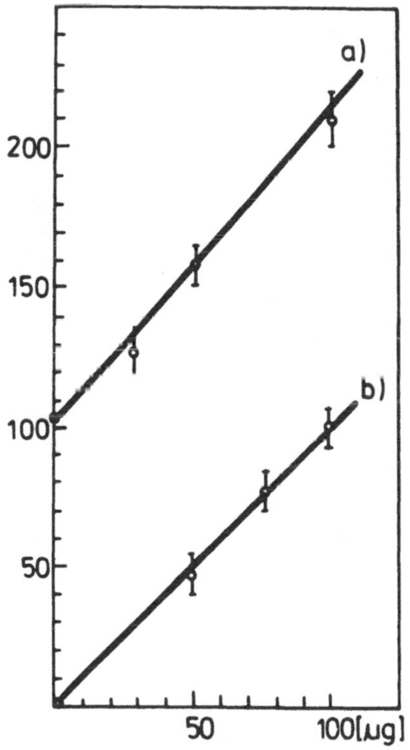

FIGURE 31. Recovery curve of Be^{2+}. (a) Ore sample; (b) ore sample free of Be^{2+}. (From Le Van So and Do Minh Vung, *Radiochem. Radioanal. Lett.*, 56, 343, 1982. With permission.)

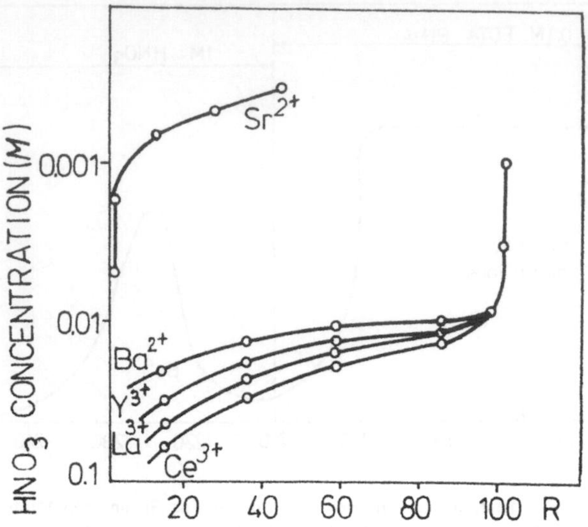

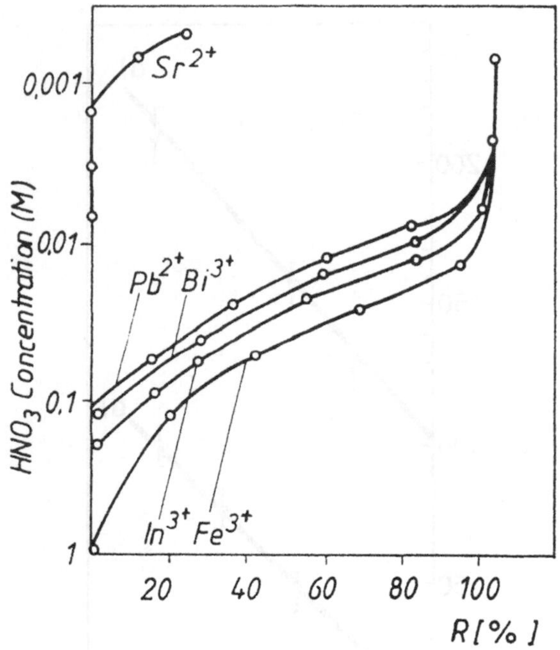

FIGURE 32. Effect of HNO_3 concentration on the retention of metal ions on Si-MDOH. (From Le Van So, Ph.D. thesis, Budapest, 1985. With permission.)

When the concentration of the HNO_3 solution was higher than 10^{-2} *M*, retention of the metal ions reached 90% (Figure 33). Based on the breakthrough capacity data (Table 10), if the total Sr^{2+} content in the sample is less than 250 mg, good separations take place if:

$$\frac{1}{2.5}(m_{Bi} + m_{Pb} + m_{In} + m_{Ba} + m_{Y} + m_{La} + m_{Ce}) + \frac{1}{3.5} m_{Fe} < M \qquad (4)$$

where *M* is the necessary weight of Si-MDOH in grams and m_{Me} is the metal ion content

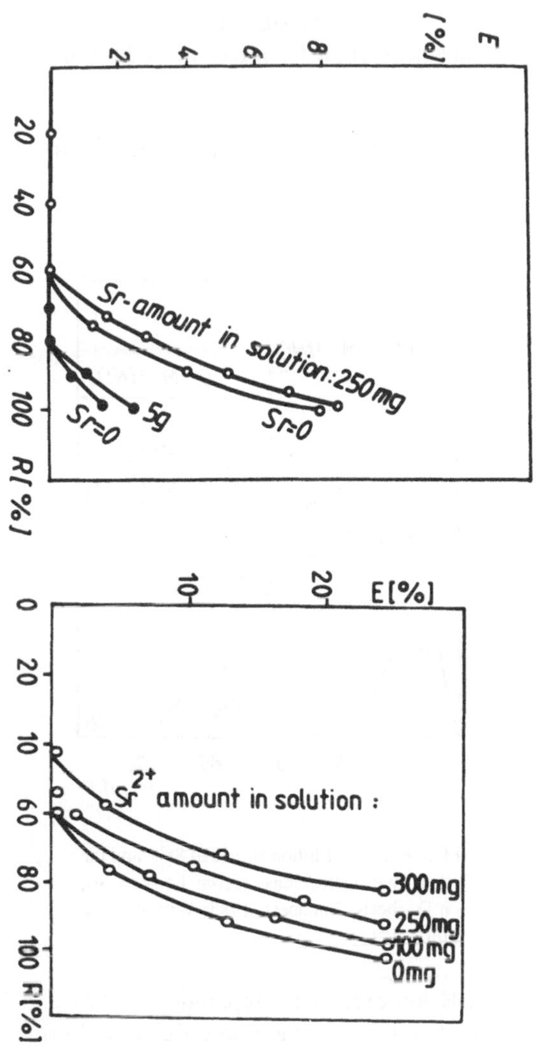

FIGURE 33. Effect of Sr^{2+} loading on the retention on Si-MDOH. (From Le Van So, Ph.D. thesis, Budapest, 1985. With permission.)

of the examined sample in milligrams. It was found that the retention of Sr^{2+} changed as a function of the metal ion content (Figure 34).

Taking these results into consideration, 250 mg of Sr^{2+} was separated from a solution containing 1-1 g of other metal ions. Also 5 g of Sr^{2+} was separated from the other solution containing 50 mg of Fe^{3+}. As shown in the elution curves, the separations were carried out with good efficiency. These successful separations suggest the possibility of determining the Sr^{2+} content of real samples. The determination may be fulfilled in two ways, due to the Sr^{2+} + Ca^{2+} content of the investigated sample.

When the Sr^{2+} + Ca^{2+} content is less than 250 mg per sample, the described separation system can be used. When the Sr^{2+} + Ca^{2+} content is higher than 250 mg per sample, the separation is executed as follows: first, the Fe^{3+} ions are adsorbed on the Si-MDOH column, then, the eluate is evaporated. From the residue, the $Ca(NO_3)_2$ is dissolved in acetone, after which the Sr^{2+} is also dissolved in 30 ml of 0.005 *M* HNO_3 solution. If this solution is passed through another Si-MDOH column, the Sr^{2+} can be separated from the other possible metal ions. Figure 35 shows the efficiency of the separation.

TABLE 10
Breakthrough Capacity of Si-MDOH

	Ion							
	Y^{3+}	Ce^{3+}	La^{3+}	Ba^{2+}	Pb^{2+}	Bi^{3+}	In^{2+}	Fe^{3+}
Capacity (mg/g)	4	7	7	9	4	6	7	5

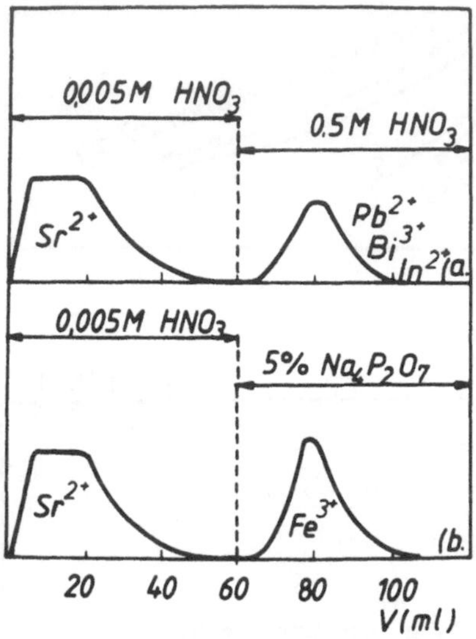

FIGURE 34. Elution curves of Sr^{2+} and the other investigated ions. (From Le Van So, Ph.D. thesis, Budapest, 1985. With permission.)

With the same Si-MDOH ion exchanger, separation of ^{90}Sr and ^{90}Y radionuclides was also possible (Figure 36), with a 99.9% separation efficiency. Further, it was observed that if the NH_4^+ ion concentration is less than 5% and the total amount of Ca^{2+} + NH_4^+ + Co^{2+} + Ni^{2+} + Zn^{2+} + Cd^{2+} is less than 15 mg per sample, these metal ions have no influence on separation efficiency. By means of this method, the ^{90}Sr content of a cabbage sample and that of a water sample (Xuanhoung Lake, Vietnam) were determined. The results showed that ^{90}Sr in Cabbage was 0.7×10^{-12} Ci/g and that in water samples was 0.2×10^{-12} Ci/1. These values are in agreement with those determined by a standard method.[182]

Amphlett[183] and Wilding[184] used Si-AMP to separate Cs^+ from fission products in a wastewater sample. They obtained good results (practically no loss in ion exchanger) in acidic media. All fission products, with the exception of Rb^+, Zr^{4+}, and Nb^{5+} ions, were eluted with 1 *M* HC1 solution. After this Nb^{5+} was eluted with formic acid.

The ^{137}Cs radiocesium content of sea-water samples has also been determined by the ion-exchange separation method.[176] For this purpose, Si-AMP was used with success. The column operations were effected under the following conditions: column length = 100 cm, diameter = 30 mm, 100 g of ion exchanger, and flow-rate = 100 ml/min.

The pH of the sea-water sample, before its treatment, was adjusted to 4 to 5 with a 0.1 *M* HNO_3 solution. Fifty liters of sample was passed through the column, and the adsorbed radiocesium ($^{137}Cs^+$) eluted with 5 *M* NH_4NO_3 solution. Radiometric determination showed that the total amount of $^{137}Cs^+$ was separated in one cycle. Ahrland and co-workers[63] and Cvjeticanin[64] described the separation of U and Pu from fission products using silica gel as the ion exchanger.

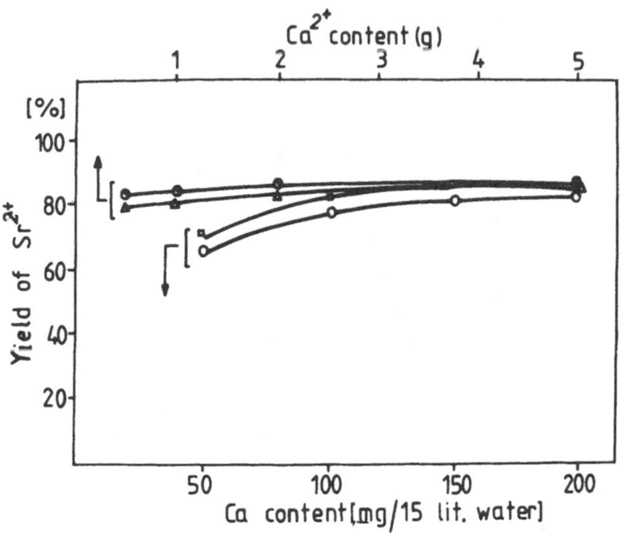

FIGURE 35. Yield of Sr^{2+} as a function of Sr^{2+} and Ca^{2+} content of the solution. (○) 50 mg Sr^{2+}; (□) 100 mg Sr^{2+} added to water sample; (●) 50 mg Sr^{2+}; (△) 100 mg Sr^{2+}. (From Le Van So, Ph.D. thesis, Budapest, 1985. With permission.)

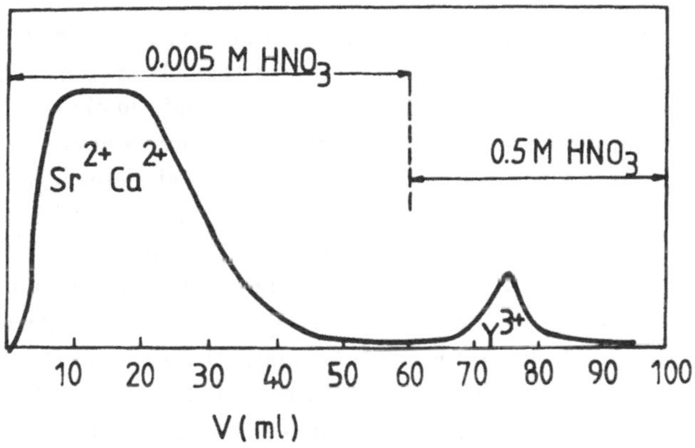

FIGURE 36. Elution curves of Sr^{2+} and Y^{3+} ions. (From Le Van So, Ph.D. thesis, Budapest, 1985. With permission.)

An inorganic ion exchanger having SiO_2 in the matrix was successfully used by Whitehurst and co-workers[185] for the isolation of heavy-metal ions from petrol and other hydrocarbons.

In parallel with other inorganic ion exchangers Na^+ form Si-ZPc was also tested in a ^{99}Mo-^{99m}Tc generator system.[176] The system consisted of two columns, one of which was filled with irradiated amorphous zirconium molybdate (ZMa). Apart from working as an ion exchanger it also contained the mother element (^{99}Mo). The second column contained Si-ZPc in a Na^+ form ion exchanger, having a particle size of ~ 100 mesh, for sorption of the possible quantity of $^{99}MoO_4^+$ and $^{95}Zr^{4+}$ in the eluant (Figure 37).

The ^{99m}Tc radionuclide arising from the mother ^{99}Mo nuclide was successfully eluted with 0.9% NaCl solution. Gamma-spectrometric results showed that 85% of the total originated ^{99m}Tc was eluted in a relatively low eluant volume. This efficiency of separation remained constant during repeated elutions (five times). The system is now successfully used in everyday practice for production of the ^{99m}Tc radionuclide.

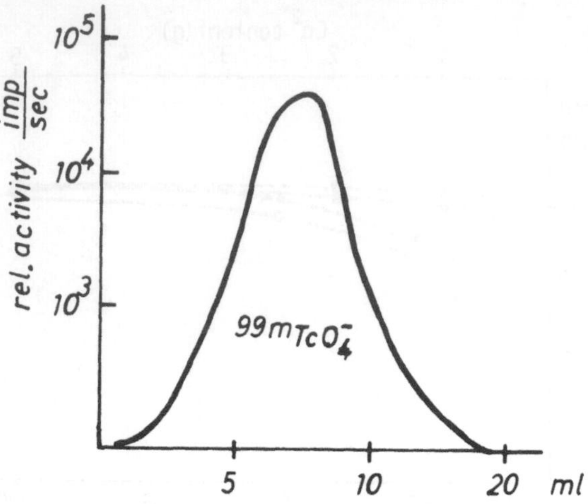

FIGURE 37. Elution curve of $^{99m}TcO_4^-$. (From Le Van So, Ph.D. thesis, Budapest, 1985. With permission.)

V. SPECIAL APPLICATIONS OF INORGANIC ION EXCHANGERS

A. ION-EXCHANGE MEMBRANES

The good resistance of inorganic ion exchangers against radiation effect and heat means that they are very effective as semipermeable membranes. Dravnieks and Bregman[186] constructed a heavy membrane (with a diameter of ~ 25 mm and a thickness of 1 mm) consisting of pressurized zirconium phosphate in a carbon plate, which contained the electrodes. The membrane was used for the transition of H^+ ions during fuel-element production. Alberti[187] described another method: the author deposited the zirconium phosphate on fibrous glass and, after washing, dried it at 50°C and pressurized it until the plate thickness reached 0.5 cm. He found that the permeability of such a membrane in a 0.1 *M* solution for Li^+, Na^+, K^+, and Cs^+ was 35, 45, 64, and 67% respectively. However, Alberti stated that this type of membrane is not suitable for use in the production of fuel elements.

Hydrous zirconium oxide (HZO) has recently become important in the water desalination process due to hyperfiltration properties. Dynamically formed membranes have led to this material showing promising self-rejections by virtue of Donnan electrolyte exclusion.[188] For the same purposes, exchange membranes were prepared from hydrous thorium oxide by Rajan and co-workers.[189]

Ciciscilli and Krupennyikova[190] presented data on the separation of crystalline alumina silicona (type 4A) in a Na-form molecular filter and described the results of Sr^{2+} sorption on it. As described, the filter contained:

Loss under heat	H_2O	SiO_2	Al_2O_3	Na_2O
11.70	9.10	32.97	29.95	16.01

in percentage.

The distribution coefficients were determined using 0.5 g of the adsorber and 50 ml of $Sr(NO_3)_2$solutions of various concentrations (Table 11).

TABLE 11
Distribution Coefficients (K_d) at Various Solution Concentrations

Initial solution conc. (*M*)	Absorp. (mval/g)	Equilibrium concentration (mval/ml)	K_d
0.75	5.25	0.70	8
0.50	4.80	0.45	11
0.30	3.39	0.27	13
0.10	4.34	0.06	77
0.05	3.94	0.01	371
0.001	0.10	5×10^{-6}	2×10^4

TABLE 12
R_f Values at Various Zr Concentrations

Temperature (°C)	R_f values at Zr concentrations (%) 5	10	15	20	25	30	Ion
20	0.71	0.63	0.46	0.44	0.38	0.30	UO_2^{2+}
50	0.80	0.62	0.52	0.40	0.37	0.34	
20	0.74	0.68	0.58	0.57	0.52	0.47	Pb^{2+}
50	0.86	0.85	0.69	0.66	0.65	0.48	

B. INORGANIC ION EXCHANGERS IN PAPER AND THIN-LAYER CHROMATOGRAPHIC SEPARATIONS

Paper impregnated with inorganic ion exchangers is generally prepared by moisturizing it with the acidic solution of one of the components of the ion exchanger. After a short time, the wet paper is brought into contact with the other component necessary for precipitation, then washed with distilled water and dried at room temperature (in air). Alberti and coworkers[191-193] made papers treated with ZPa by this method. Grassini[194] reported the following method for preparing paper. Whatman No. 1 paper was drawn through $ZrOCl_2 \cdot 8H_2O$ in a 4 *M* HCl solution. The $ZrOCl_2 \cdot 8H_2O$ concentrations were 5, 10, 15, 20, 25, and 30%. After that, the paper was dried for 12 h. Then the strips were dipped into a 60% solution of H_3PO_4 in a 4 *M* HCl solution. The strips were dried for 6 h and then washed with a 2 *M* HCl solution and distilled water for 30 min each.

Chromatographic separation of UO_2^{2+} and Pb^{2+} was carried out using these paper strips. UO_2^{2+} and Pb^{2+} were separated from a 1% solution of uranyl and lead nitrates in $HClO_4$ solution on paper strips 21 cm long (the start point was 2 cm from the end).

During the process, UO_2^{2+} was eluted with a 3 *M* $HClO_4$ solution and Pb^{2+} with a 0.4 *M* $HClO_4$ solution. The R_f values given in Table 12 were found by the author. Using a paper pretreated with ZPa, Adloff[195] demonstrated that the following separations may be carried out with success:

1. Co^{2+}, Cu^{2+}, Fe^{3+}
   Cr^{3+}, Cu^{2+}, Fe^{3+} eluting with 0.035 *M* HCl—0.045 *M* NaCl
2. UO_2^{2+}, Fe^{3+}, Th^{4+}, or Ti^{4+}, eluting with 2 *M* HCl—4 *M* NH_4Cl
3. Mn^{2+} or Cr^{3+}, Al^{3+}, and Fe^{3+}, eluting with 0.5 *M* HCl.

Sastri and Rao[196] reported that the same paper is also capable of separating from each other the heavy radionuclides (transuranium elements), such as AcB-AcC″, ThB-ThC, and

TABLE 13
R_f Values for Various Elements

	Eluant		
Element	**0.1 *M* HCl**	**0.5 *M* HCl**	**1.0 *M* HCl**
Li^+	0.82	—	—
Na^+	0.68	0.77	0.83
K^+	0.53	—	—
Rb^+	0.12	—	—
Cs^+	0	—	—
Ca^{2+}	0.81	—	—
Sr^{2+}	0.81	—	—
Ba^{2+}	0.60	—	—
Ra^{2+}	0.44	—	—
Co^{2+}	0.67	—	—
Ni^{2+}	0.61	0.80	0.85
Zn^{2+}	0.69	—	—
Cu^{2+}	0.59	—	—
Cd^{2+}	0.60	—	—
Hg^{2+}	0.67	—	—
Fe^{3+}	0	0	0.40
Al^{3+}	0.13	0.40	0.90
Cr^{3+}	0.75	0.80	—
Ce^{3+}	0.19	—	—
La^{3+}	0.31	0.73	0.80
Ti^{4+}	0	0	0
Th^{4+}	0	0	0
UO_2^{2+}	0	—	—

From Alberti, C. and Grassini, G., *J. Chromatogr.*, 4, 83, 1960. With permission.

AcX-Fr, and even they can, in turn, be used to separate as parts of the same elements having different valency states.

Similar results were found by other authors using paper pretreated with AMP. Schroeder[197] described the separation of natural radioactive elements using the paper chromatographic method or simple filtration. On an AMP-treated paper using a 6 *M* HNO_3 solution as the chromatographic agent, ThC″ (^{208}Tl)-ThB (^{212}Pb), ThC (^{212}Bi)-AcK (^{233}Fe), and AcC″ (^{207}Tl)-^{227}Ac were separated.[198]

The high K_d values found in acidic medium (pH = 2) for Cs^+ (k_d^{Cs} = 5500) and Tl (k_d^{Tl} = 4300) reflected their good separation under these conditions (eluant, 0.1 *M* NH_4NO_3 solution). Schroeder[197] observed that the Tl^+ ion is less sorbed on paper treated with ZPa than on those treated with HZO.

Cabral[199] presented a method for separating the alkaline earth metal ions from Cs^+ using paper treated with ZPa. The first step is to separate the Cs^+ using 0.5 *M* HCl solution as eluant. Then the other ions may be separated from each other with a 0.9 *M* NH_4Cl solution. Nunes da Costa and Jeronimo[153] reported that under the same conditions, using a 0.1 *M* HCl solution as eluant, Ag^+-Cu^{2+} ions may be separated (R_f = 0; R_f^{Cu} = 0.7 to 0.8). Alberti and Grassini[191] used paper strips treated with ZPa to determine the R_f values for many elements (Table 13).

As described, the alkali metals were separated using a paper treated with AMP. Eluting the alkali metals with a mixture of 0.1 *M* NH_4NO_3 and 0.2 *M* HNO_3 solutions, they separated the metals in three groups by cutting the paper in three parts Cs^+ + Rb^+, K^+, and Na^+ + Li^+. The first piece was eluted with a mixture of 0.2 *M* HNO_3 and 3.5 *M* NH_4NO_3 solutions to separate Cs^+ and Rb^+ from each other. Then, by washing the third part with 96% ethanol, Li^+ was separated from Na^+.

Zabin and Rollins[200] described separations using thin-layer chromatography (with layers made from inorganic ion exchangers). Kornyei and co-workers[201] presented a method for determining different chemical forms of Ga, and Shimomura and Walton[202] used thin layers of inorganic materials for the separation of amines.

C. INORGANIC ION EXCHANGERS IN GAS CHROMATOGRAPHY

Some gas chromatographic separations have been performed successfully on well-defined crystalline inorganic ion exchangers, such as ZPc and zirconium arsenate (ZAc).

Lykourghiotis and co-workers[203] have reported that rapid determinations can be achieved with selective retention of some organic compounds on an HAP column. As noted by the authors, at the same temperature, the logarithm of the corrected retention volume of compounds belonging to the same homologous series is a linear function of the number of C atoms, as is usually the case in gas-liquid chromatography. In addition, they found that the eluted chromatographic peaks are generally symmetrical; because of this, good separations were achieved for various pairs with similar chromatographic properties. Making use of this specific property, HAP is able to retain completely the alkenes and alkynes. Thus, it is possible to separate saturated hydrocarbons from unsaturated ones. The hydrodynamic properties of Si-ZPc are being investigated by the author in order to use the latter in HPLC practice.[204]

REFERENCES

1. **Adams, B. A. and Holmes, E. L.,** *J. Soc. Chem. Ind. London,* 54, 1T, 1935.
2. **Adams, B. A. and Holmes, E. L.,** *English Patent,* 450308, 1935.
3. **Kiseleva, E. D.,** *Z. Phys. Chim.,* 36, 2457, 1962.
4. **Kiseleva, E. D.,** Trudii 2-go Vsesoiusn. Soves.po Radiacionnoi Chim., *Izv. Acad. Sci.,* USSR, Moscow, 27, 1962
5. **Kiseleva, E. D.,** *Izv. Acad. Sci. USSR Ser. Chim.* 6, 990, 1964.
6. **Szirtes, L.,** Investigation on Zirconium Phosphate, Ph.D thesis, 1968.
7. **Ahrland, S.,** The use of inorganic ion exchangers for reprocessing purposes, paper presented at the OECD Symp., Brussels, 1963.
8. **Pekarek, V. and Vesely, V.,** Synthetic inorganic ion exchangers. Hydrous oxides and acidic salts of monovalent metals, *Talanta,* 19, 219, 1972.
9. **Marinsky, J. A. and Marcus, Y., Eds.,** *Ion Exchange and Solvent Extraction,* Vol. 5, Marcel Dekker, New York, 1973, 92.
10. **Clearfield, A., Ed.,** *Inorganic Ion Exchange Materials,* CRC Press, Boca Raton, FL 1982, 141.
11. **Kraus, K. A., Phillips, H. O., Carlson, T. A., and Johnson, J. S.,** Ion exchange properties of hydrous oxides, in Proc. 2nd Int. Conf. P. U. A. E., Vol. 3, Geneva, 1958.
12. **Cannon, P.,** The reaction of alkali tungstate solutions with hydrochloric acid, *J. Inorg. Nucl. Chem.,* 13, 269, 1960.
13. **Smit, J. van R., Jacobs, J. J., and Robb, W.,** Cation exchange properties of the ammonium heteropolyacid salts, *J. Inorg. Nucl. Chem.,* 12, 95, 1959.
14. **Smit, J. van R. and Robb, W.,** Ion exchange on ammonium molybdophosphate. II. Bivalent and trivalent ions, *J. Inorg. Nucl. Chem.,* 26, 509, 1964.
15. **Cziboly, Cs., Szirtes, L., and Zsinka, L.,** Investigation of some heteropolyacid ion exchangers, *Radiochem. Radioanal. Lett.,* 8, 11, 1971.
16. **Amphlett, C. B.,** *Inorganic Ion Exchangers,* Elsevier, Amsterdam 1964, 97.
17. **Szirtes, L. and Zsinka, L.,** Synthetic inorganic ion exchangers: preparation and some uses of them in radiochemistry, *Izotoptechnika,* 17, 545, 1974.
18. **Clearfield, A. Ed.,** *Inorganic Ion Exchange Materials,* CRC Press, Boca Raton, FL, 1982, 152.
19. **Cathers, C. I., Blanco, R. E., Ferguson, D. E., Higgins, I. R., Kibbey, A. H., Mansfield, R. G., and Wischow, R. P.,** Radiation damage to radiochemical processing reagents, *Proc. Int. Conf. P U A E* 7, 490, 1955.

20. **Amphlett, C. B.,** *Inorganic Ion Exchangers,* Elsevier, Amsterdam, 1964.
21. **Materova, E. A., Belinskaya, Militsina, E. A., and Skabichevskii, P. A.,** *Ionnyiobmen, Leningr. Gos. Univ.*, 1965.
22. **Tachimori, S. and Amano, H.,** Inorganic ion exchangers, *Rep. JAERI-13726,* Japan Tokyo, 1969.
23. **Pai, K. R., Krishnaswamy, N., and Datar, D. S.,** Properties of hydrous oxide inorganic ion exchangers, in Papers Conf. *Ion Exch.*, Process Ind., Soc. Chem. Ind., London, 1970, 322.
24. **Price, J. W.,** Inorganic tin compounds, in Conf. Tin Consumption, London, 1972, 199.
25. **Roslyakova, N. G., Dushina, A. P. and Aleskovskii, V. B.,** Determination of the composition of cations sorbed on silica gel, *Ionny Obmen, Ionity, Izd. Nauka,* Leningrad, 1970, 137.
26. **Pekarek, V. and Vesely, V.,** Synthetic inorganic ion exchangers, hydrous oxides and acidic salts of multivalent elements, *Talanta,* 19, 56, 1972.
27. **Materova, E. A.,** Ion exchangers based on the elements from groups IV-VI., *Neorg. Ionoobmen. Mater* 1, 56, 1974.
28. **Nikolskii, B. P., Ed.,** *Neorg. Ionoobmen, Mater. Vyp. I. Izd. Leningr.* Gos. Univ. Lenigrad, 1974.
29. **Qureshi, M.,** Structural aspects and some novel features of synthetic inorganic ion exchangers, in Proc. 2nd Chem. Symp. Indian Bombay, 1972, 1.
30. **Belinskaya, F. A.,** Structure of inorganic ion exchangers, *Vestn. Leningr. Univ. Fiz. Khim.*, 1, 94, 1974.
31. **Michael, N. and Fletcher, W. D.,** Some performance characteristics of zirconium phosphate and zirconium oxide ion exchange materials, *Trans. Am. Nucl. Soc.*, 3, 46, 1960.
32. **Michael, N. Sterling, P. F., and Cohen, P.,** Inorganic ion exchange resins could purify hotter reactor water, *Nucleonics,* 62, 1963.
33. **Ruvarac, A. and Tolic, A.,** Purification of Nuclear Reactor Cooling Water at Higher Temperatures and Pressures, Nucl. Sci. *Rep. IBK-452,* Boris Kidric Institute of Nuclear Science, Belgrade, 1966.
34. **Ahrland, S. and Carleson, G.,** Inorganic ion exchangers: Purification of water on elevated temperatures by a combination of zirconium phosphate and zirconium oxide gels, *J. Inorg. Nucl. Chem.*, 33, 2229, 1977.
35. **Amphlett, C. B.,** Synthetic inorganic ion exchangers and their applications in atomic energy, in *Proc. 2nd Int. Conf. P U A E,* 28, 17, 1976.
36. **Kraus, K. A. and Phillips, H. O.,** Anion exchange studies. XIX. Anion exchange properties of hydrous zirconium oxide, *J. Am. Chem. Soc.*, 78, 249, 1965.
37. **Korshunov, I. A., Chernorukov, N. G., and Prokofieva, T. V.,** Sorption of Np and Pu on several difficulty soluble compounds, *Radiokhimiya,* 18, 5, 1976.
38. **Kennedy, J., Peckett, J. W., and Perkins, R.,** The Removal of Pu and Certain Fisson Products from Alkaline Media by Hydrated Titanium Oxide, Rep. AERE-R-4516 U. K., Atomic Energy Research Establishment, Harwell, Didcot, Berkshire, 1964.
39. **Myasoedov, B. F.,** Use of higher valence states during isolation and determination of Am and Bk, *Usp. Anal. Khim.*, 148, 1974.
40. **Brandona, A., Meloni, S., Girardi, F. and Sabbioni, E.,** Radiochemical separation by adsorption on tin dioxide, *Analysis,* 2, 300, 1973.
41. **Eschrich, H., Herrera-Huertas, M., and Tallberg, K.,** Sorption of Antimony by Silica Gel from HNO_3 and HC1 solutions, Rep. NP-18738, U.S. Atomic Energy Commission, 1970.
42. Anon., Reprocessing nuclear fuel, *Res. Discl.* 179, 104, 1979.
43. **Kraus, K. A., Phillips, H. O., and Carlson, T. A.,** Rep. ORNL-2195, 1956, 41.
44. **Keen, N. J.,** Extraction of U from sea water, *J. Br. Nucl. Energy Soc.*, 7, 178, 1968.
45. **Kanno, M.,** Extraction of U from Sea Water, Rep. IAEA-CN-36/161, IAEA, Vienna, 1977.
46. **Takesute, E. and Miyamatsu, T.,** Uranium collector, Jn Kokai, 77, 114, 511, 1977.
47. **Yamashita, H., Ozawa, Y., Nakajima, F. and Murata, T.,** *Nippon Kagaku Kaishi,* 8, 1057, 1978.
48. **Bettinali, C. and Panetti, F.,** U from sea water, possibilities of recovery exploiting slow coastal currents, in *Proc. Advisory Group Meeting, 1975,* IAEA, Vienna, 1976.
49. **Tustakowski, S.,** Separation of halide anions on hydrous zirconium oxide, *J. Chromatogr.*, 31, 268, 1967.
50. **Tucher, W. D.,** Rep. BNL-3746, 1958.
51. **Szirtes, L.,** Trends of isotope production in Europe, *Izotoptechnika,* 18, 5, 1975.
52. **Shishkov, D. A.,** *God* Minno. Geol. Institute, 1960, 213.
53. **Johnson, R. S. and Vosburgh, W. C.,** *J. Electrochem. Soc.*, 99, 317, 1952.
54. **Sasaki, K.,** *Mem. Fac. Eng. Nagoya Univ.*, 3, 81, 1951.
55. **Kozawa, A.,** *J. Electrochem. Soc.,* 106, 552, 1959.
56. **Gerevini, T. and Somigliana, R.,** *Energ Nucl. (Milan),* 6, 339, 1959.
57. **Puskariev, V. V.,** Investigations on synthetized manganese dioxide, *Radiokhimiya,* 4, 49, 1960.
58. **Benton, D. P. and Horsfall, G. A.,** *J. Chem. Soc.* 3899, 1962.
59. **Renault, N.,** Fixation of elements of the first transition series on tin dioxide in hydroorganic media, *Anal. Chem. Acta,* 70, 469, 1974.

60. **Renault, N. and Deschamps, N.,** Retention of some transition elements on tin dioxide in water-acetone, nitric-acid-acetone and hydrochloric acid-acetone media, *Radiochem. Radioanal. Lett.*, 13, 207, 1973.
61. **Cleyrergue, Ch. and Deschamps, N.,** Automation of a systematic analysis scheme for high purity iron, *J. Radioanal. Chem.*, 17, 139, 1973.
62. **Ahrland, S., Grenthe, I., and Noren, B.,** The ion exchange properties of silica gel. I. The sorption of Na^+, Ca^{2+}, Bn^+, UO^{2+}, Gd^{3+}, Zr^{4+}, Nb^{5+}, U^{4+} and Pu^{4+}, *Acta Chem. Scand.*, 14, 1059, 1960.
63. **Ahrland, S., Grenthe, I., and Noren, B.,** The ion exchange properties of silica gel II. Separation of Pu and fission products from irradiated U, *Acta Chem. Scand.*, 14, 1077, 1960.
64. **Cvjeticanin, D. and Cvjeticanin, N.,** JENER Rep. 1958, 54.
65. **Milone, M., Cetini, G., and Ricca, F.,** in *Proc. 2nd Int. Conf. P U A E.*, 18, 133, 1958.
66. **Laskorin, B. N., Bondarenko, L. I., Strelko, V. V., Kulbich, T. S., and Denisov, V. I.,** Characteristics of the sorption of divalent cations by ion exchangers based on antimonic acid, *Dokl. Akad. Nauk SSSR*, 229, 1411, 1976.
67. **Jander, G. and Simon, A.,** Formation kinetics of antimony pentoxide hydrate, *Z. Anorg. Allg. Chem.*, 127, 68, 1923.
68. **Jander, G.,** Antimony and antimonates, *Kolloid Z.*, 23, 122, 1919.
69. **Ghosh, S. and Dhar, N. R.,** Investigation on the colloidal behaviour of antimony pentoxide, *J. Ind. Chem. Soc.*, 6, 17, 1929.
70. **Abe, M. and Ito, T.,** Ion adsorptive properties of insoluble quadri-, quinque-, and sexi-valent metal hydrous oxides, *Nippon Kagaku Zasshi*, 86, 817, 1259, 1965.
71. **Ito, T. and Abe, M.,** Ion exchange properties of "so-called" antimonic acid, in 15th Nat. Meet. Japan Chemical Society Tokyo, 1962, 58.
72. **Abe, M. and Ito, T.,** Cation exchange properties of antimonic (V) acid, *Nippon Kagaku Zasshi*, 87, 1174, 1966.
73. **Lefebvre, J.,** Antimonic acid as ion exchanger, detection and character of exchange, *C. R. Acad. Sci. (Paris)*, 260, 5575, 1965.
74. **Lefebvre, J. and Gaymard, F.,** Antimonic acid as ion exchanger, capacities and selectivities for alkali and alkaline earth metals, *C. R. Acad. Sci. (Paris)*, 260, 6911, 1965.
75. **Abe, M. and Ito, T.,** Preparation and properties of "so-called" antimonic (V) acid, *Bull. Chem. Soc. Jpn.*, 41, 333, 1968.
76. **Abe, M.,** A study on ion-exchange properties of crystalline antimonic (V) acid and hydrated antimony pentoxide for various metal ions in nitric acid media, *Sep. Sci. Technol.*, 15, 23, 1980.
77. **Girardi, F., Pietra, R., and Sabbioni, E.,** Radiochemical separations by retention on ionic precipitate. Adsorption tests on 11 materials, *J. Radioanal. Chem.*, 5, 141, 1970.
78. **Krishnan, S. S. and Crapper, D. R.,** Sodium removal by hydrated antimony pentoxide in neutron activation analysis, *Radiochem. Radioanal. Lett.*, 20, 287, 1975.
79. **Baetsle, L. H., van Deyck, D., Huys, D., and Guery, A.,** The Use of Inorganic Exchanger in Acid Media for Recovery of Cs and Sr from Reprocessing Solution, AEC Accession No. 7613, Rep. BLG. 267, U.S. Atomic Energy Commission, 1964.
80. **Baetsle, L. H., van Deyck, D., Huys, D., and Guery, A.,** Separation of ^{137}Cs and ^{90}Sr from Fission Products in an acid Medium on Mineral Exchangers, AEC Accession No. 3871, Rep. EUR 24970, U.S. Atomic Energy Commission, 1975.
81. **Abe, M. and Ito, T.,** Mutual separation of alkali metals with antimonic (V) acid, *Bull. Chem. Soc. Jpn.*, 40, 1013, 1967.
82. **Abe, M.,** The mutual separation of alkali metals with three different antimonic (V) acid, *Bull. Chem. Soc. Jpn.*, 42, 2683, 1969.
83. **Novikov, B. G., Materova, E. A., and Belinskaiya, F. A.,** Nature and stability of precipitated polyantimonic acid, *Vestn. Leningr. Univ. Fiz. Khim.*, 22, 97, 1976.
84. **Abe, M. and Uno, K.,** Ion exchange behaviour and separation of alkaline earth metals on crystalline antimonic (V) acid., *Sep. Sci. Technol.*, 14, 355, 1979.
85. **Konecny, C. and Kourim, V.,** Recovery of Sr by means of a polyantimonic (V) acid cation exchanger, *Radioanal. Radiochem. Lett.*, 2, 47, 1969.
86. **Abe, M.,** Selective separation of Cd from Zn and Cu(II) with crystalline antimonic (V) acid as a cation-exchanger, *Chem. Lett.*, 561, 1979.
87. **Abe, M. and Kasai, K.,** Distribution coefficients and possible separation of transition metals on crystalline antimonic (V) acid, *Sep. Sci. Technol.*, 14, 895, 1979.
88. **Abe, M. and Akimoto, M.,** Ion exchange properties of crystalline antimonic (V) acid towards noble metals in nitric acid media, *Bull. Chem. Soc. Jpn.*, 53, 121, 1980.
89. **Girardi, F. and Sabbioni, E.,** Selective removal of radio sodium from neutron-activated materials by retention on hydrated antimony pentoxide, *J. Radioanal. Chem.*, 1, 169, 1968.
90. **Caletka, R., Konecny, C., and Simkova, M.,** Removal of Na from mineral acid solutions by adsorption on modified polyantimonic (V) acid, *J. Radioanal. Chem.*, 10, 5, 1972.

91. **Abe, M. and Tsuji, M.,** The ion exchange properties of Na^+ and K^+ ions on crystalline and amorphous antimonic (V) acids treated thermally at elevated temperatures, *J. Radioanal. Chem.*, 54, 137, 1979.
92. **Higuchi, H. and Takehira, T.,** Removal of Na^+ from HAP, *Bunseki Kagaku,* 21, 808, 1972.
93. **Abe, M. and Sudoh, K.,** Ion exchange equilibria of transition metals and H^+ ions in crystalline antimonic (V) acid, *J. Inorg. Nucl. Chem.*, 42, 1051, 1980.
94. **Baetsle, L. H. and Huys, D.,** Structure and ion exchange characteristics of polyantimonic acid, *J. Inorg. Nucl. Chem.*, 30, 639, 1968.
95. **Aubertin, C., Lefebvre, J. and Galand, G.,** Separating Sr from fission product solutions, *Chem. Abstr.*, 70, 16481, 1969.
96. **Belinskaiya, F. A., Materova, E. A., Militsina, E. A., Karmanova, L. A., and Novikov, B. G.,** *Chem. Abstr.*, 85, 96498x, 1976.
97. **Nagy, L. G., Torok, G., Foti, G., Toth, T., and Feuer, L.,** Investigation on the structure and sorption properties of removal of radiosodium, in *Proc. Int. Conf. Colloid. Surf. Sci.*, 1, 33, 1975.
98. **Meinke, W. W.,** L'Analyse par activation et ses applications aux sciences biologiques, Presses Univ. de France, 1964, 149.
99. **Lyon, W. S.,** in *Proc. Int. Conf. Act. Techn.* paper SM-91/68, IAEA, Vienna, 1967.
100. **Cuypers, J., Girardi, F., and Mousty, F.,** The application of inorganic exchangers to radiochemical separation on neutron-activated high-purity materials, *J. Radioanal. Chem.*, 17, 115, 1973.
101. **Pietra, R., Sabbioni, E., and Girardi, F.,** Determination of Ca, Mg, Ni and Si in biological materials by neutron activation and Cerenkov counting, *Radiochem. Radioanal. Lett.*, 22, 243, 1975.
102. **Torok, G., Schelenz, R., Fischer, E., and Diehl, J. F.,** Separation of Na, K and P by means of inorganic separators in the neutron activation analysis of biological materials, *Fresenius Z. Anal. Chem.*, 263, 110, 1973.
103. **Clearfield, A., Ed.,** *Inorganic Ion Exchangers,* CRC Press Boca Raton, FL, 1982, 243.
104. **Tjioe, P. S., CeGoeij, J. J. M., and Houtman, J. P. W.,** Automated chemical separation in routine activation analysis, *J. Radioanal. Chem.*, 16, 153, 1973.
105. **Ralson, H. R. and Sato, E. S.,** Na removal as an aid to neutron activation analysis, *Anal. Chem.*, 43, 129, 1971.
106. **Ching-Wang, H., Higuchi, H., and Hamaguchi, H.,** Multielment neutron activation analysis of human hair samples, *Bunseki Kagaku,* 22, 1586, 1973.
107. **Buenafama, H. D. and Rudelli, M. D.,** Determination of Cs, Rb and Ba in neutorn activated biological materials, *J. Radioanal. Chem.*, 16, 269, 1973.
108. **Platin, L. O.,** A method for the determination of Mn, Cu, Zn and Na in small tissue biopsies by neutron activation analysis, *J. Radioanal. Chem.*, 12, 441, 1972.
109. **Velandia, J. A. and Perkons, A. K.,** Survey of 33 constitutent elements in heart tissue by instrumental activation analysis, *J. Radional. Chem.*, 14, 171, 1973.
110. **Livens, P., Cornelis, R. and Hoste, J.,** A separation scheme for the determination of trace elements in biological materials by neutron activation analysis, *Anal. Chim. Acta,* 80, 97, 1975.
111. **Maziere, B., Gaudray, A., Stanilewicz, W., and Comer, D.,** Possibilities and limits of the multielemental determinations in biological samples by neutron activation analysis with and without chemical separations, *J. Radioanal. Chem.*, 16, 281, 1973.
112. **Treuil, M., Jaffrezic, H., Deschamps, N., Derre, C., Guichard, F., Joron, J. L., Pelletier, B., and Courfois, C.,** Analysis of lanthanides, Hf, Sc, Cr, Mn, Co, cu and Zn in the mineral rocks by neutron activation, *J. Radioanal. Chem.*, 18, 55, 1973.
113. **Tshiashala, M. D. and DeMicheli, F. O.,** Separation of K from rock samples in neutron activation analysis, *Radiochem. Radioanal. Lett.*, 25, 101, 1976.
114. **Kudo, K., Kobayashi, K., and Shigematsu, T.,** Substoichiometric and non-destructive determinations of trace impurities in high-purity optical glasses by neutron activation analysis, *J. Radioanal. Chem.*, 27, 329, 1975.
115. **Gills, T. E., Marlow, W. F., and Thompson, B. A.,** Determination of trace elements in glass by activation analysis using HAP, for sodium removal, *Anal. Chem.*, 42, 1831, 1970.
116. **Schelhorn, H., Pfrepeer, G. and Geisler, M.,** Determination of trace element in biological materials with neutron activation analysis, *J. Radioanal. Chem.*, 33, 187, 1976.
117. **Yoshida, H. and Yonezawa, C.,** Determination of Na in high purity Zn and Se by neutron activation, *Bunseki Kagaku,* 22, 929, 1973.
118. **Mitchell, J. W., Riley, J. E., and Northover, W. R.,** Determination of trace transition elements in ultra high purity Na and Ca-carbonates by activation analysis, *J. Radioanal. Chem.*, 18, 133, 1973.
119. **Higuchi, H., Nonaka, N., Hamaguchi, H. and Tomura, K.,** Determination of Na and Cl in pure water by neutron activation analysis, *J. Radioanal. Chem.*, 36, 457, 1977.
120. **Casella, V. R., Grant, P. M., and O'Brien, H. A., Jr.,** The quantitative recovery and purification of spallogenic ^{43}K for nuclear medicine, *J. Radioanal. Chem.*, 36, 337, 1977.

121. **Massart, D. L.,** The elimination of K activity in activation analysis by isotopic exchange on column, *J. Radioanal. Chem.*, 4, 265, 1970.
122. **Schumacher, J., Maier-Borst, W., and Hauser, H.,** A half automated, non time consuming radiochemical separation scheme for determination of 25 trace elements in biological specimens, *J. Radioanal. Chem.*, 37, 503, 1977.
123. **Abe, M.,** Chromatographic separation of microamounts of Na and K from a large quantity of LiCl by using crystalline antimonic(V) acid as a cation exchanger, *Sep. Sci. Technol.*, 13, 347, 1978.
124. **Abe, M., Ichsan, E. A. W. A., and Hayashi, K.,** Ion exchange separation of Li from large amount of Na, Ca and other elements by a double column of Dowex 50W-X8 and crystalline antimonic (V) acid, *Anal. Chem.*, 52, 524, 1980.
125. **Arino, H., Skraba, W. J., and Kramer, H. H.,** A new^{68}Ge/^{68}Ga radioisotope generator system, *Int. J. Appl. Radiat. Isot.*, 29, 117, 1978.
126. **Vandenborre, H. and Leysen, R.,** On Inorganic membrane-electrolyte wqater electrolysis, *Electrochim. Acta*, 23, 803, 1978.
127. **Amphlett, C. B., McDonald, L. A., and Redman, M. J.,** Synthetic inorganic ion exchange materials. II. Hydrous zirconium oxide and other oxides, *J. Inorg. Nucl. Chem.*, 6, 236, 1958.
128. **De, A. K. and Chowdhury, K.,** Synthetic ion exchangers. X. Hydrous tungsten (VI) oxides (synthesis, physico-chemical properties and ion exchange separations), *Chromatographia*, 11, 586, 1978.
129. **Shannon, R. D. and Prewitt, C. T.,** Effective ionic radii in oxides and fluorides, *Acta Crystallogr. B*, 25, 925, 1969.
130. **Szalay, T. and and Bartha, L.,** Ion exchange adsorption process on tungsten oxides, *Mag. Kem. Foly.*, 81, 67, 1975.
131. **Tsuboya, T., Kaya, A., and Hoshino, T.,** Np separation from uranium nitrate-nitric acid solutions with tungstic acid exchanger, in *Tokai Works Semi annual Progress Rep. No. N 831-70-02*, 1970, 72.
132. **Kraus, K. A. and Phillips, H. O.,** Adsorption on inorganic materials, I. Cation exchange properties of zirconium phosphate, *J. Am. Chem. Soc.*, 78, 644, 1956.
133. **Baetsle, L. H. and Huys, D.,** Ion exchange equilibrium of zirconium phosphate, *J. Inorg. Nucl. Chem.*, 21, 133, 1961; 25, 271, 1963.
134. **Amphlett, C. B., McDonald, L. A., and Redman, M. J.,** Cation exchange properties of zirconium phosphate, *Chem. Ind.* London, 1314, 1956.
135. **Maeck, W. J., Kussy, M. E., and Rein, J. E.,** *Anal. Chem.*, 35, 2086, 1963.
136. **Amphlett, C. B.,** *Inorganic Ion Exchangers*, Elsevier, Amsterdam, 1964, chap. 5.
137. **Kraus, K. A. and Phillips, H. O.,** Adsorption in inorganic materials, *J. Am. Chem. Soc.*, 78, 694, 1956.
138. **Phillips, H. O., Nelson, F., and Kraus, K. a.,** *Rep. ORNL-2159*, 1956, 37.
139. **Amphlett, C. B., McDonald, L. A., and Redman, M. J.,** Synthetic inorganic ion exchange materials, *J. Inorg. Nucl. Chem.*, 6, 220, 1958.
140. **Gal, I. and Ruvarac, A.,** The separation of Pu and fission products on zirconium phosphate column, *J. Chromatogr.*, 13, 549, 1964.
141. **Zsinka, L. and Szirtes, L.,** Separation of microamount of Rb from macroamount of Sr, *Radiokhimiya*, 12, 774, 1970.
142. **Szirtes, L. and Zsinka, L.,** Study of ion exchange equilibria using radioactive isotopes, *J. Radioanal. Chem.*, 30, 131, 1976.
143. **Iofa, B. Z., Szirtes, L., and Zsinka, L.,** Separation of Fe and Co on synthetic inorganic ion exchangers, *Radiokhimiya*, 10, 491, 1968.
144. **Zsinka, L. and Szirtes, L.,** Use of synthetic inorganic ion exchangers for the separation of Ca and Sc, *Mag. Kem. Lapja*, 21, 536, 1966.
145. **Kornyei, J. and Szirtes, L.,** Separation of Sr from radioactive waste water on crystalline zirconium phosphate, *Izotoptechnika*, 23, 243, 1980.
146. **Alberti, G., Bertrami, R., Casciola, M., Costantino, U. and Gupta, J. P.,** Crystalline insoluble acid salts of tetravalent metals. XXI. Ion exchange mechanism of alkaline earth metal ions on crystalline ZrHNa $(PO_4)_2 \cdot 5\ H_2O$, *J. Inorg. Nucl. Chem.*, 38, 843, 1976.
147. **Gal, I. and Gal, O.,** Separation of uranium and plutonium using amorphous zirconium and titanium phosphates as ion exchangers, in *Proc. 2nd Int. Conf. P. U. A. E.* 28, 24, 1958.
148. **Ruvarac, A.,** Purification of cooling water from nuclear reactor RA, Rep. IBK-560, Inst. Bk., Boris Kidric Institute of Nuclear Science. Belgrade, 1967.
149. **Kraus, K. A. and Raridon, R. J.,** Temperature dependence of some cation exchange equilibria in the range 0—200°, *J. Phys. Chem.*, 63, 190, 1959.
150. **Grover, J. R. and Chidley, B. E.,** *Ind. Chem.* 39, 31, 1963.
151. **Cziboly, Cs., Szirtes, L., and Zsinka, L.,** Possibilities of the separation of alkali and alkaline earth metals on heteropoly acid type ion exchangers, *Mag. Kem. Lapja*, 24, 470, 1969.
152. **Kraus, K. A., Carlson, T. A., and Johnson, J. S.,** Cation exchange properties of Zr(IV)-W(VI) precipitates, *Nature*, 177, 1128, 1956.

153. **Nunes da Costa, M. J. and Jeronimo, M. A. S.,** *J. Chromatogr.*, 5, 456, 1961.
154. **Crouch, E. A. C., Corbett, J. A., and Willis, H. H.,** AERE-C/R 2325, U. K. A. E. A., 1957.
155. **Szirtes, L. and Zsinka, L.,** Use of zirconium tellurate in ion exchange separation technique, *Radiochem. Radioanal. Lett.*, 7, 61, 1971.
156. **Bresler, S. E., Sinotskin, J. D., Egorov, A. I., and Perumov, D. A.,** Separation of alkaline earth metals on zirconium tungstate ion exchanger, *Radiokhimiya*, 1, 507, 1959.
157. **Sinotskin, I. D., Perumov, D. A., Patin, A. C., and Azaza, E. G.,** Zirconium base ion exchangers. Application of zirconium oxalate for determination of $^{90}Sr/^{90}Y$ in sea water samples, *Radiokhimiya*, 4, 198, 1962.
158. **Zsinka, L. and Szirtes, L.,** Investigation on a chromium phosphate ion exchanger, *Acta Chim. Hung.*, 69, 249, 1971.
159. **Zsinka, L. and Szirtes, L.,** Use of chromium phosphate in ion exchange separation technique, *Radiochem. Radioanal. Lett.*, 2, 257, 1969.
160. **Fussy, E. and Szirtes, L.,** Technetium-^{99m} and indium-^{133m} generators using chromium phosphate, *Radiochem. Radioanal. Lett.*, 8, 19, 1971.
161. **Smit, J. van R.,** Ion exchange on the ammonium molybdophosphate. III. Preparation and behaviour of NH_4 salts of heteropoly acids, *J. Inorg. Nucl. Chem.*, 27, 227, 1965.
162. **Smit, J. van R.,** Ammonium salts of the heteropolyacids as cation exchangers, *Nature*, 181, 1530, 1958.
163. **Smit, J. van R., Robb, W., and Jacobs, J. J.,** AMP-effective ion exchanger for treating fission waste, *Nucleonics*, 17 (g), 116, 1959.
164. **Krtil, I. and Krivy, I.,** Exchange behaviour of NH_4-salts of heteropoly acids. V. Sorption of alkali metals on AMWP, *J. Inorg. Nucl. Chem.*, 25, 1191, 1963.
165. **Kourim, V., Lavrukhina, A. K., and Rodin, S. S.,** Coprecipitation of Fr and Cs with heteropoly acids, *J. Inorg. Nucl. Chem.*, 21, 375, 1961.
166. **Kourim, V., Lavrukhina, A. K., and Rodin, S. S.,** *Dokl. Acad. Nauk SSSR*, 140, 832, 1961.
167. **Lavrukhina, A. K., Kourim, V., and Rodin, S. S.,** New method for separation of radioactive Rb and Cs, *Acta Chim. Hung.*, 33, 309, 1962.
168. **Krtil, J.,** Exchange properties of ammonium salts of 12-heteropolyacids. III. The sorption of strontium and yttrium on ammonium phosphotungstate and their separation from caesium, *J. Inorg. Nucl. Chem.*, 22, 247, 1961.
169. **Rodinskii, S. Z.,** Preparation of concentrates of radioactive Cs^+ with ferrocyanate of heavy metals from a solution with high content of other salts, *Radiokhimiya*, 2, 438, 1960.
170. **Baetsle, L. H. and Dejonghe, P.,** Treatment and Storage of High Level Radioactive Wastes, IAEA, Vienna, 1963, 553.
171. **Szirtes, L.,** unpublished data.
172. **Ames, L. L.,** *Am. Mineral.*, 45, 689, 1960.
173. **Herr, W. and Riedel, H. J.,** *Radiochim. Acta*, 1, 32, 1962.
174. **Amphlett, C. B.,** *Inorganic Ion Exchangers*, Elsevier, Amsterdam, 1964, 48.
175. **Rodinskii, S. Z.,** Preparation of concentrates of radioactive Cs^+ using glauconite ion exchange column, *Radiokhimiya*, 2, 431, 1960.
176. **Le Van So,** Investigation on Various Inorganic Ion Exchangers on the Silica Matrix, Ph.D. thesis, Budapest, 1985.
177. **Szirtes, L.,** unpublished data.
178. **Le Van So and Szirtes, L.,** Investigation of silica gel supported inorganic ion exchangers, *J. Radioanal. Nucl. Chem. Art.*, 99, 45, 1986.
179. **Le Van So and Szirtes, L.,** Silica gel supported microcrystalline zirconium phosphate ion exchanger Si-ZrP and its application in chemical separation. III. Separation of alkali and transition metal ions, *J. Radioanal. Nucl. Chem. Art.*, 99, 55, 1986.
180. **Le Van So and Do Minh Vung,** Quantitative separation of Be from other elements using crystalline zirconium phosphate ion exchanger, *Radiochem. Radioanal. Lett.*, 56, 343, 1982.
181. **Le Van So,** Preparation of microcrystalline zirconium phosphate and its silica gel supported form, *Radiochem. Radioanal. Lett.*, 57, 23, 1983.
182. Standard SSR No. GOST-18913-73.
183. **Amphlett, C. B.,** *Treatment and Disposal of Radioactive Wastes*, Pergamon Press, Oxford, 1961, 29.
184. **Wilding, M. W.,** *Rep. IDO 14540*, 1960.
185. **Whitehurst, D. D., Butter, S. A., and Rodewald, P. G.,** U.S. Patent 3793, 185, 1974.
186. **Dravnieks, A. and Bregman, J. I.,** Inorganic membrane works in fuel cells; polymeric $Zro(H_2PO_4)_2$ may lead to more efficient fuel cells, allows high temperature operation, *Chem. Eng. News.*, 39, 40, 1961.
187. **Alberti, G.,** Perm-selectivity of inorganic ion exchange membranes consisting of amorphous ZrP supported on glass wool fibers, *Atti. Accad. Naz. Lincei, Cl. Sci. Fis. Mat. Nat. Rend.*, 31, 427, 1961.
188. **Kraus, K. A., Shoz, A. J., and Johnson, J. S.,** Hyperfiltration studies X. Hyperfiltration with dynamically formed membranes, *Desalination*, 2, 243, 1967.

189. **Rajan, K. S., Boies, D. B., Casolo, A. J., and Bregman, J. I.,** Inorganic ion exchange membranes and their application to electrodialysis, *Desalination,* 7, 237, 1966.
190. **Ciciscilli, G. V. and Krupennykova, A. I.,** Sorption of the Sr^{2+} on a Na-formed molecule filter, *Radiokhimiya,* 5, 656, 1963.
191. **Alberti, G. and Grassini, G.,** Chroimatography on paper impregnated with zirconium phosphate, *J. Chromatogr.,* 4, 83, 1960.
192. **Alberti, G., Dobici, F., and Grassini, G.,** Chromatography on paper impregnated with inorganic ion exchangers. III. Chromatography of inorganic ions on zirconium phosphate paper with HC1, H_2SO_4 HNO_3, $HC1O_4$ and CH_3COOH at various concentrations, *J. Chromatogr.,* 13, 561, 1961.
193. **Alberti, G.,** Chromatographic separations on paper impregnated with inorganic ion exchangers, *Chromatogr. Rev.,* 8, 246, 1969.
194. **Grassini, G.,** The exchange capacity of papers impregnated with zirconium phosphate, *J. Chromatogr.,* 5, 365, 1964.
195. **Adloff, J. P.,** Separation de radioelements par chromatographie sur papier impregne d'un echangeur d'ion mineral, *J. Chromatogr.,* 5, 365, 1961.
196. **Sastri, M. N. and Rao, A. P.,** Separation of the valency states of some elements on paper impregnated with zirconium phosphate, *J. Chromatogr.,* 9, 250, 1962.
197. **Schroeder, M. J.,** *Radiochim. Acta,* 1, 27, 1962.
198. **Amphlett,** *Inorganic Ion Exchangers,* Elsevier, Amsterdam, 1964, 108.
199. **Cabral, J. M. P.,** Chromatography on paper impregnated with inorganic ion exchangers, *J. Chromatogr.,* 4, 86, 1960.
200. **Zabin, B. A. and Rollins, C. B.,** Inorganic ion exchangers for thin layer chromatography, *J. Chromatogr.,* 14, 534, 1964.
201. **Kornyei, J., Szirtes, L., and Kecskes, F.,** Radio-paper-chromatographic determination of different chemical forms of Ga, *Radioanal Nucl. Chem. Lett.,* 103, 313, 1986.
202. **Shimomura, K. S. and Walton, H. F.,** Thin-layer chromatography of amines by ligand exchange, *Sep. Sci.,* 3, 493, 1968.
203. **Lykourghiotis, A., Hadzistelios, I., and Katsanos, N. A.,** Chromatographic and adsorption properties of antimony pentoxide, *J. Chromatogr.,* 110, 287, 1975.
204. **Szirtes, L.,** unpublished data, 1986.

Chapter 5

INORGANIC ION EXCHANGERS FOR ION-SELECTIVE ELECTRODES

C. J. Coetzee

TABLE OF CONTENTS

I. INTRODUCTION

One of the newer and very elegant tools of analytical chemistry are ion-selective electrodes. They have been developed and designed to respond to a particular ion in solution. The development of ion-selective electrodes blew new life into potentiometry and the manufacturing of equipment for potentiometric measurement.

With a large number of commercially available ion-selective electrodes for both common cations and anions, a large number of analytical applications exist. There is, however, still a need for the development of ion-selective electrodes for certain ions, and a large number of ion-selective electrodes using inorganic ion-exchangers have been developed during the past 25 years.

Research in the field of ion-selective electrodes was enormously stimulated by the publication by Frant and Ross[1,2] of their articles on the fluoride and calcium ion-selective electrodes.

A large number of books, articles, and reviews in which the application and theory of operation of ion-selective electrodes are discussed, have been published. See, for example, References 3 to 9 and references cited herein.

Although it is not the aim of this work to discuss the theory of ion-selective electrodes, the following basic equations have to be kept in mind. When an ion-selective electrode is used in conjunction with a suitable reference electrode for an ion of activity a_A and charge z_A, the following cell potential is measured:

$$E = \text{constant} \pm \frac{2.303\ RT}{Z_A\ F} \log a_A$$

and in the presence of an interfering ion of activity a_B and charge z_B,

$$E = \text{constant} \pm \frac{2.303\ RT}{z_A\ F} \log (a_A + k_{AB}^{pot})$$

where k_{AB}^{pot} is the selectivity coefficient of the electrode for ion A in the presence of ion B.

In cases where the slope of the calibration curve is not equal to the Nernstian value, the factor $2.303\ RT/z_A\ F$ may be represented by S, indicating the slope.

II. CLASSIFICATION OF ION-SELECTIVE ELECTRODES

The older classification of ion-selective electrodes into the types (1) solid state, (2) heterogeneous, and (3) liquid ion-exchangers is no longer appropriate, as some active materials can be used in all three forms. The suggested classifications by the International Union of Pure and Applied Chemistry (IUPAC)[10] divides electrodes into the following types:

1. Primary electrodes
   a. Crystalline electrodes
      (1) Homogeneous membrane — a single compound, e.g., LaF_3, or mixture, e.g., $AgI\text{-}Ag_2S$
      (2) Heterogeneous membrane — an active substance mixed with an inert matrix, e.g., AgI in silicone rubber
   b. Noncrystalline, with or without support, such as porous glass, Millipore filter, PVC, etc.

(1) Rigid matrix, e.g., Na^+-glass electrodes
(2) With a mobile carrier which may be positively charged, negatively charged, or neutral; e.g., quaternary ammonium cations, tetraphenylborate anions, and valinomycin

c. Sensitized electrodes
(1) Gas-sensing electrodes
(2) Enzyme-substrate electrodes

Although the above classification for different ion-selective electrodes is good, active materials for ion-selective electrodes can be broadly classified into the following categories:[7]

1. Glass
2. Insoluble inorganic salts
3. Long-chain ion-exchange materials, often those used in solvent extraction

All these materials share a common property when brought into contact with an electrolyte solution containing the suitable ions — their ability to rapidly set up an exchange equilibrium or ion-exchange process across the phase boundary. A second requisite property is the ability of the active material to conduct electricity.

By using inorganic ion-exchange materials, the above types are combined, and they do shape quite well as prospective membrane materials for use in ion-selective electrodes. The possibility of using this type of membrane in potentiometric measurements for the determination of ion activities was envisaged in an article by Geyer and Syring.[11] Membrane electrodes have been tried for many ions; the only setback at present seems to be the question of selectivity.

III. CLASSIFICATION OF INORGANIC ION EXCHANGERS AND MATERIALS

Inorganic ion exchangers have played a very important role in analytical chemistry as well as industrial applications. The first synthetic ion exchangers were, in fact, inorganic materials. They were the synthetic zeolites or permutites developed by Gans and other workers for use in water-softening processes.[12]

Due to their diverse properties, not all the so-called inorganic ion exchangers are suitable for use as sensors in the construction of ion-selective electrodes. The major limiting factor in most cases may be a lack of a high degree of selectivity for single cations. As many of these compounds are crystalline salts and also have a diversity of physical properties, some are not even suitable for use in column chromatography. Some of these materials are slow to react or are soft and powdery. Various classifications for these compounds have been given, but a very convenient one is that by Vesely and Pekarek,[13] which groups the compounds together as follows:

1. Hydrated oxides of metals
2. Acidic salts of polyvalent metals
3. Insoluble salts of heteropolyacids
4. Insoluble hydrated metal hexacyanoferrate (II) and (III)
5. Synthetic zeolites
6. Other substances with weak exchange properties

IV. REQUIREMENTS OF MATERIALS TO BE USED IN HETEROGENEOUS, SOLID-MEMBRANE ION-SELECTIVE ELECTRODES

The inorganic ion exchangers are all solids which are relatively insoluble. They cannot easily be pressed into pellets or discs and used as such in electrodes. They have to be incorporated into a membrane matrix to form a heterogeneous membrane. In general, heterogeneous electrodes include the cases where the active materials are dispersed in silicone rubber, polyvinylchloride (PVC), epoxy resin, and paraffin. Individual properties and transport mechanisms may vary, but, in any case, presumably involve the transfer of ions between crystals that are in physical contact within the matrix.

The properties of the support and suitable solids to be incorporated into heterogeneous solid-state membranes are the following. A suitable matrix support must:

1. Be chemically inert and provide good adhesion for the sensor particles
2. Be hydrophobic
3. Be tough, flexible, yet nonporous and crack-resistant to prevent leakage of internal solutions
4. Not swell in sample solutions

In turn, the active sensor material should:

1. Be physically compatible with the matrix
2. Have a low solubility product
3. Be of the right grain size (1 to 15 nm), which is a function of the precipitation technique
4. Be mixed with the matrix support in the right proportion, usually 50 mass percent
5. Undergo rapid ion exchange at the membrane-sample interface

The insoluble hydrated metahexacyanoferrates (II) and (III) and the insoluble salts of the heteropolyacids seem to be the most successful inorganic ion exchangers to have been used in heterogeneous ion-selective membrane electrodes.

In his book on analysis with ion-selective electrodes, Bailey[14] lists a number of characteristics against which the performance of an electrode may be assessed. These include:

1. Response range and slope
2. Selectivity
3. Stability and reproducibility
4. Response time
5. Sensitivity to temperature, pressure, and light
6. Frequency and ease of maintenance
7. Mechanical design
8. Availability
9. Cost and life-time

The above general criteria shall be used in the following discussions, although some of them are not applicable to many of the electrode systems using inorganic ion exchangers.

V. CONSTRUCTION METHODS

Heterogeneous membranes have a solid compound dispersed in an inert binding material and thus consist of more than a single phase. The binding material which forms the inert matrix of the membrane must be chemically inert and insoluble in water. It must provide correct adhesion properties to effectively bind the particles of the active compound in such a manner as to ensure electrical contact between them. The binder should be hydrophobic and produce a membrane with good mechanical properties, such as flexibility and resistance to cracking and swelling in solution. The active reagent, in this case the inorganic ion exchanger, should have a low solubility, and must be finely divided in order to ensure good contact between the particles throughout the membrane. Optimum grain size appears to be between 1 and 15 nm. The active compound should constitute about 50% or more by mass of the membrane.

Various matrix binders have been used. Examples are paraffin, polystyrene, agar, polypropylene, methylmethacrylate, and thermoplastics such as polythene, silicone rubber, and epoxy resins. The most popular one seems to be the epoxy resin Araldite, which was first used by Coetzee and Basson.[15]

In the preparation of the heterogeneous membrane electrodes, the active material is mixed with the binder and spread out to a thin layer of about 1 mm thickness. It is then left to cure or dry. Small discs are cut from these sheets and are fixed to the end of a glass tube. In some cases, the binder and the active compound mixture are pressed into a disc under very high pressure, after which it is fixed to the end of a glass tube with an appropriate adhesive. The electrode is completed with an appropriate internal reference electrode, usually a silver-silver chloride or a saturated calomel electrode and a filling solution, which usually is a 0.1 *M* salt solution of the ion in question. The electrode is conditioned before use in a solution (usually 0.1 *M*) of the ion to be determined until stable potential readings are found. The prepared electrode is then used in an electrochemical cell with a suitable reference electrode.

The usual direct potentiometric studies, such as:

1. Calibration curves
2. Effect of pH on electrode potential
3. Effect of temperature on potential
4. Dynamic response characteristics
5. Selectivity

are carried out on the prepared electrode. These studies are followed by some analytical application of the prepared electrode.

VI. ELECTRODES BASED ON INORGANIC ION EXCHANGERS

A. INSOLUBLE SALTS OF COMPLEX CYANIDES

The ion-exchange properties of these compounds are well known. Not many cases of the use of this class of compounds as materials for the preparation of possible ion sensors have been reported. The compound, potassium zinc hexacyanoferrate(II), ($K_2Zn_3[Fe(CN)_6]_3 \cdot 8H_2O$), appears to have been used most often. Vesely and Pekarek[13] discussed the ion-exchange properties of the compound. This compound has been used to coprecipitate traces of cesium ion[16] as well as organic compounds.[17]

Finely ground commercial solid potassium zinc hexacyanoferrate(II) was used by Fogg et al.[18] to prepare two different heterogeneous membrane electrodes: one was a silicone rubber matrix electrode and the other a PVC matrix electrode. The membranes were prepared as follows: A 4:4:1 mixture of potassium zinc hexacyanoferrate(II), Silastomer 70, and catalyst BC was used to prepare the silicone rubber membrane, and the PVC electrode membrane by mixing 0.45 g of the salt with a solution of PVC powder (0.23 g) in 8 cm^3

TABLE 1
Selectivity Coefficients for a Potassium Electrode

	Selectivity coefficients k_{KC}^{pot}	
C^+ species	Silicone rubber electrode	PVC electrode
Cs^+	9.5	11.6
Rb^+	3.3	4.3
K^+	1.0	1.0
Na^+	0.025	0.021
Li^+	0.003	0.017
NH_4^+	1.8	2.2

of tetrahydrofuran. Potentiometric measurements were made at 25°C in a flow-through thermostated cell. In order to exclude leakage of potassium ions into the measuring cell, an Orion double-junction reference electrode was used, with the outer compartment filled with a 0.1 *M* solution of lithium trichloroacetate. Both electrodes gave a Nernstian response to potassium ion activities in the concentration range of 5×10^{-4} to 10^{-1} *M*. However, they also gave a Nernstian response to Cs^+, Rb^+, and NH_4^+ ion activities. The effect of pH on the response of the silicone rubber electrode was studied, and between pH 4 and 9.5, a constant response was found. Selectivity coefficients, based on potential measurements on single cation solutions at 0.1 *M* concentration, are indicated in Table 1.

The selectivity of these electrodes for potassium over sodium is of the same order as that for a good potassium ion-selective glass electrode. Greater selectivity can, however, be obtained with the valinomycin electrode. As expected from ion-exchange considerations, the selectivity coefficients indicate that the electrodes are actually more selective for cesium ion activities than for potassium ions. Response times were rather slow, and the final potential reading at each concentration of potassium ion was reached within 1 min and fluctuated by about 2 mV from the mean. The potentials were reproducible over a period of at least 1 month.

Three years later, the same salt ($K_2Zn_3[Fe(CN)_6]_3 \cdot 8H_2O$) in a heterogeneous salt-plus-silicone rubber membrane was investigated as an ion sensor by Rock et al.[19] Although this electrode gave a Nernstian response in potassium ion solutions over the concentration range 10^{-3} to 1 *M*, it took much longer to attain stable potentials (3 to 5 min). Calibration curves with the above-mentioned electrodes compared favorably with those of an Orion potassium ion-selective electrode and a Beckman 39137 glass electrode under similar experimental conditions. It is presumed that the commercial electrodes gave stable potential readings much quicker. Only one selectivity coefficient, that of the electrode for the K^+ ion over the Na^+ ion, was determined by the extrapolation intersection method. This value is given as approximately 0.10, which is about the same value as that for the cationic glass electrode with which it was compared. No potentiometric titrations in which the electrode was used were done. This work suggested that an electrode prepared from potassium zinc hexacyanoferrate(II) would be more selective to cesium ion than to potassium ion activities. D'Olieslager and Heerman[20] employed this compound in a PVC matrix as a cesium ion-selective electrode. Previous studies[21,22] demonstrated a high selectivity of this compound for the cesium ion over other alkali metal ions when used as an ion exchanger. The membranes were prepared by mixing the dry materials (60 to 70% precipitate content) and cold pressing. Thereafter, the membranes were conditioned overnight in a 0.1 *M* cesium chloride solution. Electrodes prepared from the above-mentioned membranes gave a Nernstian response in aqueous cesium ion solutions, and calibration curves were linear in the concentration range 1.5×10^{-4} to

0.10 *M*. Selectivity coefficients were estimated from experiments in which the concentration of cesium chloride was kept constant (0.001 *M*), and the concentration of potassium, sodium, or lithium ions, added as their fluorides, was varied. Representative values of the selectivity coefficients are

$$k_{Csk}^{pot} \simeq 0.09;\ k_{CsNa}^{pot} \simeq 0.02;\ k_{CsLi}^{pot} \simeq 0.006$$

Ammonium and rubidium ions were not included in the selectivity study. According to the authors, the selectivity coefficients were not as low as was hoped for on the basis of ion-exchange data. This is attributed to the lower mobility of cesium ions with respect to potassium or sodium ions in the matrix. No potentiometric titrations were done during which the electrode was used as indicator electrode.

Epoxy-resin-based membranes of chromium ferrocyanide gel, saturated with $[Co(en)_3]^{3+}$ ions, have been prepared and used as an ion-selective electrode for the measurement of the activity of $[Co(en)_3]^{3+}$ ions by Saraswat et al.[23] The response time of the membranes is less than 1 min, and cell potentials are reproducible and remain constant for more than 4 h. The slope of the calibration curve is not mentioned, but the electrode can be used within the concentration range 10^{-4} to 10^{-1} *M* at pH 3.5 to 6.5.

The properties of a solid-membrane copper hexacyanoferrate(II) in Araldite as a thallium(I)-sensitive electrode have been reported by Jain et al.[24] The salt was prepared by precipitation from aqueous solution and a number of membranes containing 40% Araldite were prepared in the usual way.[25] Membranes prepared by using Araldite as a binder exhibited no swelling and could be used over a period of 6 months. Before use, membranes were equilibrated in 0.1 *M* thallium(I) nitrate solution for 5 to 6 d. The prepared membrane electrode gave a linear near-Nernstian response to the thallium(I) ion activities in the concentration range of 5×10^{-4} to 10^{-1} *M*. It can be used to determine concentrations down to 10^{-4} *M*. Electrode response is fast, and steady potentials were obtained in less than 1 min. The only cations that showed some interference on electrode response were lead and copper. This electrode functions well at all pH values above 4, and is in this respect superior to all other thallium ion-selective electrodes based on salts of heteropolyacids, as they do not function above pH 6 due to depolymerization of the salt.[25] The above-mentioned electrode was successfully used as an indicator electrode during the potentiometric titration of about 41 mg of thallium(I), using potassium chromate as titrant.

A novel type of potassium-selective electrode was introduced by Engel and Graber.[26] Chemically modified electrodes were prepared by deposition of copper hexacyanoferrate on glassy carbon. This material has been described recently by Kuwana and Siperco.[27] These electrodes were characterized by voltametric and potentiometric measurements. After suitable conditioning, they exhibit a nearly Nernstian response to changes in the activity of potassium or ammonium ions, as well as good selectivity with respect to sodium ions. Potassium and ammonium ions are reversibly exchanged between film and aqueous solutions; sodium ions are not. This is a consequence of the zeolitic structure of the copper hexacyanoferrate film with channels through which weakly hydrated potassium and ammonium ions can pass. Strongly hydrated sodium ions cannot. The behavior of the above-mentioned electrodes can be explained by the following reaction:

$$\underset{\text{film}}{KCuFe^{III}(CN)_6} + \underset{\text{electrode}}{e^-} + \underset{\text{solution}}{K^+} \rightleftharpoons \underset{\text{film}}{K_2CuFe^{II}(CN)_6}$$

for which the Nernst equation holds:

$$E = E^{\circ} + \frac{RT}{F} \ln \frac{a_{Fe^{III}} \cdot a_{k}}{a_{Fe^{II}}}$$

A distinct potentiometric Nernstian response to K^+ in the solution can be expected if conditions can be set up where the FeIII/FeII activity ratio is constant. Slopes for potential versus log activity plots were 50 mV/pK^+ and 41 mV/pNH_4^+ after potentiostatic redox conditioning, and 42 mV/pK^+ after chemical redox conditioning. Selectivity coefficients with respect to sodium were determined, and k^{pot}_{KNa} values ranged from 7×10^{-3} to 1.8×10^{-1}, depending on K^+ activity (ca. 10^{-3} *M*). Only two electrodes of the complex cyanide type were examined as anionic ion sensors. Jain and co-workers[28] used potassium chromium hexacyanoferrate (II) ($K_6Cr_2[Fe(CN)_6]_3 \cdot 16H_2O$), as active material, together with Araldite in a 7 to 3 ratio, for the preparation of membranes. In order to ensure steady potentials, these membranes had to be soaked for at least 15 d in potassium hexacyanoferrate(II) solutions (0.10 *M*).

Preliminary investigations showed that although the membranes were meant to be hexacyanoferrate(II) sensitive, they also gave a linear response to different chromium(III) activities after equilibration in chromium(III) nitrate solutions, the slope being 50 mV/decade. When equilibrated with hexacyanoferrate(II) solutions, the membranes exhibited a linear response to hexacyanoferrate(II) concentrations with a 50 mV/decade slope, the direction of potential change with concentration being opposite to that for chromium(III) ions, showing that the membrane was responding to anionic species. A steady potential was attained within 1 min and remained constant for about 1 h. As in the case of other epoxy membranes, their lifetimes were about 6 months. The inert nature of the epoxy-resin matrix and the insolubility probably account for the longevity of the membranes. No selectivity determinations were done.

These membranes were used as endpoint indicators in two types of potentiometric titrations. First, small amounts of potassium hexacyanoferrate(II) (20 cm^3 of a 10^{-3} *M* solution) with Hg^{2+}-, UO_2^{2+}-, and Th^{4+}-containing solutions, and 20 cm^3 of 10^{-4} *M* solution with Pb^{2+}-, Zn^{2+}-, and Cu^{2+}-containing solutions. Second, 20 cm^3 of 10^{-3} *M* solutions of these cations were titrated with standard potassium hexacyanoferrate(II) solutions. Compositions of the precipitates formed were a function of the manner of addition of the reagents. The titration curves did not have the expected traditional shape. They were straight lines intersecting at different points, and their shapes correspond to those previously reported by Coetzee and Basson[29,30] for heteropolyacid salt membranes, which will later be discussed in detail. If the last breaks in the titration curves were taken as the stoichiometric endpoints, the compositions of the various compounds are in agreement with previously reported values.[31,32] Electrodes sensitive to hexacyanoferrate(III) ions were reported by Ihn.[33] These electrodes were prepared with Ag_2S and $Ag_3Fe(CN)_6$ in mole ratios varying from 3:1 to 7:1. The 5:1 Ag_2S-to-$Ag_3Fe(CN)_6$ composition is superior to others in terms of potentiometric response, rapidity of response, and reproducibility. Testing was done over the concentration range of 10^{-6} to 10^{-1} *M* at pH 6.8 with constant ionic strength. The calibration curve was linear and coincided with a Nernstian slope of 19.7 mV/decade. The lifetime of these electrodes was 3 weeks, and the principal interfering ions were I^-, Br^-, SCN^-, and $Fe(CN)_6^{4-}$.

Jain and co-workers[34] studied the use of chromium hexacyanoferrate(II) epoxy resin-matrix ion-exchange membranes as indicator electrodes in potentiometric titrations. Membranes were prepared by mixing 0.3 g of Araldite with 0.7 g of chromium hexacyanoferrate(II). These membranes also had to be equilibrated for at least 15 d with a 0.1 *M* potassium hexacyanoferrate(II) solution.

Samples (20 cm^3) of 10^{-3} *M* potassium hexacyanoferrate(II) solutions were titrated with 10^{-2} *M* solutions of Bi^{3+}, La^{3+}, Fe^{3+}, Cr^{3+}, and Ce^{3+}. These titrations were represented by straight-line plots intersecting one another at the equivalence point. This is due to the fact that the membrane electrode also responds to some of the ions added during the titration. The trivalent metal hexacyanoferrate(II) that precipitated during titrations corresponds to the composition KM(III) $Fe(CN)_6$.

B. HETEROPOLYACID SALTS

The heteropolyacid salts of large inorganic and organic cations form insoluble salts. A number of these salts have found application in a variety of analytical procedures. Several of these compounds have also been used as ion exchangers for the concentration of cesium as well as the separation of various alkali metal cations.[35,36]

1. Cesium 12-Molybdophosphate

Coetzee and Basson[15,37] first investigated the possibility of using some of these compounds as ion sensors in electrodes. In the first communication,[15] the construction of a cesium-sensitive electrode based on cesium 12-molybdophosphate was discussed. The salt was embedded in silicone rubber to form a membrane which was fixed to one end of a glass tube. Saturated calomel electrodes were used as both the internal and the reference electrode. The electrode was then used as indicating electrode during potentiometric titrations of the cesium ion. The only reagent type that could be used as titrant in this case had to be anions that form slightly soluble compounds with cesium. In this case, standard solutions of 12-molybdophosphoric acid and sodium tetraphenylborate were used as titrants.

Gran plots[38] of the experimental data had to be used to indicate the endpoints. It was shown that in the case of a titration using 12-molybdophosphoric acid as titrant, the precipitation took place in three stages:

$$Cs^+ + H_3PMo_{12}O_{40} \rightleftharpoons CsH_2PMo_{12}O_{40} + H^+$$

$$Cs^+ + CsH_2PMo_{12}O_{40} \rightleftharpoons Cs_2HPMo_{12}O_{40} + H^+$$

$$Cs^+ + Cs_2HPMo_{12}O_{40} \rightleftharpoons Cs_3PMo_{12}O_{40} + H^+$$

As little as 38.5 mg of Cs^+ ion in 25-cm^3 samples could be determined by applying the standard addition potentiometric technique. A 0.1 *M* lithium nitrate solution was used to keep the ionic strength constant. The general properties of the above-mentioned electrode were described in a subsequent paper.[37] The membrane was prepared by mixing equal masses of cesium 12-molybdophosphate and GE silicone XA12x80 and pressing them into a circular disc at a pressure of 1140 kg cm^{-2}. The membrane was left in a 0.1 *M* cesium nitrate solution for equilibrium for 24 h. The slope of the calibration curve of an electrode employing such a membrane was 24.5 mV/pCs, which is very much smaller than the expected Nernstian value, but the electrode could be used to determine cesium concentrations as low as 10^{-5} *M*. The membrane lasted for about 4 months. The pH dependence of the electrode was studied for the pH range 1 to 6. Higher pH values could not be studied as the heteropoly compounds begin to depolymerize and dissolve. The cell potential increased with pH in the range 1 to 4 and then remained constant. The working range of the electrode is between pH 4 and 6. This study is one of the few in which a temperature coefficient is mentioned. The temperature coefficient of the electrode was determined by measuring the cell potential in 10^{-2} and 10^{-3} *M* cesium nitrate solutions between 10 and 50°C, and the observed temperature coefficient was 0.33 mV per degree. The response times of the electrode were rather long and all measurements had to be taken 2 min after insertion of the electrode in sample solutions.

Selectivity coefficients which were determined in the presence of a fixed cesium ion concentration of 0.01 *M* for various B species were: Tl^+, 5623; NH_4^+, 6.3; Rb^+, 427; K^+, 0.18; Na^+, 0.0072; and Li^+, 0.12. The various values also show a sharp dependence on concentration and are much larger than those expected from normal ion-exchange equilibria. The extent of interference shows a qualitative relationship to results obtained in ion-exchange studies with the exchange of monovalent ions on ammonium 12-molybdophosphate.[39] The electrode is not selective in the presence of the heavier monovalent alkali cations and thallium.

TABLE 2
Selectivity Coefficient Values for Thallium 12-Molybdophosphate and Thallium 12-Tungstophosphate Membrane Electrodes

Interfering cation	TlMP	TlWP
Li^+	10^{-2}	0.28
Na^+	10^{-2}	0.53
K^+	10^{-2}	0.66
Rb^+	0.065	0.80
Cs^+	10^{-2}	0.29
NH_4^+	10^{-2}	0.57
Ag	—	0.65
Mg^{2+}	10^{-3}	2×10^{-2}
Ca^{2+}	10^{-3}	2×10^{-2}
Sr^{2+}	10^{-3}	1.3×10^{-2}
Ba^{2+}	10^{-3}	1.2×10^{-2}

2. Tungstophosphate and Molybdophosphate of Thallium(I)

Thallium(I) salts of molybdophosphoric and tungstophosphoric acids in epoxy-resin-matrix supports were used by Coetzee and Basson[25] to construct electrodes sensitive to thallium(I) ion activities. This was the first investigation during which Araldite was used as a binder in this type of heteropolyacid salt electrode, and it proved to be the most convenient way of preparing membranes of this type, as this became the standard method of preparation. Equal masses of the dry salts and Araldite were mixed and thinly spread out on filter paper. After hardening, they were left in 0.1 *M* thallium nitrate solutions to equilibrate for 6 d. Both the reference and inner electrodes were fiber junction-saturated calomel electrodes, and no problems were reported from the possible leakage of potassium chloride into thallium nitrate solutions.

Although both electrodes gave calibration curves with sub-Nernstian slopes — 41 mV/pTl for thallium(I) tungstophate and 36 mV/pTl for thallium(I) molybdophosphate — both were found suitable to determine thallium concentrations as low as 10^{-4} *M*. The optimum pH range of the previously reported cesium electrode,[37] namely 4 to 6, was also operative for these electrodes. A positive temperature coefficient of 0.27 mV per degree was reported for both thallium(I) electrodes. Constant potential readings were obtained after about 2 min.

Selectivity coefficients for both electrodes were determined by a mixed solution method and are quoted in Table 2. The values show that the thallium(I) 12-molybdophosphate membrane functions more selectively for thallium(I) ions than does the thallium(I) 12-tungstophosphate membrane.

The response of the membrane electrodes in thallium solutions containing various concentrations of methanol, ethanol, *n*-propanol, and acetone was studied. The slopes of the calibration curves are summarized in Table 3. The results show no exceptional change in the behavior of the electrodes in these solvents, and the response times were the same as in water.

The above-mentioned electrodes were successfully used as indicator electrodes during the potentiometric titration of thallium(I) ions with potassium bromide, potassium chromate, and sodium tetraphenylborate as precipitating agents. These potentiometric titrations will be discussed in detail later in the text.

3. Zirconium 12-Molybdophosphate

An epoxy-resin-impregnated solid membrane electrode for the estimation of thallium(I) was reported by Malik et al.[40] Heterogeneous membranes were prepared by mixing powdered

TABLE 3
Effect of Solvent Composition on Calibration Curves

Solvent composition (% v/v in water)	Slope mV per pTl	
	TlMP	TlWP
water	36.0	40.0
10% methanol	38.0	39.0
25% methanol	39.0	40.0
10% ethanol	37.0	40.7
25% ethanol	39.0	42.0
10% n-propanol	36.3	42.0
25% n-propanol	40.0	44.7
10% acetone	37.0	40.3
25% acetone	38.0	46.3

zirconium molybdophosphate with Araldite in a 70 to 30% salt-epoxy resin ratio by mass. Equilibration of 4 to 5 d in a 0.1 *M* thallium solution was necessary. The response of this electrode to thallium (I) ions was non-Nernstian, but a linear plot of cell potential against log thallium concentration was linear in the concentration range 2.5×10^{-4} to 10^{-1} *M*, with a slope of 55.2 mV per decade change in concentration. The potentials generated in the concentration range 10^{-4} to 10^{-1} *M* are reproducible, and the slope of the calibration curve is large enough to permit the use of the membrane electrode for ion-activity measurements in this concentration range. The electrode response in a 25% ethanolic medium was almost identical to that in water. Similar behavior has been observed for acetone-water media. Response time was found to be 25 s, and the electrode system could be used for 6 months. The pH range in which the electrode could be used was 3 to 6.

Selectivity coefficients were determined at two different interfering cation concentrations for 15 different cations. The values were all less than one. About 31 mg thallium(I) ion was successfully titrated with a standard molybdophosphoric acid solution. The titration curve had a break indicating the final step in the reaction, i.e., complete precipitation of $Tl_3PMo_{12}O_{40}$.

4. Cesium Tungstophosphate and Tungstosilicate

Membranes prepared from cesium 12-tungstophosphate and an epoxy resin (50% m/m composition) were used by Coetzee and Basson[29] as ion sensors for cesium ions. As in previously reported cases,[37] the slopes of the calibration curves were sub-Nernstian. For the electrode employing cesium tungstophosphate, the slope was 36.0 mV/pCs, and for the cesium tungstosilicate membrane electrode, it was 31.5 mV/pCs. The calibration curves were linear, in the case of cesium tungstophosphate, between 10^{-1} and 10^{-4} *M*, and in the case of cesium tungstosilicate, between 10^{-1} and 10^{-3} *M*. The observed temperature coefficients for both electrodes were 0.62 mV per degree. Both these electrodes proved not to be entirely selective to cesium ions, as can be seen from Table 4, which shows selectivity coefficients which were determined by the mixed-solution method.

The selectivity coefficient values for these two electrodes are strikingly similar. They seem to be more selective for rubidium ions than for cesium. Their subsequent use in potentiometric titrations will be discussed later on in the text.

5. Tungstoarsenates

Malik and and co-workers[41] reported the use of cesium and thallium(I) tungstoarsenate-Araldite membranes. The same procedure for membrane preparation as that described by Coetzee and Basson[25] was followed. These membranes gave a non-Nernstian response.

TABLE 4
k^{pot}_{CsB} Values for Cesium Tungstophosphate (CsWP) and Cesium Tungstosilicate (CsWSi) Membrane Electrodes[29]

	Selectivity coefficients	
B species	CsWP	CsWSi
Li^+	0.17	0.19
Na^+	0.21	0.24
K^+	0.37	0.41
Rb^+	1.27	1.28
NH_4^+	0.87	0.97
Tl^+	1.02	1.12
Mg^{2+}	4.7×10^{-4}	5.2×10^{-4}
Ca^{2+}	6.5×10^{-4}	7.1×10^{-4}
Sr^{2+}	6.9×10^{-4}	7.3×10^{-4}
Ba^{2+}	11.9×10^{-4}	12.0×10^{-4}

Although the slope on the cesium electrode was close to the Nernstian values (57.4 mV/pCs) and that of the thallium electrode (48.4 mV/pTl in water), the working pH range was found to be 3 to 6. It seems that most of these types of electrodes respond rather sluggishly, as constant potential in this case was attained only after about 2 min. Selectivity coefficients for the electrodes were reported with respect to the following ions: Na^+, K^+, NH_4^+, Rb^+, Ag^+, Tl^+, Sr^{2+}, and Ba^{2+}. Compared to other reported values for this type of electrode with the same interfering ions, these values seem to be remarkably small. For the cesium electrode, all values were smaller than 0.06, and for the thallium electrode, smaller than 0.04. No appreciable change in the slopes of the calibration curves was observed in solutions containing up to 25% methanol, ethanol, and acetone. Thus, it was concluded that these electrodes can be used in solutions having a nonaqueous content of up to 25%.

A membrane consisting of 70% strontium 12-tungstoarsenate and 30% Araldite content was prepared by Jain and co-workers[42] and used as a strontium ion-sensor in an electrode. The slope of the calibration curve of the electrode is Nernstian in the concentration range of 10^{-2} to 10^{-3} *M*, but in the range of 10^{-2} to 10^{-1} *M*, there is a super-Nernstian response, and between 10^{-4} and 10^{-5} *M*, there is a sub-Nernstian response. No explanation is given for these phenomena. The electrode is pH-independent between pH values 3 to 6. The electrode response is reasonably fast (20 to 50 s), and the membrane has been used over a period of 1 year. The response of the electrode in nonaqueous media was non-Nernstian (45 mV/pSr) in the concentration range 10^{-1} to 10^{-4} *M* or Sr^{2+} in solutions containing up to 25% ethanol or acetone. Selectivity coefficients were determined by a separate solution method. These values for Li^+, Na^+, K^+, Rb^+, and Cs^+ ions are all larger than 10; for Mg^{2+}, Ca^{2+}, Ba^{2+}, Ni^{2+}, Co^{2+}, Pb^{2+}, Fe^{2+}, Ce^{3+}, and Fe^{3+}, between 1 and 10^{-1}; for Zn^{2+}, and Hg^{2+} ions, between 10^{-1} and 10^{-2}; and for Al^{3+}, it is 7.6×10^{-4}. The major interfering ions thus are the univalent alkali cations. The anions Cl^-, Br^-, I^-, NO_3^-, PO_4^{3-}, and AsO_4^{3-} do not interfere with the electrode's performance. The cationic surfactants hexadecylpyridium bromide and hexadecyltrimethylammonium bromide show interference on electrode behavior at concentrations as low as 2×10^{-4} *M*.

6. Critical Micelle Concentration

As suggested in the previous publication,[42] strontium tungstoarsenate was found to be sensitive to cationic surfactants. A strontium tungstoarsenate-Araldite membrane electrode was thus used by Srivastava and co-workers[43] for the measurement of the activity and critical micelle concentration of cationic surfactants. Membranes were prepared in the usual way with a 60% inorganic salt content. The specific conductances of membranes equilibrated with hexadecylpyridinium bromide (CPB) and hexadecyltrimethylammonium bromide (CTAB) were found to be $0.6 \times 10^{-4}\ \Omega^{-1}\ cm^{-1}$ and $0.4 \times 10^{-4}\ \Omega^{-1}\ cm^{-1}$, respectively. After equilibration for 7 d in 10^{-3} *M* solutions, potentiometric measurements were carried out with two different membranes, and the results are in close agreement, indicating reproducibility.

The plots of membrane potentials against the log of surfactant concentration showed rectilinear relationships. There is, however, an abrupt change in the slope of the line at a certain concentration which seems to be the critical micelle concentration. The surfactant solution behaves as a strong electrolyte at concentrations below this value. The slope of the graph is close to the Nernstian value for CTAB (55 mV per decade of concentration). For CPB, however, it is smaller (38 mV per decade of concentration). Above the critical micelle concentrations owing to micellization of the surfactant cations, there is very little rise in potential even when the concentration increases considerably. The determined critical micelle concentrations of the surfactants, 7.9×10^{-4} *M* for CPB and 10×10^{-4} *M* for CTAB, are in close agreement with those obtained by other methods.[44,45]

Jain and co-workers[44] investigated the silver selectivity of a membrane electrode made from rubidium 12-tungstoarsenate and Araldite. Membranes of different compositions were prepared, and it was observed that the one with 30% Araldite content gave the best performance with regard to mechanical stability, conductivity, and selective response. Freshly prepared membranes were kept in 0.1 *M* silver nitrate solution for 3 to 4 d for complete equilibration. If it is accepted that the ion-exchange properties of the salts of heteropolyacids are in a broad sense alike, it must be that during this equilibration a certain amount of the rubidium ions in the membrane will be exchanged for silver ions.[46] Although constant potentials were reached after only 40 s, this electrode calibration curve, like several others of this type,[25,37,41] also indicated a sub-Nernstian response (34 mV per decade activity change). The useful pH range was between 4 and 6.

Selectivity coefficients (constants in the test) were determined by a separate solution method for the following ions: Li^+, Na^+, K^+, Rb^+, Cs^+, Tl^+, Mg^{2+}, Ca^{2+}, Zn^{2+}, Cd^{2+}, and Hg^{2+}. The values reported were all for thallium(I). The values are rather small for this type of electrode, and it seems that selectivity coefficients determined in this way, with this type of electrode, tend to be on the small side. The following anions are also listed as noninterfering: Cl^-, NO_3^-, NO_2^-, S^{2-}, SO_3^{2-}, SO_4^{2-}, $Fe(CN)_6^{4-}$, $Fe(CN)_6^{3-}$, WO_4^{2-}, MoO_4^{2-}, AsO_4^{3-}, and PO_4^{3-}. The disadvantage of the separate solution method is that anions which are listed as causing no interference will actually seriously interfere in the presence of silver ions, as many of them form insoluble silver salts. A better measure of selectivity might be by selectivity coefficients determined by the mixed-solution method. The membrane has also been used as an endpoint indicator electrode in the potentiometric titrations involving silver ions.

An Araldite-based titanium tungstoarsenate solid-membrane electrode has been developed by Srivastava and co-workers[47] to estimate rubidium ion concentration in the range 4×10^{-5} to 10^{-1} *M*. Titanium tungstoarsenate was prepared by adding 0.25 *M* titanic chloride solution at pH 1 to a mixture of 0.25 *M* sodium arsenate and sodium tungstate in the ratio 2:1:1 (v/v). The obtained precipitate was left for 24 h and washed with distilled water, and finally with 2 *M* nitric acid. The gel, dried at 50°C, broke down to small particles when immersed in water. The Ti:W:As ratio was found to be 1:4:2, and the exchange capacity of the material was 0.86 mol/g (univalent ion). Membranes were prepared as previously described[15,25] with a 40% Araldite content. Membranes were equilibrated for 1 week in a 0.1 *M* rubidium chloride solution and had a sub-Nernstian response in the concentration range 10^{-4} to 10^{-1} *M*, the slope of the calibration being 53.3 mV per decade of concentration change. Response times were less than 1 min.

The working range for this electrode is 4×10^{-5} to 10^{-1} *M* rubidium ion concentration. The pH dependence of the electrode was investigated in the pH range 1 to 8 at 10^{-3} and 10^{-2} *M* concentrations of Rb^+ ions. The potentials were constant within the pH range 3 to 8, which constitutes a much larger pH range than any of the other heteropolyacid salt membranes.

The ion-exchange capacity of the membrane was found to be 0.399 mol/g of dry membrane in the hydrogen form, which is less than the exchange capacity of the heteropolyacid salt (0.86 mol/g), as the mass of the binder is also included in the membrane's mass. The specific conductances of the membranes which had been equilibrated with 10^{-3}, 10^{-2}, and 10^{-1} *M* aqueous RbCl solutions were observed to be 1.14×10^{-4}, 1.26×10^{-4}, and $1.7 \times 10^{-4}\ \Omega^{-1}\ cm^{-1}$, respectively. A similar value of the membrane conductance with change in the external solution concentration implies that the Donnan exclusion is fully effective in this concentration range. The specific conductance of the membranes in the H^+, Li^+, Na^+, K^+, and Cs^+ forms were found to be in the range 0.52×10^{-4} to $1.4 \times 10^{-4}\ \Omega^{-1}\ cm^{-1}$, except for the membrane in the H^+ form (0.22 to $1.13 \times 10^{-4}\ \Omega^{-1}\ cm^{-1}$), which indicates that the Donnan exclusion is not fully effective in this case. Selectivity coefficients were determined under a fixed concentration of an interferent ion, the concentration of the reference RbCl solution being 10^{-1} *M* at pH 5.0. The values are given in Table 5.

The electrode exhibits fairly good selectivity in its response to rubidium ion activity over a large number of cations. The effect of the presence of surfactants on electrode performance was also studied. When a membrane was treated with the anionic surfactant sodium dodecylsulfate (10^{-4} *M* for 48 h), it became immune to the effect of anionic surfactants. The membrane also behaves as a better sensor for rubidium ions and has a larger validity range. Cationic and nonionic surfactants do not interfere if they are present in concentrations $\leq 10^{-4}$ *M*. The response of the electrode was linear in the concentration range 10^{-4} to 10^{-1} *M* in 25% ethanol in water.

An electrode which responds selectively to copper(II) ions was reported by Srivastava and co-workers.[48] The membrane used was prepared by mixing copper tungstoarsenate and Araldite in a 7:3 mass ratio. The hardened membrane was equilibrated with a 1 *M* copper nitrate solution for 4 d. The membrane electrode could be used up to 6 months and exhibited a near-Nernstian response of 28 mV/pCu in the concentration range 10^{-5} to 10^{-1} copper(II) solutions. Response times were reasonably fast (20 to 30 s). The response of the membrane electrode in partially nonaqueous media is non-Nernstian, although a linear dependence of potentials with log activity of copper ions in the concentration range 10^{-5} to 10^{-1} *M* is observed in solutions containing up to 30% ethanol or acetone (the value of the slope is not mentioned). Selectivity studies showed that, except for Ba^{2+} and Fe^{3+}, the electrode is more selective for copper ions, compared to other uni-, bi-, and trivalent cations. Although nitrate and acetate did not interfere in the electrode performance, sulfate and chloride did, even when present in small concentrations. The addition of a small amount of a cationic surfactant like cetylpyridinium bromide causes large shifts in membrane potentials, but after treatment with this surfactant, a membrane gets desensitized to potential shifts and shows response to copper(II) ions even in the presence of the surfactant cation. The electrode has also been used successfully to estimate copper in metal processing and plating wastes.

Polystyrene-based zirconium tungstoarsenate membranes were described by Srivastava and co-workers[49] for the determination of zirconyl ions. Zirconium tungstoarsenate[50] and the polystyrene-supported membranes[51] were prepared. The plot of potential against log concentration was found to be linear in the concentration range 5×10^{-4} to 10^{-1} *M*, with a slope of 53.3 mV per decade of concentration. Response time was less than 1 min, and the electrode could be used in the pH range 3 to 6. The electrode may also be used in

TABLE 5
Selectivity Coefficients for a Rubidium-Selective Titanium Tungstoarsenate Membrane Electrode

Ion	Concentration of B_n^+ $10^3k^{pot}_{RbB}$ 10^{-3} *M*	10^{-2} *M*
Cs^+	632	642
Tl^+	522	532
Li^+	363	390
NH_4^+	198	200
K^+	218	231
Na^+	393	390
Hg^{2+}	9.91	9.99
Mg^{2+}	1.94	2.04
Zn^{2+}	1.84	3.15
Pb^{2+}	3.33	3.31
Ca^{2+}	6.33	8.99
Ba^{2+}	8.91	9.66
Co^{2+}	2.45	2.45
Cu^{2+}	8.59	8.69
Ni^{2+}	5.32	6.51
Al^{3+}	3.98	3.99
Bi^{3+}	1.99	1.99
Fe^{3+}	1.99	1.99
Th^{4+}	0.52	0.52

From Srivastava, S. K., Kumar, S., and Kumar, S., *J. Electroanal. Chem.*, 161, 345, 1984. With permission.

partially aqueous media (e.g., 20% ethanol or acetone). There seems to be no serious interference from the 22 cations listed. The functioning of the electrode is, however, disturbed by molybdate, phosphate, and vanadate ions.

7. Molybdoarsenates

Due to the need for an electrode for thallium ion determination because of its toxicity, and also to the similarity of some of the reactions of thallium and silver, interest in the preparation of membrane electrodes selective for thallium(I) ions exists. Jain[52] and co-workers used thallium(I) 12-molybdoarsenate in a 60 to 40 ratio with Araldite as a membrane to prepare a thallium-ion-selective electrode. The solubility of the thallium salt in the membrane was determined by labeling the thallium with ^{204}Tl and shaking the membrane in water. The solubility was found to be 3×10^{-6} mol dm^{-3}.

The electrode gave a near-Nernstian response (54 mV per decade of activity change) in the concentration range of 10^{-3} to 10^{-1} *M*. Response time for the electrode was about 40 s and the useful pH range was between 4 and 6. Electrode response in partly nonaqueous media (ethanol-water and acetone-water mixtures of 10% and 25%, respectively) showed no drastic changes in electrode behavior. In acetone media, there was a 4 mV increase in the value of the slope of the calibration curve.

The selectivity of the electrode for thallium(I) ions over the cations Li^+, Na^+, K^+, Ag^+, Mg^{2+}, Ca^{2+}, Sr^{2+}, Ba^{2+}, Hg^{2+}, Co^{2+}, Ni^{2+}, Cu^{2+}, and Cd^{2+} was investigated by the mixed-solution method, and was smaller than 1.35×10^{-2} for the monovalent and smaller than 2.5×10^{-3} for the divalent ions. Unfortunately, rubidium and cesium ions were not included in the study. The interference of anions was investigated by a separate-solutions method. The following anions do not interfere: Cl^-, Br^-, I^-, NO_3^-, NO_2^-, SO_4^{2-}, PO_4^{3-}, MoO_4^{2-}, WO_4^{2-}, AsO_4^{3-}, $Fe(CN)_6^{3-}$, and $Fe(CN)_6^{4-}$. Many of these anions would, however, interfere with thallium(I) determinations, as they form insoluble salts with thallium(I).

The use of a pyridinium molybdoarsenate-Araldite membrane prepared in a 65 to 35 mass ratio as a pyridinium-selective electrode has been investigated by Srivastava and co-workers.[53] The electrode response in aqueous solution was sub-Nernstian (40 mV per decade change in concentration). The membranes, however, are suitable for determining pyridinium ion concentrations from 10^{-3} to 1 *M*. Response time is about 1 min and the useful pH range is 3 to 6. Selectivity coefficients were determined by a mixed-solution method. Concentrations of the interfering and pyridinium ions were 10^{-2} and 10^{-4} *M*, respectively. Reported values for the interfering ions Na^+, K^+, Rb^+, NH_4^+, Ag^+, Tl^+, Sr^{2+}, Ca^{2+}, PO_4^{3-}, AsO_4^{3-}, MoO_4^{2-} were all $<10^{-2}$. At pyridinium ion concentrations above 10^{-3} *M*, the organic cations picolinium, lutidinium, and collidinium do not interfere. The potential versus log activity plots in partially nonaqueous media are linear. Other than methanol, the slope in ethanol-water mixtures and acetone-water mixtures are equal to or larger than the slope of a calibration curve in water. In 25% acetone or ethanol solutions, the calibration curves are linear between 10^{-4} and 1 *M*. This "better response" is attributed by the authors to the low colloidal dispersion of molybdo- and tungstoarsenate in aqueous media. The dispersion effect vanishes in nonaqueous media of low dielectric constant. Addition of a small amount of the cationic surfactant hexadecyltrimethylammonium bromide causes a shift in the membrane potential in the presence of low concentrations of pyridinium ion to move negative values and extends the linear response range. This effect, which can be made permanent by treating the membrane with a surfactant solution (10^{-4} *M*) for 1 h, is attributed to adsorption of surfactant on the membrane.

Another membrane that was evaluated as a thallium(I)-selective electrode by Jain and co-workers[53,54] was prepared out of α-picolinium molybdoarsenate and Araldite in a 7 to 3 mass ratio. In the concentration range 10^{-3} to 10^{-1} *M*, the linear calibration curve was linear, with a slope of 48 mV per decade of concentration. Deviation from linearity was observed in the range 10^{-3} to 10^{-5} *M*, but the potentials were reproducible, permitting the use of the electrode in this region for thallium(I) estimation. Response times were found to be about 20 s for the range down to 10^{-3} *M*, but increased to 1 min for the most dilute solutions. The useful pH range was between 3 and 6. Cation selectivity coefficients were determined by a mixed-solution method. Selectivity coefficient values for Li^+, Na^+, K^+, Rb^+, and Cs^+ were between 4.5×10^{-3} and 10^{-2}, and for Mg^{2+}, Ca^{2+}, Sr^{2+}, Ba^{2+}, Co^{2+}, Ni^{2+}, and Mn^{2+}, between 1.1×10^{-3} and 1.6×10^{-3}. For Ag^+, Cd^{2+}, and Hg^{2+}, the values were 1.3×10^{-2}, 2.1×10^{-4}, and 5.6×10^{-5}, respectively. The anions chloride, nitrate, sulfite, sulfate, phosphate, molybdate, tungstate, arsenate, hexacyanoferrate(II), and hexacyanoferrate(III) did not interfere with electrode performance. The response of the electrode was also evaluated after pretreatment with a cationic surfactant, cetyltrimethylammonium bromide, but there was no improvement in the performance of the electrode as a whole.

8. Ammonium 12-Molybdophosphate

An ammonium-ion-sensing electrode was reported by Longhi and co-workers.[55] Membrane discs were prepared by mixing ammonium molybdophosphate in a 50 mass percent

TABLE 6
Selectivity Coefficient Values ($k^{pot}_{A,B}$) for the Molybdate Hydrous Zirconium Oxide Membrane Electrode[78]

Ion	$k^{pot}_{A,B}$ 10^{-4} *M*	$k^{pot}_{A,B}$ 10^{-2} *M*
Cl^-	12	15
SCN^-	14	18
ClC_4^-	14	18
NO_3^-	12	13
$C_2O_4^{2-}$	0.11	0.14
WO_4^{2-}	0.21	0.23
SO_4^{2-}	0.12	0.14
AsO_4^{3-}	0.029	0.034
PO_4^{3-}	0.016	0.020
$Fe(CN)_6^{4-}$	0.021	0.023

From Srivastava, S. K., Sharma, A. K., and Jain, A. K., *Talanta*, 30, 285, 1983. With permission.

ratio with methacrylate powder, followed by sintering at 130°C. Such a membrane was used as an ion-selective electrode for ammonium ions. The linear portion of the calibration curve was between 3×10^{-1} and 3×10^{-4} *M*. The slope was a near-Nernstian 57 mV per decade of activity change. The authors claim an operational pH range of 2 to 9. The electrode gave stable and reproducible results over a period longer than 6 months.

Selectivity coefficients for the electrode were determined by the mixed-solution method, and are as follows:

Ions (B^+)	H^+	Li^+	Na^+	K^+	Rb^+	Ca^+	$(CH_3)_4N^+$
k^{pot}_{NH4B}	1.84	0.16	0.30	0.63	0.12	0.19	0.41

These values compare very favorably with those for a commercial NH_4^+-ion glass electrode.[56] The low cost and ease of construction of the electrode are its main advantages. The electrode can be advantageously utilized in the construction of ammonia gas-sensing electrodes. It can also be used for enzyme-membrane probes, and for sensing any nitrogen-bearing species where there are available specific enzyme-mediated reactions that finally produce ammonia.

9. Crown Ether Tungstophosphate

Crown ethers have high complexing ability with various metal ions[57-60] and are expected to be suitable neutral carriers in ion-selective electrodes. Some crown ethers have been successfully applied as neutral carriers in some alkali metal ion electrodes.[61-64] Fernando[65] and co-workers prepared precipitates of some crown ethers with phosphomolybdic acid, which showed characteristics similar to a crown ether. A sodium ion-selective PVC membrane electrode, based on 12-crown-4-phosphotungstic acid precipitates with dipentyl phthalate as plasticizer, has been described by Jeng and Shih.[66] During the work, both the 12-crown-4- and 15-crown-5 compounds were tried as membrane materials.

The precipitate of the crown ether-phosphotungstic acid was prepared by mixing 50 cm^3 of an aqueous 0.02 *M* phosphotungstic acid, which was 2 *M* in nitric acid, with 50 cm^3 of dichloromethane solution that contained about 1 g of crown ether. The precipitate was dried under vacuum. The molecular structures of the precipitates were found to be 3,2(15-crown-5)PW and 5,6(12-crown-4)PW (PW = phosphotungstate). Electrodes were prepared as follows. A mixture of 100 mg of PVC and 50 mg of dibutyl phthalate was dissolved in 15 cm^3 of tetrahydrofuran. This solution was mixed with 120 mg of the precipitate of crown ether-phosphotungstic acid. The mixture was poured into a glass dish of 9.5 cm diameter, and a nontransparent membrane 0.04 mm thick was obtained on evaporation of the tetrahydrofuran. The electrochemical system for the study was

Ag-AgCl/internal solution/PVC membrane/test solution/satd KCl/AgCl-Ag
(10^{-3} *M* NaCl)

The PVC electrode, based on 12-crown-4-PW, exhibits a linear response to sodium ion activities within the concentration range 10^{-1} to $10^{-3.5}$ *M* NaCl with a Nernstian slope. A linear response in the concentration range 10^{-1} to 10^{-4} *M* NaCl was found for the electrode based on 15-crown-5-PW, but the slope was only 30 mV per decade of concentration.

Crown ethers are sensitive to ion size. The sodium ion can be fitted into the cavity of 15-crown-5 very well and therefore this complex may be more stable than that of 12-crown-4-PW with the sodium ion. This effect leads to a higher concentration of complexed species in the membrane with 15-crown-5. Boles and Buc[67] reported that at high concentrations of complexed species in membranes, an anion response can be observed which results in the reduction of the cation response. The non-Nernstian response when 15-crown-5 is used can be attributed to the low mobility of the 15-crown-5-Na^+ complex in the membrane, or to the high mobility of the free anion. Using 12-crown-4, the Nernstian slope response seems to indicate a much higher cation complex mobility than that of the free anion in the membrane phase. The 12-crown-4-PW precipitate was also studied as a carrier for K^+ and Li^+ ions. The slopes of the response curves to these ions were 34 and 42 mV per decade concentration, respectively. The selectivity of the sodium electrode was determined at a fixed sodium ion concentration of 5×10^{-3} *M* with varying interferent ion concentrations. Potassium and rubidium ions were the most important interferents with the sodium electrode. The working pH range of the electrode was between 3 and 5, and the response time was less than 10 s.

In a second paper, Wang and Shih[68] described a cesium-ion-selective electrode based on 15-crown-5 phosphotungstic acid precipitates. The electrode was prepared in the same manner as the previously described sodium-ion-selective electrode.[66]

The response characteristics of cesium ion-selective electrodes based on dibenzo-18-crown-6-PW were studied. The 15-crown-5 phosphotungstate appeared to give the best sensitivity to cesium ion activities, with a Nernstian slope of 60 ± 1 mV per decade of concentration change. The calibration plot was linear within the concentration range 10^{-4} to 10^{-1} *M*. It is well known that crown ethers are very sensitive to the size of ions, and as expected, the 15-crown-5-PW system is the best for the neutral carrier of the cesium-ion-selective electrode. Membranes with the ratio of PVC to 15-crown-5-PW to dibutyl phthalate of 100:50:150 gave the best performance, and the electrode with 10^{-3} *M* $CsNO_3$ as the internal solution showed the best response. Response times were shorter than 1 min.

The selectivity of the cesium ion electrode was determined from potential measurements in solutions containing a fixed amount of cesium ion (10^{-3} *M*) and different amounts of the interfering ion (M^{z+}). Alkaline earth ions and some transition metal ions such as Ni^{2+}, Fe^{3+}, Zn^{2+}, and Hg^{2+} showed negligible interference at concentrations $<10^{-2}$ *M*. Alkali metal ions such as Na^+, K^+, and Rb^+ show some interference, the selectivity coefficient for K^+ being the largest (0.49). The working pH range was between 3.5 and 5.5. The addition of ethanol to aqueous test solutions resulted in a reduction of the slope of the potential response when the water-to-ethanol ratio was less than 6:1. The electrode was used to indicate the endpoint during the potentiometric titration of 50 cm^3 of a 0.01 *M* cesium ion solution with a 0.1 *M* sodium tetraphenylborate solution.

10. Liquid Membrane Electrodes

Liquid membrane types of ion-selective electrodes may have conventional liquid membranes supported by an inert phase such as a Millipore filter, and may be replenished from a reservoir of the ion exchanger in a solvent mediator (the liquid ion exchanger), or the liquid ion exchanger may be immobilized in a polymer matrix, such as PVC.

Preliminary work by Fogg and Yoo[69] on liquid-state electrodes based on crystal violet tetraphenylborate and crystal violet 12-tungstosilicate showed that sharper potential jumps can be obtained during titrations of crystal violet with tetraphenylborate than with other types of electrodes. In a later communication, Fogg and Yoo[70] describe a PVC-tetraphenylphosphonium 12-tungstosilicate electrode. This membrane was prepared as follows. A paste was made from 0.15 g of tetraphenylphosphonium 12-tungstosilicate and 0.2 cm^3 of 1-nitrophenyl-*n*-butyrate (solvent mediator); 0.2 cm^3 of diisooctylphthalate (plasticizer) was then mixed well with a solution of 0.15 g of PVC powder in 6 cm^3 of tetrahydrofuran. The complete electrode was prepared in the normal manner. This electrode was tried out in solutions of both tetraphenylphosphonium and basic dyes.

a. Response to Tetraphenylphosphonium Ion

After conditioning in 10^{-3} *M* solutions for less than 1 d, the response was Nernstian to the tetraphenylphosphonium ion in the 10^{-6} to 10^{-2} *M* range. The response fell to 35 mV per decade ion concentration change after electrode storage in the same solution for 37 d. Initially, the electrode also gave a Nernstian response to the tetraphenylborate ion, but this response also decreased with time.

b. Response to Basic Dyes

The response to basic dyes was less than Nernstian. For example, the response to crystal violet was 25 to 30 mV per decade in the range 10^{-5} to 10^{-3} *M*. Satisfactory titration curves were, nevertheless, obtained for several basic dyes when they were titrated with standard sodium tetraphenylborate solution. Acid dyes were titrated with a standard solution of crystal violet using the tetraphenylphosphonium 12-tungstosilicate electrode as indicator electrode. As expected from the higher water solubility of the crystal violet acid-dye salts, the potential jumps were smaller and the endpoints were less sharp than those obtained for the tetraphenylborate ion.

The only recorded attempt to use conventional heteropoly compounds in liquid-liquid ion exchange-membrane electrodes is that of Guilbault and Brignac.[71] The insolubility of all the heteropolyacid salts in question creates problems in successfully developing liquid-liquid electrodes. Furthermore, the liquid ion exchanger must also be in electrolytic contact with the sample solutions.

Phosphotungstic and phosphomolybdic acids are very soluble in water and in oxygen-containing organic solvents. For these electrodes, *n*-pentanol was chosen as solvent. Since the free heteropolyacids are soluble in both the aqueous and organic layers, a porous membrane was placed between the two layers, which could restrict the passage of the complex from the organic layer to the aqueous layer. This is based on the theory that a high-molecular-mass compound, such as the heteropolyacid, could not diffuse through, but that smaller ions could penetrate quite easily. Guilbault and Brignac[71] used Orion model 92-07 electrode barrels with an *n*-pentanol solution of phosphotungstic acid (1.25 g/10 cm^3) as nonaqueous filling solution. The electrodes were conditioned in 0.1 *M* phosphoric acid solution for 24 h. In solutions of sodium dihydrogenphosphate (pH 4, 6, and 7), a linear calibration curve down to a concentration of 10^{-4} *M* with a slope of about 28 mV per decade of concentration change was found. The selectivity over anions was poor. The response was leveled by the presence of a high electrolyte concentration of other anions. With varying phosphoric acid concentrations, a rectilinear calibration curve in the concentration range 10^{-1} to 10^{-5} *M* was found with a slope of 42 mV per decade. Selectivity was still poor.

With phosphomolybdic acid as the active ion exchanger, the electrode response in sodium dihydrogenphosphate solutions was poor (nonlinear calibration curve). In phosphoric acid solutions, the response was as reported above for phosphotungstic acid. The phosphomolybdic acid system was reduced to a blue solution within 24 h. The response characteristics of the electrode changed, indicating a nonreproducible system.

C. CLAY MEMBRANES AND SYNTHETIC ZEOLITES

Ion-exchange processes of zeolite molecular sieves are interesting due to unusual mechanisms for providing ion selectivity. Cation selectivities do not follow the same trends that typify other inorganic and organic exchangers. Marshall and co-workers[72-75] made extensive studies of membranes prepared from natural zeolites. Initially, membranes were ground from large, single crystals of chabazite or apophyllite. Later, they used films made from colloidal clays such as montmorillionite and beidellite. The properties could be changed by heat treatment. Membranes were prepared with which ionic activities could be measured. Good permselectivity was obtained, resulting in a near-Nernstian slope when a single cation was present. Membrane supports like polystyrene, polyethylene, polymethyl methacrylate, and phenolic resins all resulted in less than ideal wetting of the crystals.[76] The research reported on the above work was, however, carried out during the days of real interest in ion-selective electrodes, and although near-Nernstian response for some cations has been demonstrated, no other data usually associated with the characteristics of an ion-selective electrode are available.

In two publications by Johansson and co-workers,[77,78] the characteristics and preparation of ion-selective electrodes prepared from a crystalline synthetic zeolite of the mordenite type is described. Zeolites are crystalline, hydrated aluminosilicates made up from a three-dimensional network of $A1O_4$ and SiO_4 tetrahedra linked to each other by sharing of all the oxygens. The framework contains channels and interconnected voids which are occupied by exchangeable cations and water molecules. The water can be removed at high temperatures from several of the zeolites and it will be replaced reversibly at lower temperatures. Synthetic zeolites are better suited for research purposes because of their greater purity and uniformity in composition.

The zeolite was prepared in the laboratory and the product had a Si to Al ratio of 6.4:1. The crystals were packed in a column and converted to the cesium form by passing through a 1 *M* cesium chloride solution. The total ion-exchange capacity was 1.65 meq/g (dry mass). The chemical composition of the material used was given as $CsAl\ Si_{6.4}O_{14.8} \cdot nH_2O$.

An epoxy resin mixture with low viscosity and high chemical resistance was selected as membrane material. Membranes were fixed to a glass tube by means of a special embedding solution. The response of the electrode toward cesium ions was linear from about 3×10^{-5} to 10^{-1} *M*, with a slope of 57.7 mV per decade of activity change. The limit of detection for cesium ions was 2×10^{-5} *M*. The response and selectivity were independent of the anion when activity changes were taken into account. The working pH range was between 3.5 and 9. The order of selectivity was Cs > Ag, K > Na > Li and Cs >> Ba > Ca > Cu. Heat treatment of the zeolites decreased the ion selectivity. Membranes prepared from zeolite A showed much less ion selectivity. Whereas the selectivity order for mordenite follows that of the ion-exchange selectivity, this is not the case for zeolite A.

D. HYDRATED OXIDES OF METALS

A polystyrene-based zirconium oxide membrane electrode has been described by Srivastava and co-workers[79] for the determination of molybdate ion concentrations in the range 0.5 to 10^{-3} *M*. Hydrous zirconium oxide (HZO) was prepared by the method described by Shor et al.[80] Membranes supported with a polystyrene binder were found to be quite stable and were made by mixing finely divided polystyrene and HZO in a 1:9 m/m ratio and heating the homogeneous mass in a die of 2.5 cm diameter at 120°C under a pressure of 5600 to 7000 psi. Membranes were equilibrated in 0.01 *M* sodium molybdate solutions for 2 to 3 d.

The electrode responded fast (within a few seconds). The response was linear and near-Nernstian (30 mV/pMoO_4^{2-}) over the concentration range 0.5 to 10^{-3} *M*. The working pH range was between 7 and 11, and the electrode was usable for at least 6 months.

The electrode could also be used in partially nonaqueous media up to a maximum of 25% v/v nonaqueous content. The response in 25% v/v methanol, ethanol, and acetone was linear, but not quite Nernstian, being 32, 31, and 33 mV per decade concentration change, respectively. Selectivity coefficients for electrode performance in the presence of various anions were determined by the fixed-interference method and are tabulated below. The electrode has been used as an endpoint indicator in the titration of molybdate ions with thorium nitrate solutions. A membrane based on hydrated thorium oxide and 15% powdered polystyrene was used by Strivastava and Jain[81] for the preparation of a strontium ion-selective electrode. Electrode response was rectilinear over the concentration range 10^{-3} to 10^{-1} *M*, with a slope of 31.5 mV per decade of concentration change at pH 2.5 to 5.5. Selectivity coefficients toward a range of cations were satisfactory. The cationic surfactants hexadecyltrimethylammonium bromide and hexadecylpyridinium bromide interfered seriously at a concentration of about 10^{-3} *M*. The membrane could be rendered immune by treatment for 24 h with 10^{-3} *M* surfactant and subsequently showed rectilinear response down to 50 μM strontium ion concentration. The electrode was used satisfactorily to detect the endpoint in the titration of 0.01 *M* strontium chloride solution with 0.1 *M* $NH_4H_2PO_4$.

A solid membrane electrode selective to sulfate ions has also been prepared from hydrous thorium oxide gel with polystyrene (15%) as binder.[100] Membranes were equilibrated with 0.1 *M* potassium sulfate solution for 2 to 3 d. Electrode response was rectilinear over the concentration range $10^{-3.5}$ to 10^{-1} *M*, with a slope of 17 mV per decade of concentration change. The working pH range for the electrode was between 6 and 10. The electrode behaved satisfactorily in solutions having a nonaqueous content up to 25% (acetone, methanol, and ethanol). The selectivity coefficients listed are small and show that most common anions would cause no serious interference. The electrode system was also tolerant of the presence of small amounts of cationic and nonionic surfactants. The membrane electrode has been successfully used to estimate sulfate ions in wastes from the tannery, paper, and pulp industries.

A new sensor for the determination of sodium ions has been developed by Srivastava and co-workers[82] using hydrous zirconium oxide (HZO) as active membrane material. The HZO was prepared as described by Amphlett and co-workers.[83] Heterogeneous membranes were obtained by embedding polystyrene in zirconium oxide gel (ratio 1:9 by mass). After equilibration in 0.1 *M* sodium chloride for 2 to 3 d, the membrane was used in an electrode assembly, as reported earlier.[51] When tested in solutions of sodium chloride, bromide, nitrate, and sulfate, and at constant ionic strength, the electrode gave a linear response in the concentration range 10^{-5} to 10^{-1} *M*. The slope of the calibration plot was only 38.5 mV per decade of concentration change. The response time was found to be 20 to 30 s and the operational pH range was between 7 and 10.

The selectivity of this membrane electrode toward sodium ions over other cations was assessed by the fixed-interference method. The silver ion does not interfere, and the interference due to the potassium ion is significantly lower than that reported for glass membranes. Selectivity coefficients listed are $H^+ \simeq 200$, $NH_4^+ \simeq 10.5$, $Tl^+ \simeq 10^{-2}$, $K^+ \simeq 10^{-2}$, $Li^+ \simeq 10^{-3}$, and $Ag^+ \simeq 10^{-4}$. The effects of surfactants on electrode behavior were also investigated. Anionic surfactants (sodium dodecylsulfonate or sodium dodecylbenzenesulfonate) do not cause any disturbance. Cationic surfactants (cetyltrimethylammonium bromide or cetylpyridinium bromide) cause appreciable disturbance and shift potentials to more

TABLE 7
Characteristics of the Electrodes with Various Solvents

Solvent	pH range	ΔE/ΔpH mV	Response time (s)
1-Octanol	0—5	47	<15
1-Decanol	0—5	56	<10
1-Dodecanol	0—5	54	<15
Cyclohexanol	0—5	46	<30

From Wakida, S., Tanaka, T., Kawahara, A., and Hiiro, K., *Fresenius Z. Anal. Chem.*, 323, 142, 1986. With permission.

negative values even when present in small concentrations (10^{-5} to 10^{-6} *M*). This effect can be overcome by treating membranes for 6 to 8 h with a 10^{-4} *M* solution of the surfactant, after which the membrane behaves in a normal way in the presence of surfactants. The membrane functions well in partially nonaqueous solvents (50% methanol, 50% ethanol, and 40% acetone). The membrane electrode was successfully used to measure sodium ion activity in the discharge obtained from the electrolytic cell for the manufacture of sodium metal and in the discharge from a soda ash manufacturing plant using the Solvay process.

A new hydrogen ion-selective electrode based on an iron hydroxo complex was proposed by Wakida and co-workers.[84] The ion-sensing membrane was composed of the iron(III) hydroxo complex, membrane solvent, and PVC. The proposed electrode using 1-decanol as the most favorable solvent showed a linear pH response from 0 to 5, with a slope of 56 mV per pH unit.

Membranes were prepared as follows. Three milligrams of iron(III) chloride hexahydrate was dissolved in a mixture of 400 mg of organic solvent and 6 cm^3 of tetrahydrofuran; 200 mg of PVC powder was then dissolved in this mixture. The solution was transferred to a glass cylinder and allowed to stand for several days, giving a clear yellow membrane on complete evaporation of the tetrahydrofuran. To prepare an electrode, a disc was cut with a cork borer (10 mm diameter) and fixed to the end of a PVC tube capped with a glass body. An aqueous 10^{-2} *M* sulfuric acid solution was used as internal solution and a silver-silver chloride electrode as the internal electrode.

The performance of various organic liquids as membrane solvents was tested, and the results are summarized in Table 7. Selectivity coefficients were determined by the mixed-solution method. They were represented schematically as log k^{pot}_{HM} values and, in the case of the electrode using 1-decanol, ranged from $-1 > Cs^+ > Li^+, K^+ > Rb^+ > Na^+ > NH_4^+ > Mg^{2+}, Sr^{2+} > Ca^{2+} > -4$.

The mechanism of the pH response of the electrode is explained as follows. Iron(III) chloride is soluble in organic solvents such as alcohols because an olation reaction of iron(III) proceeds in the hydrophobic solvent. Polynuclear iron hydroxo complexes and condensed macromolecular ions are formed. It is supposed that the iron(III) hydroxo complex is formed in the membrane of the electrode. At the interface between the membrane and the solution, the surface hydroxyl group is in equilibrium with the hydrogen ion in solution.

$$\text{M–O}^- \rightleftharpoons \text{M–OH} \rightleftharpoons \text{M–OH}_2^+$$

An electrical double layer is formed between the M–O$^-$ group and H_3O^+ at the interface, and the interface potential is generated. Changes in the interfacial potential correspond to changes in the pH of the solution, the latter resulting from changes in the surface density of the M–O$^-$ group, according to the above equation. It is also assumed that the hydrogen ion is transported in the membrane through movement of the hydroxy group of the iron(III)

TABLE 8
Selectivity Coefficients k_{ij}^{pot} for Zirconium Molybdate Membrane Electrode in 10^{-4}, 5×10^{-4}, and 10^{-3} *M* Solutions of Interfering Ions[85]

	k_{ij}^{pot}		
Ion	10^{-4} *M*	5×10^{-4} *M*	10^{-3} *M*
VO_3^-	1.8	1.8	1.2
Cl^-	0.8	0.8	0.9
Br^-	0.6	0.6	0.1
I^-	0.6	0.6	0.1
CNS^-	1.2	1.4	1.1
NO_3^-	1.4	1.6	0.8
SO_4^{2-}	0.012	0.010	0.028
WO_4^{2-}	0.018	0.014	0.022
$S_2O_3^{2-}$	0.011	0.011	0.018
CrO_4^{2-}	0.016	0.014	0.024
PO_4^{3-}	0.010	0.008	0.015
AsO_4^{3-}	0.009	0.008	0.021
$Fe(CN)_6^{4-}$	0.12	0.11	0.16

From Malik, W. V., Srivastava, S. K., and Bansal, A., *Anal. Chem.*, 54, 1399, 1982. With permission.

hydroxo complex. This is the first report on a hydrogen ion-selective electrode based on a metal hydroxo complex membrane. Preliminary studies[84] showed that the membrane electrode based on an aluminium(III) hydroxo complex also worked as a hydrogen-ion-selective electrode.

E. VARIOUS OTHER MEMBRANES

The preparation and electrochemical properties of a number of inorganic ion-exchanger membranes have been studied by Alberti and co-workers.[85] Although these membranes were not employed as electrodes, they must be cited. The membranes examined were ammonium 12-molybdophosphatepolyethylene and zirconium phosphate-, zirconium antimonate-, and zirconium oxide-glass fiber membranes. The potentials of these membranes were measured in 0.1 and 0.5 *M* potassium chloride solutions at pH 6 (except for zirconium oxide, which was used at pH 3). All of these showed a near-Nernstian response.

A solid-membrane electrode for the measurement of molybdate ions has been reported by Malik and co-workers.[86] Heterogeneous membranes obtained by embedding polystyrene in zirconium molybdate gel were used for electrode preparation. The ratio used was 15:85 by mass of polystyrene to gel. The calibration curve was linear in the concentration range 10^{-5} to 5×10^{-3} *M*, with a slope of 24.5 mV per decade of concentration change. The operational pH range was between 6.5 and 9.0. The response of the electrode in ethanol-water (25% ethanol) medium was also linear in the same concentration range, with an almost identical slope.

Selectivity coefficients were determined for a large variety of anions by the fixed-interference method and are listed in Table 8. The electrode exhibits fairly good selectivity in its behavior toward molybdate over several other anions. The ions VO_3^-, NO_3^-, and CNS^- are the only ones that show serious interference. It was also observed that the surfactants cetylpyridinium bromide and sodium dodecylsulfate did not cause any disturbance in the concentration range 10^{-6} to 10^{-4} *M*. The time response was fast (30 to 40 s). The electrode was used to indicate endpoints during the titration of 20 cm^3 of 0.01 *M* molybdate solutions with titanium tetrachloride and ceric ammonium nitrate, respectively.

The behavior of an Araldite-based membrane of crystalline antimonic(V) acid as a nitrate ion-selective electrode was studied by Agrawal and Abe.[87] The membrane consisting of a 7:3 ratio by mass of crystalline antimonic(V) acid to Araldite was prepared by the method of Coetzee and Basson.[15] When acting as a nitrate ion-selective electrode, a near-Nernstian response was found for nitrate ion concentrations between 10^{-5} and 10^{-1} *M*. Stable potentials were observed within 10 and 30 s, and the useful pH range was 3.5 to 11 at 5×10^{-3} *M* nitrate and 4.5 to 9 at 5×10^{-4} *M* nitrate. The membrane responds to nitrate ions in a solution containing 25% ethanol, and the working curve in 10% ethanol is the same as that in aqueous solution. Membranes can be used for a period of 2 months without any significant change in potentials.

Selectivity coefficients were determined for 21 anions and a number of cations (not listed) by means of mixed-solution and separate-solution methods. The membrane is selective for nitrate ions with respect to many anions except iodide. The results also indicate that this electrode can determine NO_3^- concentrations in the presence of Br^-, ClO_3^-, and SO_4^{2-}, which is a great advantage over other electrodes. Ion-selectivity electrodes based on ion exchangers involve ion-exchange processes at the electrode interface. The crystalline antimonic(V) acid (C-SbA) is, however, a cation exchanger and has no adsorptive properties toward anions in the pH range studied. It is known that fast conduction of protons occurs within the pyrochlore framework in C-SbA in the H^+ form. The Araldite used as binder contains polyamide. The functional group $=N_=^+$ may remain in the polymer after polymerization and may behave as an anion exchanger. The response of this membrane may be due to the anion-exchange contribution of the polymer, and the C-SbA acts as an electroconductive material.

A solid-membrane electrode, prepared from titanium arsenate gel as membrane material and Araldite as binder, was described by Srivastava et al.[88] Membranes with 40% Araldite content showed the best electrochemical performance and were selected for the study. The membranes were equilibrated with 0.1 *M* lead nitrate for 1 week and then used in electrodes for lead ion determination.

These electrodes gave a linear response over the concentration range 10^{-5} to 10^{-1} *M*, with a non-Nernstian slope of 33.3 mV/pPb, but were used to measure lead ion concentrations down to 5×10^{-6} *M*. Response times were less than 1 min, and the electrodes had a lifespan of about 4 months. Between pH values of 2 and 5, electrode response was constant.

Selectivity coefficients were determined by the fixed-interference method for a large number of ions. Univalent ions seem to be the only ions that do interfere to a large extent, selectivity coefficient values ranging from 0.11 for the sodium ion to 0.33 and 0.32 for potassium and thallium ions, respectively. If the membrane is modified by soaking in 10^{-4} *M* cetylpyridinium chloride for 4 d, the selectivity with respect to univalent cations (except ammonium) is improved. The slope of the electrode response curve is the same in ethanol-water and acetone-water mixtures (up to 30% v/v organic component) as in water. The linear range is, however, shorter (5×10^{-4} to 5×10^{-2} *M*). The electrode was successfully used as an endpoint indicator in the potentiometric titration of lead ions with molybdate solutions.

A heterogeneous membrane in which zirconium tellurite gel was embedded in polystyrene was used to prepare an electrode which was used by Srivastava at el.[89] for the estimation of chromium(VI) in water, and in tannery and plating wastes. Membranes were converted to the desired ionic form by immersing them in a solution of electrolyte (0.1 *M*) for 2 to 3 d. The electrode assembly was the same as reported earlier.[86] Linear plots over the concentration range 10^{-4} to 10^{-1} *M* indicate that the electrode can be used for the estimation of chromium as chromate (pH 9) and dichromate (pH 4). The slope of the plots varies from

22 to 23 mV per decade change in concentration. The working pH range of the electrode assembly is 3 to 6 for chromate and 8 to 11 for dichromate. The electrode is also usable in ethanolic water mixtures with an ethanolic content of less than 30%. Selectivity coefficients in the presence of ten common anions were determined. The only major interfering anion was molybdate, with a value of 0.1.

VII. POTENTIOMETRIC TITRATIONS

The most common way of employing ion-selective electrodes in analytical procedures is by using them as endpoint indicators during the potentiometric titration of ions for which they are selective. Due to the specific character of electrodes employing inorganic ion exchangers as active membranes, most of these electrodes are sensitive to large monovalent cations.

The nature of these cations is such that the large majority of titrations, except for the cases where standard addition potentiometry is used, are based on precipitation reactions. When titrating large monovalent alkali cations, the titrants are limited to solutions of sodium tetraphenylborate, derivatives thereof, or heteropolyacid solutions. In the case of thallium and silver ions, additional titrants such as solutions of halides or chromates may be used.

In the following paragraphs, some typical titrations in which a number of these electrodes were used will be discussed.

A. ELECTRODES BASED ON COMPLEX CYANIDE SALTS

The electrode hexacyanoferrate(III)-Araldite membrane reported by Jain and co-workers[24] was used in the titration of 20 cm^3 of 0.01 *M* thallium nitrate with 0.1 *M* K_2CrO_4. The usual potential against titrant plot shows two inflection points; the second, which is the largest, corresponds to the formation of $TlKCrO_4$. This phenomenon was also reported by Coetzee and Basson[25] and Malik and co-workers.[41]

A ferricyanide-sensing electrode reported by Ihn[33] was used to indicate endpoints during separate titrations of 20 ml each of 10^{-2} and 10^{-3} *M* ferricyanide with silver nitrate solutions. A mixture containing 20 ml each of ferricyanide and ferrocyanide was also titrated successfully with a silver nitrate solution, precipitation of silver ferricyanide occurring after precipitation of silver ferrocyanide. All of these titration curves had the normal shape.

The use of chromium hexacyanoferrate(II) epoxy-resin matrix ion-exchange membranes as indicator electrodes in potentiometric titrations was studied by Jain and co-workers.[28] Basically, two types of titrations were performed in which these membranes were used as endpoint indicators. First, there were the titrations of small amounts of potassium hexacyanoferrate(II) (20 cm^3 of a 10^{-3} *M* solution) with Hg^{2+}-, UO_2^{2+}-, and Th^{4+}-containing solutions and 20 cm^3 of 10^{-4} *M* solution with Pb^{2+}-, Zn^{2+}-, and Cu^{2+}-containing solutions. Second, 20 cm^3 of 10^{-3} *M* solutions of these cations were titrated with standard potassium hexacyanoferrate(II) solutions.

The compositions of the hexacyanoferrate(II) precipitates formed were a function of the manner of addition of the reagents. The titration curves did not have the expected traditional shape, as can be seen from Figures 1 to 4. They were straight lines intersecting at different points, and their shapes correspond to those previously reported by Coetzee and Basson[29,30] for heteropolyacid salt membranes, which will later be discussed in detail. If the last breaks in the titration curves are taken as the stoichiometric endpoints, the compositions of the various compounds are in agreement with previously reported values.[31,32]

In another publication by Jain and co-workers,[34] the same electrodes were used in the titration of 20 cm^3 of potassium hexacyanoferrate(II) (10^{-3} *M*) with 10^{-2} *M* solutions of Bi^{3+}, La^{3+}, Fe^{3+}, Co^{3+}, and Ce^{3+}. The titration curves also are not of the classical type because the membrane electrode responds not only to hexacyanoferrate(II), but also to other

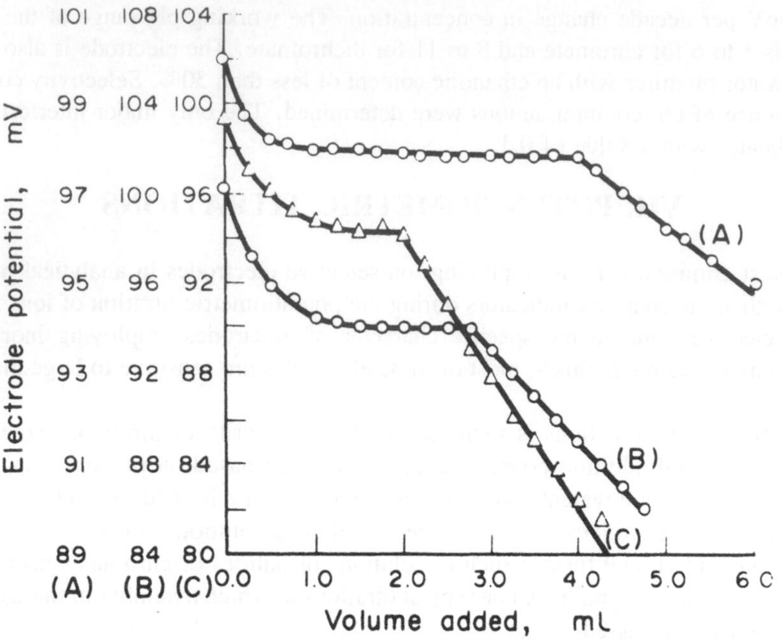

FIGURE 1. Titration of 20 ml of 10^{-3} *M* potassium ferrocyanide with 10^{-2} *M* Hg^{2+} (A), UO_2^{2+} (B), and Th^{4+} (C). (From Jain, A. K., Srivastava, S. K., Agrawal, S., and Singh, R. P., *Talanta,* 25, 531, 1978. With permission.)

ions added during the titration. The plots represent straight lines intersecting one another at the equivalence point. The compositions for the insoluble trivalent metal hexacyanoferrate(II) were calculated from the data, and corresponds to KM(III) $Fe(CN)_6$.

B. ELECTRODES BASED ON HETEROPOLYACID SALTS

Several membrane electrodes employing insoluble salts of heteropolyacids as active membrane substances have been used as indicator electrodes during potentiometric titrations. The general shapes of the resulting titration curves, that is, a plot of cell potential versus volume of titrant added, can be divided into two main groups: those which did not have the classical "S" shape and those which did have the normal form. The standard addition potentiometric technique was reported only by Coetzee and Basson,[15] using a cesium ion-selective electrode for the determination of 38.6 mg cesium, with a 0.100 *M* $CsNO_3$ solution as titrant and 0.1 *M* $LiNO_3$ as ionic-strength adjustor. In order to indicate endpoints clearly during titrations, Coetzee and Basson[15] had to use Gran plots[38] of titration data for the titration of 72.4 mg of cesium with a 0.245 *M* 12-molybdophosphoric acid solution and the titration of 6.55 mg of cesium with a 0.0176 *M* sodium tetraphenylborate solution. This method was also used for the titration of 17.5 mg of strontium with 0.067 *M* diammonium-hydrogen phosphate, as reported by Jain et al.,[42] employing a strontium tungstoarsenate membrane electrode.

Electrodes made from the tungstophosphate and molybdophosphate of thallium(I) were used as indicator electrodes during the potentiometric titration of thallium(I) ions with potassium bromide, potassium chromate, and sodium tetraphenylborate as precipitating agents.[25] Titration curves of these titrations had the shapes shown in Figures 5, 6 and 7, respectively.

This type of potentiometric titration curve was also noted by Coetzee and Basson[29,30] when they reported the use of cesium- and thallium-sensitive epoxy-resin membranes. An-

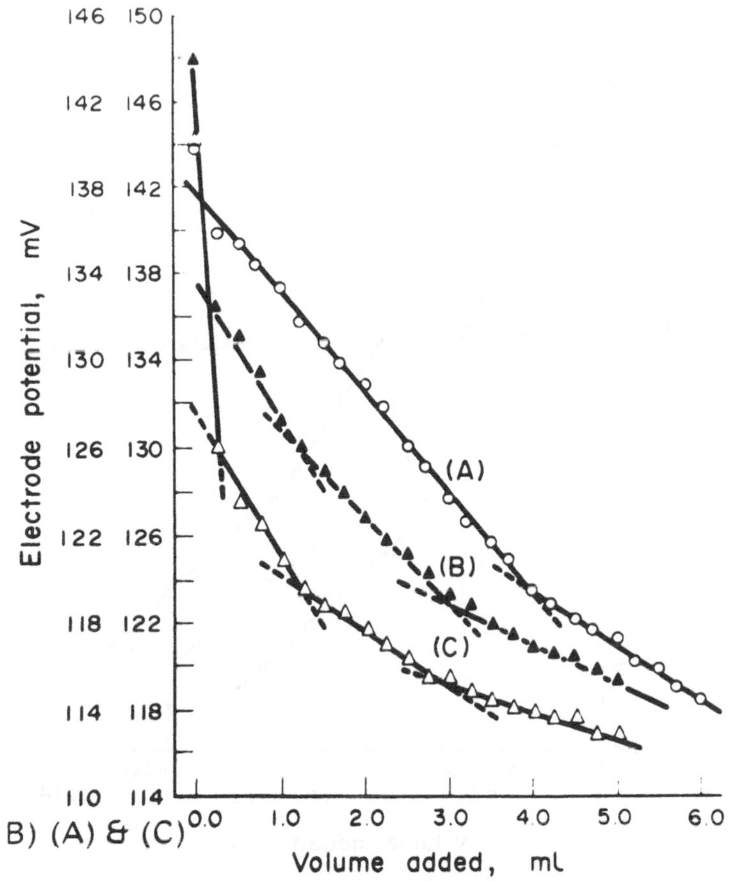

FIGURE 2. Titration of 20 ml of 10^{-4} *M* potassium ferrocyanide with 10^{-3} *M* Pb^{2+} (A), Zn^{2+} (B), and Cu^{2+} (C). (From Jain, A. K., Srivastava, S. K., Agrawal, S., and Singh, R. P., *Talanta*, 25, 531, 1978. With permission.)

other interesting fact indicated by these titrations is that when thallium ion was titrated with a K_2CrO_4 solution, the stoichiometric endpoint indicated the formation of $KTlCrO_4$. This was confirmed by Malik and co-workers,[41] using a thallium tungstoarsenate membrane electrode. The abnormally shaped titration curves are not only restricted to the titration of monovalent cations. Srivastava and co-workers[48] observed the same phenomena when using a copper tungstoarsenate membrane electrode during the titration of copper solutions with EDTA. The shape of the plots might be due to the fact that these membranes are also sensitive to changes in the activities of other monovalent or divalent cations. During titration, the primary cation is effectively removed from solution, but is replaced by another cation to which the electrode responds to a certain extent.

A type of titrant which was quite often used for the titration of the large monovalent cations was a standard solution of a heteropolyacid.

Coetzee and Basson[15] showed that when cesium was titrated with 12-molybdophosphoric acid, the precipitation took place in three stages:

$$Cs^+ + H_3PMo_{12}O_{40} \rightleftharpoons CsH_2PMo_{12}O_{40} + H^+$$

$$Cs^+ + CsH_2PMo_{12}O_{40} \rightleftharpoons Cs_2HPMo_{12}O_{40} + H^+$$

$$Cs^+ + Cs_2HPMo_{12}O_{40} \rightleftharpoons Cs_3PMo_{12}O_{40} + H^+$$

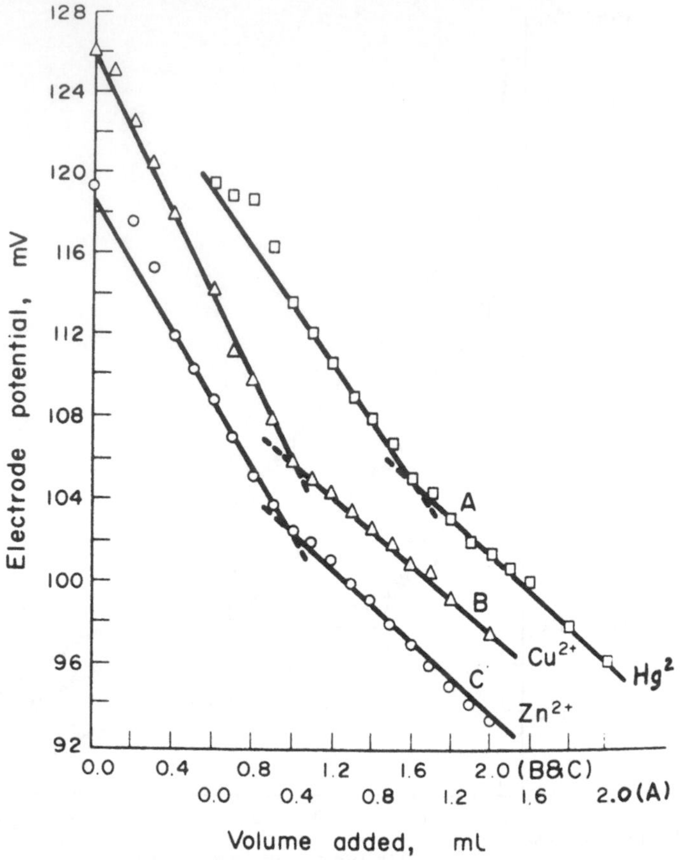

FIGURE 3. Titration of 20 ml of 10^{-3} *M* Hg^{2+} (A), Cu^{2+} (B), and Zn^{2+} (C) with 10^{-2} *M* potassium ferrocyanide. (From Jain, A. K., Srivastava, S. K., Agrawal, S., and Singh, R. P., *Talanta,* 25, 531, 1978. With permission.)

For the precipitation of thallium[40] as well as cesium,[41] Malik and co-workers reported that they found three easily detectable endpoints. They could not, however, detect exact stoichiometric ratios for the first two steps; the third step gave the correct stoichiometric value for complete precipitation. The same phenomenon was found by Srivastava and co-workers[47,53] when titrating rubidium ion with 12-tungstophosphoric acid[47] and pyridinium ion with 12-molybdophosphoric acid.[53]

A rubidium 12-tungstoarsenate-Araldite membrane electrode was used to monitor the titration of 2.16 mg silver with 10^{-2} *M* sodium ferrocyanide and 10^{-2} *M* potassium chloride.[90] Normal titration curves resulted.

In another publication by Jain and co-workers,[52] a thallium-selective electrode was successfully used to titrate 20 cm^3 of a 0.01 *M* thallium ion solution with potassium chromate (0.1 *M*), again confirming the formation of $TlKCrO_4$. The titration of 0.01 *M* cesium solution with a 0.1 *M* sodium tetraphenylborate solution, using a cesium ion-selective electrode based on a crown ether phosphotungstate precipitate, was reported by Wang and Shih.[68]

A PVC tetraphenylphosphonium 12-tungstosilicate electrode was described by Fogg and Yoo.[70] Satisfactory titration curves were obtained for several basic dyes when they were titrated with standard sodium tetraphenylborate solution. Acid dyes were titrated with a standard solution of crystal violet, using the tetraphenylphosphonium 12-tungstosilicate electrode as indicator electrode. As expected from the higher water solubility of the crystal violet

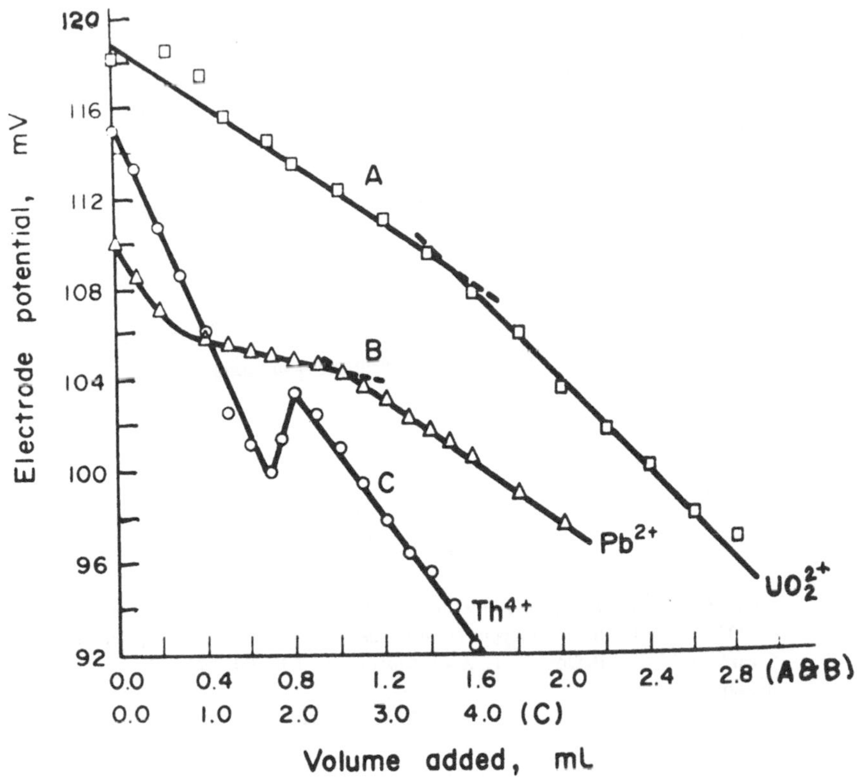

FIGURE 4. Titration of 20 ml of 10^{-3} *M* UO_2^{2+} (A), Pb^{2+} (B), and Th^{4+} (C) with 10^{-2} *M* potassium ferrocyanide. (From Jain, A. K., Srivastava, S. K., Agrawal, S., and Singh, R. P., *Talanta,* 25, 531, 1978. With permission.)

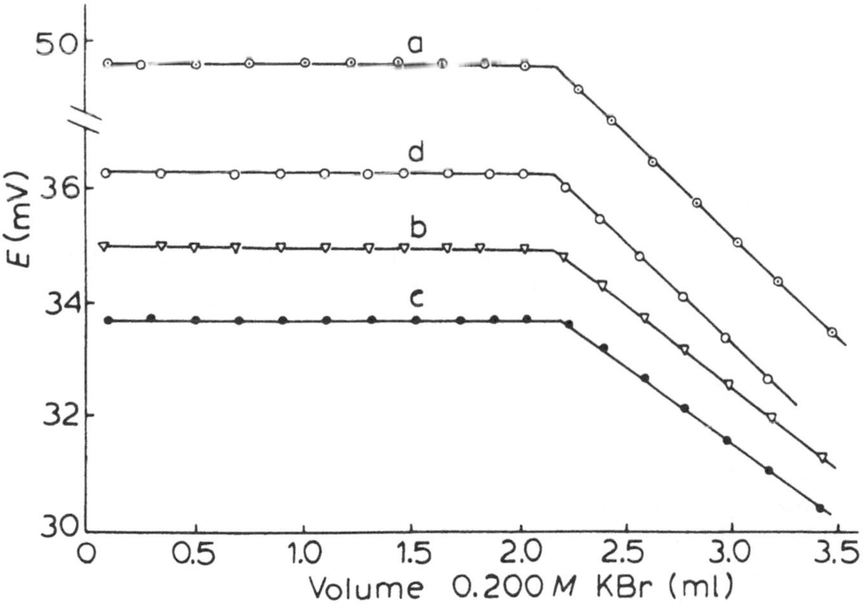

FIGURE 5. The titration of 100.9 mg of thallium(I) with a 0.2008 *M* KBr solution. In 55 ml of water (a); in 25% (v/v) ethanolic solution (b); in presence of 19.6 mg of potassium (c); in presence of 11.5 mg of sodium (d). (From Coetzee, C. J. and Basson, A. J., *Anal. Chim. Acta,* 64, 300, 1973. With permission.)

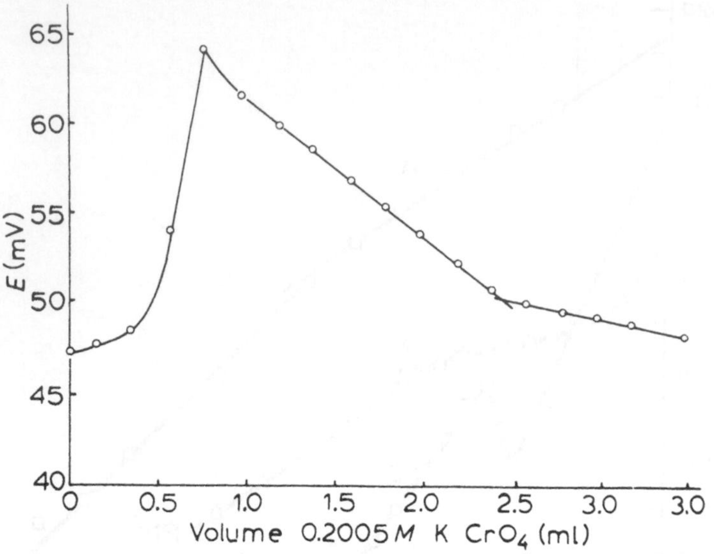

FIGURE 6. The titration of 100.9 mg of thallium(I) in 55 ml of water with a 0.2005 *M* K_2CrO_4 solution. Endpoint corresponds to formation of TlK CrO_4. (From Coetzee, C. J. and Basson, A. J., *Anal. Chim. Acta*, 64, 300, 1973. With permission.)

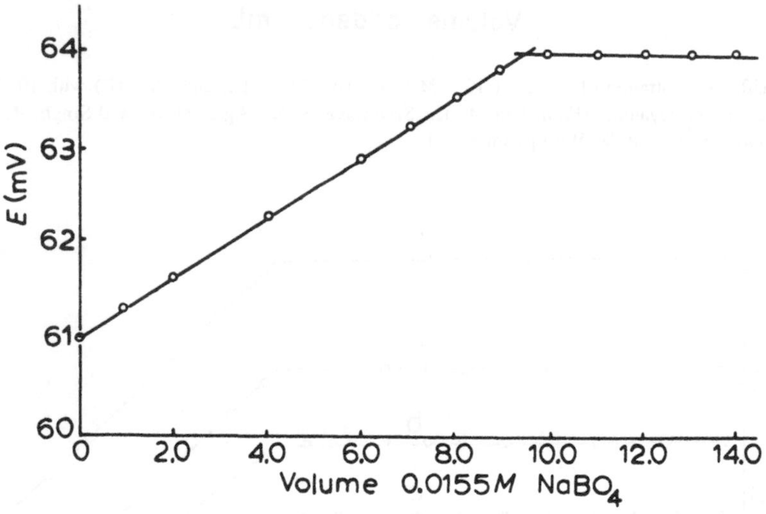

FIGURE 7. The titration of 30.27 mg of thallium(I) in 50 ml of water with a 0.0155 *M* sodium tetraphenylborate solution. (From Coetzee, C. J. and Basson, A. J., *Anal. Chim. Acta*, 64, 300, 1973. With permission.)

and dye salts, the potential jumps were smaller then those obtained for the tetraphenylborate titration of basic dyes, and the endpoints were less sharp.

Several other electrodes which were used in potentiometric titrations were reported by Srivastava and co-workers.[79,81,86] A polystyrene-based zirconium oxide membrane[79] was used during titration of various amounts of sodium molybdate, using thorium nitrate as titrant. Similarly, a polystyrene thorium oxide membrane[81] electrode was used during titration of 0.01 *M* $SrCl_2$ with 0.1 *M* $NH_4H_2PO_4$, and a zirconium molybdate polystyrene membrane electrode[86] was used during the titration of molybdate with cerium(IV) and titanium(IV) solutions.

The titration of nitrate with nitron was monitored by Agrawal,[87] using an Araldite-based membrane of crystalline antimonic(V) acid.

VIII. CONCLUSIONS AND DISCUSSION

The use of inorganic ion exchangers in liquid-liquid type of membrane electrodes is not very successful or popular. The selectivity over other anions, for example, of the anionic liquid-liquid membrane reported by Guilbault and Brignac[71] was very poor. Over 95% of the reported electrodes belong to the heterogeneous solid membrane type of electrodes. One of the main problems in constructing heterogeneous solid membrane electrodes is finding a suitable material that can be used as an inert binder in the membrane. The binding material must be chemically inert, insoluble in water, hydrophobic, flexible, effectively bind the particles, and also must resist cracking and swelling in aqueous solution. An epoxy resin first used by Coetzee and Basson[25] proved to be the most suitable and widely used material.

Probably the major disadvantage of many of these electrodes is that they are not truly selective. Most of these salts could be used as ion-exchange materials for the separation of cations. This implies that many of these cations would interfere with electrode response. The narrow pH range within which these salts are stable is also a slight disadvantage when they are applied to the manufacture of membrane electrodes.

Electrodes based on potassium zinc hyxacyanoferrate(II)[17,18,20] showed near-Nernstian behavior. In the case of ones prepared from various heteropolyacid salts, the behavior was varied. The slopes of calibration curves varied from sub-Nernstian to super-Nernstian.[37,43] In general, with ion-exchange membranes, deviations at high solution activities are caused by co-ion transference, and at lower activities, by competing hydrogen or hydroxyl ions.[9] Deviation from Nernstian behavior may be attributed to incomplete permselectivity of the membrane,[91] a factor which Helfferich[92] covered by an integral term in the function for membrane potential to account for the co-ion flux. The non-Nernstian behavior of many of the reported electrodes did not, however, prevent them from being successfully employed to determine the ions in question. A somewhat longer response time of some of the reported electrodes did not seem to pose any problems. It has been mentioned that perhaps too much emphasis has been placed on response times of electrodes, since, for direct potentiometric determinations of activity, response times longer than 1 to 2 min are quite tolerable.

Selectivity coefficient values of comparable electrodes vary widely. They also are not always related to the values expected from ion-exchange data. Many of these values were, of course, determined by different methods and not too much importance should be given to the deviations. As already mentioned, when using ion exchangers as ion sensors in electrodes, they often cannot be expected to be very highly selective to specific ions. These electrodes are of relatively low cost, easy to prepare, and may be successfully used in analytical procedures.

The ion-exchange properties of many inorganic crystalline compounds have been studied during the past decade. Not many of these salts have been used in membranes as ion sensors. Examples of these compounds are cerium(IV) phosphate, thorium arsenate, zirconium phosphate, cerium(IV) arsenate, and a fibrous thorium phosphate.[93-98] Some of these compounds have specific preferences for certain ions. Fibrous cerium phosphate,[99] for example, shows a very high selectivity for Pb^{2+} and Ag^{+} cations. Cerium(IV) phosphate[93] shows a strong preference for sodium ions at pH 4 and for lithium ions at pH 8. If the various properties of the above-mentioned materials are carefully considered, it is likely that some of them could successfully be applied as ion sensors in heterogeneous precipitate membrane electrodes.

At present, the majority of ion-selective electrodes employing inorganic ion exchangers as ion sensors can be divided into only two main groups: (1) salts of heavy metal ferrocyanides, and (2) salts of heteropolyacids. The electrodes may be used analytically mainly for the determination of monovalent cation concentrations. Although not highly selective, they do have a part to play in modern-day analytical chemistry as ion sensors.

REFERENCES

1. **Frant, M. S. and Ross, J. W.,** *Science,* 154, 1553, 1966.
2. **Ross, J. W.,** *Science,* 156, 1378, 1967.
3. **Freiser, H., Ed.,** *Ion-selective Electrodes in Analytical Chemistry,* Vol. 1 and 2, Plenum Press, New York, 1978.
4. **Morf, W. E.,** *The Principles of Ion-selective Electrodes and of Membrane Transport,* Elsevier, Budapest, 1981.
5. **Koryta, J.,** *Ion-selective Electrodes,* Cambridge University Press, Cambridge, England, 1975.
6. **Koryta, J.,** *Ions, Electrodes and Membranes,* John Wiley & Sons, New York, 1982.
7. **Covington, A. K., Ed.,** *Ion-selective Electrode Methodology,* Vols. 1 and 2, CRC Press, Boca Raton, FL, 1979.
8. **Cammann, K.,** *Working with Ion-selective Electrodes,* Springer-Verlag, Berlin, 1979.
9. **Moody, G. J. and Thomas, J. D. R.,** *Selective Ion Sensitive Electrodes,* Merrow Publishing, Watford, 1971.
10. Recommendations for Nomenclature of Ion-selective Electrodes, *Pure Appl. Chem.,* 48, 129, 1976.
11. **Geyer, R. and Syring, W.,** *Z. Chemie,* 6, 92, 1966.
12. **Gans, R.,** *Jahrb. Preuss. Geol. Landesanst.,* 26, 179, 1905.
13. **Vesely, V. and Pekarek, V.,** *Talanta,* 19, 219, 1245, 1972.
14. **Bailey, P. L.,** *Analysis with Ion-selective Electrodes,* Heyden, 1976.
15. **Coetzee, C. J. and Basson, A. J.,** *Anal. Chim. Acta,* 56, 321, 1971.
16. **Pushkarev, V. V., Skrylev, L. D., and Bagretsov, V. F.,** *Z. Prik. Khim.,* 33, 81, 1960.
17. **Fogg, A. G. and Reynolds, G. F.,** *Anal. Chim. Acta,* 32, 582, 1965.
18. **Fogg, A. G., Pathan, A. S., and Burns, D. T.,** *Anal. Lett.,* 7, 539, 1974.
19. **Rock, P. A., Eyrich, T. L., and Styer, S.,** *J. Electrochem. Soc.,* 124, 531, 1978.
20. **D'Olieslager, W. and Heerman, L.,** *J. Electrochem. Soc.,* 126, 347, 1979.
21. **Vlasselaer, S., D'Olieslager, W., and D'hont, M.,** *J. Inorg. Nucl. Chem.,* 38, 327, 1976.
22. **Vlasselaer, S., D'Olieslager, W., and D'hont, M.,** *J. Radioanal. Chem.,* 35, 211, 1977.
23. **Saraswat, I. P., Srivastava, S. K., and Sharma, A. K.,** *Fresenius Z. Anal. Chem.,* 305, 410, 1981.
24. **Jain, A. K., Singh, R. P., and Bala, C.,** *Anal. Lett.,* 15, 1557, 1982.
25. **Coetzee, C. J. and Basson, A. J.,** *Anal. Chim. Acta,* 64, 300, 1973.
26. **Engel, D. and Grabner, E. W.,** *Ber. Bunsenges. Phys. Chem.,* 89, 982, 1985.
27. **Kuwana, T. and Siperco, L.,** *J. Electrochem. Soc.,* 130, 396, 1983.
28. **Jain, A. K., Srivastava, S. K., Agrawal, S., and Singh, R. P.,** *Talanta,* 25, 531, 1978.
29. **Coetzee, C. J. and Basson, A. J.,** *J. S.A. Chem. Inst.,* 26, 39, 1973.
30. **Coetzee, C. J. and Basson, A. J.,** *Tydskr. Natuurwet.,* 13, 20, 1973.
31. **Bellomo, A.,** *Talanta,* 17, 1109, 1970.
32. **Bellomo, A., De Marco, D., and Casale, A.,** *Talanta,* 22, 197, 1975.
33. **Ihn, G. S.,** *Bull. Korean Chem. Soc.,* 7, 36, 1986.
34. **Jain, A. K., Agrawal, S., and Singh, R. P.,** *J. Indian Chem. Soc.,* 57, 343, 1980.
35. **Broadbank, R. W. C., Dhabanandana, S., and Harding, R. D.,** *Analyst,* 85, 365, 1960.
36. **Smit, J. V. R.,** *Nature,* 181, 1530, 1958.
37. **Coetzee, C. J. and Basson, A. J.,** *Anal. Chim. Acta,* 57, 478, 1971.
38. **Gran, G.,** *Analyst,* 77, 661, 1952.
39. **Coetzee, C. J. and Rohwer, E. F. C. H.,** *J. Inorg. Nucl. Chem.,* 32, 1711, 1970.
40. **Malik, W. U., Srivastava, S. K., and Bansal, A.,** *Indian J. Chem.,* 22A, 221, 1983.
41. **Malik, W. U., Srivastava, S. K., Razdan, P., and Kumar, S.,** *J. Electroanal. Chem.,* 72, 111, 1976.
42. **Jain, A. K., Srivastava, S. K., Singh, R. P., and Agrawal, S.,** *J. Appl. Chem. Biotechnol.,* 27, 680, 1977.
43. **Srivastava, S. K., Jain, A. K., Agrawal, S., and Singh, R. P.,** *J. Electroanal. Chem.,* 90, 291, 1978.

44. **Hartley, G. S., Collie, B., and Samis, C. S.,** *Trans. Faraday Soc.*, 32, 795, 1936.
45. **Harkins, W. D.,** *J. Am. Chem. Soc.*, 69, 1428, 1947.
46. **Coetzee, C. J.,** D.Sc. thesis, University of Stellenbosch, 1965.
47. **Srivastava, S. K., Kumar, S., and Kumar, S.,** *J. Electroanal. Chem.*, 161, 345, 1984.
48. **Srivastava, S. K., Pal, N., Singh, R. P., and Agarwal, S.,** *Indian J. Chem.*, 22A, 1033, 1983.
49. **Srivastava, S. K., Kumar, S., Pal, N., and Agrawal, S.,** *Fresenius Z. Anal. Chem.*, 315, 353, 1983.
50. **Jain, A. K., Singh, R. P., and Agrawal, S.,** *J. Radioanal. Chem.*, 54, 171, 1979.
51. **Malik, W. U., Srivastava, S. K., and Bansal, A.,** *Anal. Chem.*, 54, 1399, 1982.
52. **Jain, A. K., Agrawal, S., and Singh, R. P.,** *Anal. Lett.*, 12, 995, 1979.
53. **Srivastava, A. K., Jain, A. K., Agrawal, S., and Singh, R. P.,** *Talanta*, 25, 157, 1978.
54. **Jain, A. K., Singh, R. P., and Agrawal, S.,** *Fresenius Z. Anal. Chem.*, 302, 407, 1980.
55. **Longhi, P., Mussini, T., Nardi, F. M., and Rondinini, S.,** *Nouv. J. Chim.*, 3, 649, 1979.
56. **Guilbault, G. G., Smith, R. K., and Montalvo, J.,** *Anal. Chem.*, 41, 600, 1969.
57. **Pedersen, C. J.,** *J. Am. Chem. Soc.*, 89, 7017, 1967.
58. **Kolthoff, I. M.,** *Anal. Chem.*, 51, 1R, 1979.
59. **Christensen, J., Eatough, D. J., and Izatt, R. M.,** *Chem. Rev.*, 74, 351, 1974.
60. **Frensdorff, H. K.,** *J. Am. Chem. Soc.*, 93, 600, 1971.
61. **Ryba, O. and Petranek, J.,** *J. Electroanal. Chem. Interfacial Electrochem.*, 44, 425, 1973.
62. **Petranek, J. and Ryba, O.,** *Anal. Chim. Acta*, 72, 375, 1974.
63. **Shono, T., Okahara, M., Ikeda, I., Kimura, K., and Tamura, H.,** *J. Electroanal. Chem.*, 132, 99, 1982.
64. **Kimura, K., Maeda, J., and Shono, J.,** *J. Electroanal. Chem.*, 95, 91, 1979.
65. **Fernando, L. A., Miles, M. L., and Brown, L. H.,** *Anal. Chem.*, 52, 1115, 1980.
66. **Jeng, J. and Shih, J. S.,** *Analyst*, 109, 641, 1984.
67. **Boles, J. H. and Buck, R. P.,** *Anal. Chem.*, 45, 2057, 1973.
68. **Wang, D. and Shih, J. S.,** *Analyst*, 110, 635, 1985.
69. **Fogg, A. G. and Yoo, K. S.,** in *Proc. Conf. Ion-Selective Electrodes*, Elsevier, Budapest, 1977, 369.
70. **Fogg, A. G. and Yoo, K. S.,** *Anal. Chim. Acta*, 113, 165, 1980.
71. **Guilbault, G. G. and Brignac, P. J.,** *Anal. Chim. Acta*, 56, 139, 1971.
72. **Marshall, C. E. and Bergman, W. E.,** *J. Am. Chem. Soc.*, 63, 1911, 1941.
73. **Marshall, C. E. and Krinbill, C. A.,** *J. Am. Chem. Soc.*, 64, 1814, 1942.
74. **Marshall, C. E. and Ayers, A. D.,** *J. Am. Chem. Soc.*, 70, 1291, 1948.
75. **Marshall, C. E. and Eime, L. O.,** *J. Am. Chem. Soc.*, 70, 1302, 1948.
76. **Barrer, R. M. and James, S. D.,** *J. Phys. Chem.*, 64, 417, 421, 1960.
77. **Johansson, G., Risinger, L., and Fälth, L.,** *Anal. Chim. Acta*, 119, 25, 1980.
78. **Johansson, G., Fälth, L., and Risinger, L.,** *Hung. Sci. Instrum.*, 49, 47, 1980.
79. **Srivastava, S. K., Sharma, A. K., and Jain, A. K.,** *Talanta*, 30, 285, 1983.
80. **Shor, A. J., Kraus, K. A., Smith, W. T., and Johnson, J. S.,** *J. Phys. Chem.*, 72, 2200, 1968.
81. **Srivastava, S. K. and Jain, C. K.,** *Bunseki Kagaku*, 33, E525, 1984.
82. **Srivastava, S. K., Kumar, S., and Jain, C. K.,** *Analyst*, 109, 667, 1984.
83. **Amphlett, C. B., McDonald, L. A., and Redman, M. J.,** *J. Inorg. Nucl. Chem.*, 6, 236, 1958.
84. **Wakida, S., Tanaka, T., Kawahara, A., and Hiiro, K.,** *Fresenius Z. Anal. Chem.*, 323, 142, 1986.
85. **Alberti, G., Conti, A., and Torracca, E.,** *Atti. Accad. Naz. Lincei Cl. Sci. Fis. Mat. Nat. Rend.*, 35, 548, 1963.
86. **Malik, W. U., Strivastava, S. K., and Bansal, A.,** *Anal. Chem.*, 54, 1399, 1982.
87. **Agrawal, S. and Abe, M.,** *Analyst*, 108, 712, 1983.
88. **Srivastava, S. K., Kumar, S., Jain, C. K., and Kumar, S.,** *Talanta*, 33, 717, 1986.
89. **Srivastava, S. K., Singh, A. K., Garg, M., and Khanna, R.,** *Mikrochim. Acta*, 3, 377, 1985.
90. **Jain, A. K., Srivastava, S. K., Singh, R. P., and Agrawal, S.,** *Anal. Chem.*, 51, 1093, 1983.
91. **Moody, G. J. and Thomas, J. D. R.,** *Talanta*, 19, 623, 1972.
92. **Helfferich, F.,** *Ion-exchange*, McGraw-Hill, New York, 1962.
93. **Alberti, G., Costantino, U., and Zsinka, L.,** *J. Inorg. Nucl. Chem.*, 34, 3549, 1972.
94. **Alberti, G. and Massucci, M. A.,** *J. Inorg. Nucl. Chem.*, 32, 1719, 1970.
95. **Alberti, G., Constantino, U., and Gupta, J. P.,** *J. Inorg. Nucl. Chem.*, 36, 2103, 1974.
96. **Alberti, G., Constantino, U., Di Gregorio, F., and Torracca, E.,** *J. Inorg. Nucl. Chem.*, 31, 3195, 1969.
97. **Alberti, G. and Torracca, E.,** *J. Inorg. Nucl. Chem.*, 30, 3075, 1968.
98. **Alberti, G. and Constantino, U.,** *J. Chromatogr.*, 50, 482, 1970.
99. **Alberti, G., Casciola, M., Constantino, U., and Luciani, M. L.,** *J. Chromatogr.*, 128, 289, 1976.
100. **Srivestava, S. K. and Jain, C. K.,** *Water Res.*, 19, 53, 1985.
101. **Srivastava, S. K., Singh, A. K., Garg, M., and Khanna, R.,** *Mikrochim. Acta (Wien)* II, 377, 1985.

Chapter 6

AMORPHOUS INORGANIC ION EXCHANGERS

K. G. Varshney and Mukhtar A. Khan

TABLE OF CONTENTS

I. INTRODUCTION

Inorganic ion exchangers are receiving increasing attention[1] due to the fact that they are resistant to heat and radiation. They can be used for the high-temperature separation of ionic components in radioactive wastes, as solid electrolytes and as catalysts.

In Chapter 1, a detailed account of the historical background of inorganic ion exchangers has been given. Their present status and future prospects in research and analysis have also been discussed. In Chapter 2, a theoretical approach based on the zeolite models has been presented to explain the ion-exchange phenomenon occurring on the surface of these materials. They can be prepared in both the amorphous and crystalline forms. Earlier studies started with the amorphous materials. However, with the synthesis of a crystalline form of zirconium phosphate in 1964 by Clearfield and Stynes by refluxing amorphous product in phosphoric acid for a long time, a spurt of activities took place in the field of crystalline materials. Crystalline materials, besides having a more definite composition, compare favorably with the corresponding amorphous ones as regards thermal degradation and stability toward hydrolysis. Moreover, several crystalline exchangers can exhibit good ion-sieve properties and, therefore, they can be employed to carry out very good separations of inorganic ions of different crystalline radii. However, ion-exchange studies with crystalline materials are often complicated by the formation of new crystalline phases and by hysteresis phenomenon; therefore, amorphous exchangers can be preferred for some particular uses.

As the literature shows, a large number of the inorganic ion exchangers prepared are amorphous in nature. Perhaps the reasons are the easy methods of preparation of such materials and their granular nature, which is suitable for column operations. They can be prepared[2] in the form of granules simply by combining two or more salts, forming gelatenous precipitates at room temperature or at a slightly higher temperature, with or without refluxing. These granules can be obtained in a range of mesh sizes. The number of possible combinations is very large, and most of these combinations have been prepared and their ion exchange properties studied,[3] particularly by the research groups of Qureshi,[4] Rawat,[5] and Varshney[6] at Aligarh (India) and of Tandon[7] at Roorkee (India). As a result, these materials have acquired a prominent position in the field of separation science at present. As a group, they are known to have a great selectivity toward the heavier alkali and alkaline earth metal ions, particularly cesium. They have an analytical potential for the recovery and concentration of

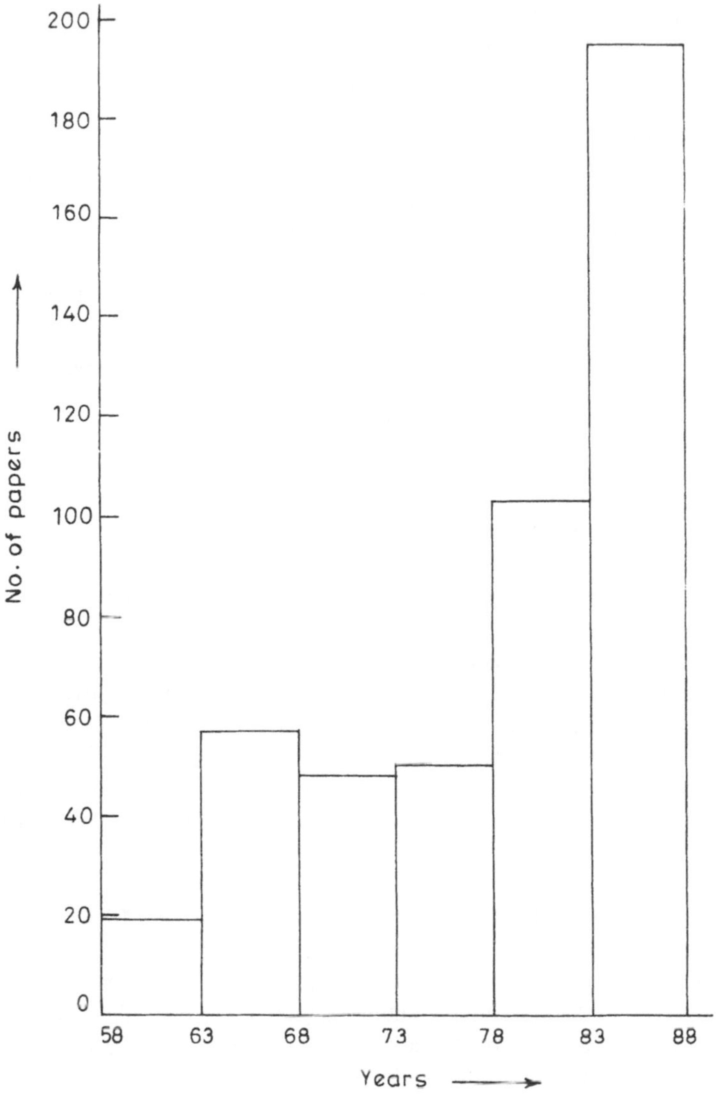

FIGURE 1. Research progress in the field of inorganic ion exchangers from the time of its infancy.

strongly adsorbed trace constituents, which has made their study more interesting. Figure 1 indicates an ever-increasing trend toward research in the field of inorganic ion exchangers from the time of its infancy. The main efforts have been on the synthesis and characterization of physically and chemically stable ion exchangers, reproducible in behavior and selective for certain metal ions.

Insoluble polybasic acid salts of multivalent metals have shown a great promise in preparative reproducibility, ion-exchange behavior, and both chemical and thermal stability. Many metals have been used for preparing materials such as aluminum, antimony, bismuth, cerium, chromium, cobalt, iron, lead, magnesium, niobium, tin, tantalum, titanium, thorium, tungsten, uranium, and zirconium. Also, a greater number of anionic species have been used to form precipitates such as phosphate, tungstate, molybdate, arsenate, antimonate, silicate, tellurate, tellurite, ferrocyanide, vanadate, arsenophosphate, arsenotungstate, arsenomolybdate, arsenosilicate, arsenovanadate, phosphotungstate, phosphomolybdate, phosphosilicate, phosphovanadate, molybdosilicate, and vanadosilicate.

The majority of the materials has been prepared by combining salts of zirconium, titanium, tin, niobium, and tantalum with any of the anionic species mentioned above. Other metals produce materials which, in general, are not very satisfactory in stability and ion-exchange characteristics.

In view of the above, the intent of the present chapter is to highlight the advances in the field of amorphous inorganic ion exchangers, with particular reference to their synthesis, characterization, and analytical applications. To be precise, only the insoluble polybasic acid salts of mutivalent metals have been selected for such a review.

II. SYNTHESIS AND CHARACTERIZATION

The insoluble acid salts of multivalent metals possessing ion-exchange properties can be produced, in general, as gelatinous precipitates by mixing rapidly the elements of groups III, IV, V, and VI of the periodic table, usually at room temperature. Sometimes, refluxing is recommended to improve their reproducibility and ion-exchange characteristics. The pH is adjusted to 0 to 1 with the help of either an acid or a base. The precipitate settles slowly to form a gelatinous cake which shrinks on drying and cracks to give a granular product. When immersed in water, the granules break down along strain lines, with the release of air bubbles which have been trapped in the gel, accompanied by heat. These granules are stable physically in water, and may be obtained in a range of mesh sizes from several millimeters downward. The method described is a general method of preparation of the amorphous ion exchangers. Only marginal differences exist in the various methods of preparation given in the literature, and they are in terms of the concentration of the mixing solutions, initial pH of the component solutions, and temperature of mixing.

Characterization of amorphous ion exchangers is mainly based on their physical and physicochemical properties, and on ion-exchange characteristics. These will be discussed individually.

A. PHYSICAL AND PHYSICOCHEMICAL CHARACTERIZATION

This includes appearance, such as the shape, size, and color of the particles produced. Granular shape is preferred over the powdered one when column operations are of primary importance. Similarly, the size of the granules plays an important role in achieving separations and in adjusting the flow rate of the column in a chromatographic process. Uniformity in size can be achieved by sieving into different mesh sizes, usually ranging from 25 to 400. The efficiency of a column is very much affected by the particle size because it governs the number of theoretical plates. The color of a material is helpful in visualizing the presence of various species on its surface, thus giving some important clues regarding the chemical properties of the substance. Hardness, opaqueness, transparency, shininess, etc. are generally observed for a material in order to ascertain its usefulness under certain specific conditions. Physicochemical methods include X-ray diffraction (in the case of crystalline materials), IR, and TGA/DTA, which give a great deal of information.

B. ION-EXCHANGE CHARACTERIZATION

The real utility of an ion exchanger depends largely on its ion-exchange characteristics. Ion-exchange capacity, concentration and elution behavior, pH titrations, and distribution behavior are some of the properties which constitute the ion-exchange characteristics of a material.

1. Ion-Exchange Capacity

Ion-exchange capacity, expressed generally as equivalents per dry gram or equivalents

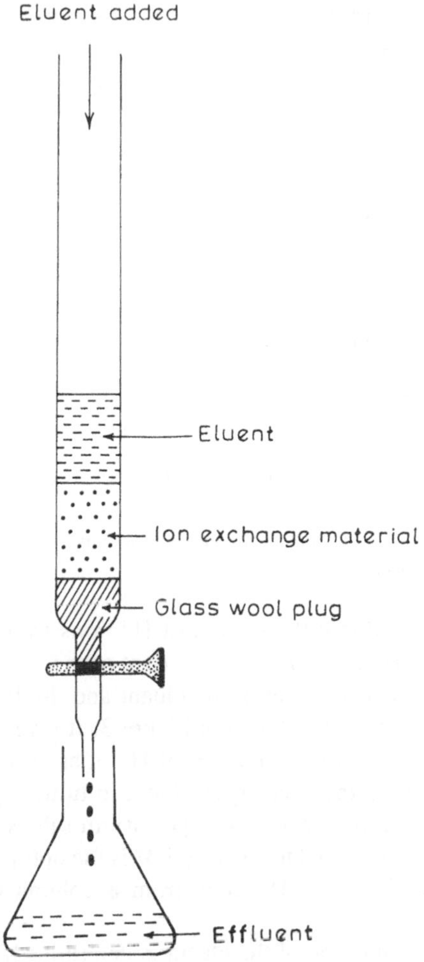

FIGURE 2. A typical ion-exchange column.

per kilogram, is the measure of the number of replaceable H^+ ions per unit mass of the exchanger. This property is determined generally by the nonequilibrium process i.e., using a column. A column is prepared by carefully packing the solid material (small particles) in a glass tube, usually by adding it to a column filled with solvent or by pouring a slurry of it into the column and allowing this to settle. The column can be mechanically vibrated or the solid material tamped with a long plunger during packing. Care must be taken to keep out air bubbles or channeling will result in the column, rendering it less effective. Figure 2 shows a typical column. A sintered glass frit or glass wool is placed in the bottom of the column to support the solid. A buret can be used as the column. A typical column may range from a few millimeters in diameter and a few centimeters in length to a few centimeters in diameter and several dozen centimeters in length. Preparative columns may have dimensions in feet and yards, but we are concerned with analytical columns. Preparative columns are used for the separation and purification of materials, which may range from a few grams or less to several pounds. Analytical columns, on the other hand, deal with much smaller quantities, usually in the milligram and submilligram range.

2. Concentration and Elution Behavior

Concentration and elution behaviors of an ion exchanger are also determined by the

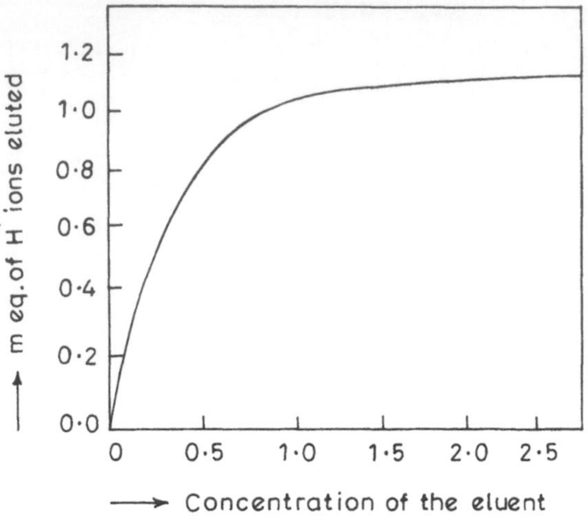

FIGURE 3. A hypothetical concentration curve of an inorganic ion exchanger.

column process. It is observed that the elution of H^+ ions from a column depends on the concentration of the eluant. Hence, for a fixed amount of the ion-exchange material, a curve can be drawn between the concentration of the eluant and the H^+ ions eluted out, keeping the rate of elution to a minimum. As shown in Figure 3, it gives an optimum concentration of the eluant necessary for a maximum elution of H^+ ions. It depends upon the nature of the ionogenic groups present in the exchanger which, in turn, depends upon the pK_a values of the acids used in its preparation. Sodium or potassium salts such as chlorides or nitrates are taken as eluants, and it is observed that nearly 1 *M* is the optimum concentration generally observed for a maximum elution of H^+ ions from a column containing 1 g of the ion exchanger.

Once the optimum concentration of the eluant is determined, it is necessary to ascertain the optimum volume of the eluant for a complete elution of the H^+ ions from a column of the exchanger. This is achieved by collecting the eluant in equal volumes in several test tubes (for example, 5-ml increments). If the flow rate is constant, then samples may be collected at equal time intervals. Automatic fraction collectors can be used for this purpose. After the various fractions are collected, they are analyzed for the H^+ ions by simple titration, and the amount or concentration in each tube is plotted as a function of the tube number or volume of solvent collected. Figure 4 represents such a curve. It gives an idea regarding the maximum volume of the eluant required for almost a complete removal of H^+ ions from a specific column, and thus the efficiency of the column.

3. pH Titrations

Acid salts of multivalent metals are acidic in nature and act as cation exchangers. They can, therefore, be titrated against an alkali, as usual, which gives the nature (weak or strong) and the number of exchange sites present in the ion exchanger. An alkali hydroxide is used to neutralize the protons and allow the reaction to go to completion, and a decinormal salt solution of the same alkali metal is used as a supporting electrolyte. A graph plotted between the number of milliequivalents of OH^- ions added and the resultant pH of the mixture at equilibrium is termed the "pH titration curve" or the "potentiometric titration curve". The functionality of an ion exchanger can also be determined with the help of these curves, as illustrated in Figure 5. Potentiometric titrations are performed by the usual batch process.[8] It is a general observation that the shape of the titration curve depends upon the time of

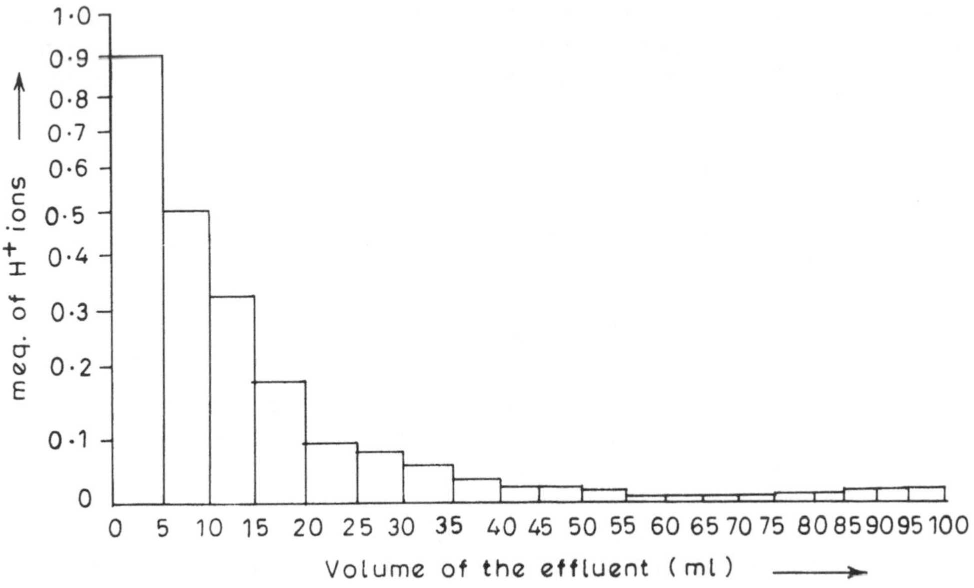

FIGURE 4. A hypothetical elution curve for an inorganic ion exchanger.

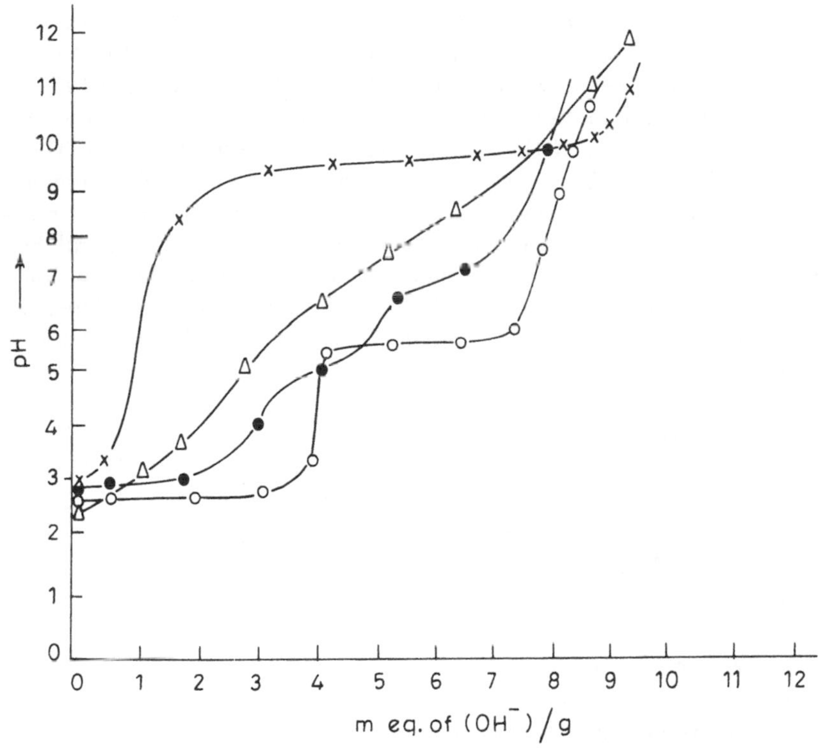

FIGURE 5. Some hypothetical potentiometric titration curves of insoluble acid salts of multivalent metals as inorganic ion exchangers. True crystalline and bifunctional (-○-○-○-); semicrystalline or poorly crystalline acid salt with a trifunctional behavior (-●-●-●-); monofunctional and weak acid salt (-△-△-△-); and an acid salt with an undefined (amorphous) structure (-x-x-x-).

contact of the exchanger beads with the solution containing OH^- ions, until an equilibrium is obtained, which sometimes takes several days if the system is put at room temperature with intermittent shaking. The equilibration time can be reduced, however, if a constant shaking device is used.

4. Distribution Behavior

The distribution coefficient (K_d) is the measure of a fractional uptake of metal ions competing for H^+ ions from a solution by an ion-exchange material. It is defined as follows:

$$Kd = \frac{I - F}{F} \times \frac{V}{W} \text{ ml/g}$$

where I is The total amount of metal ions initially in the solution, F is the amount of the metal ions left in the solution after equilibrium, V is the volume of the solution, and W is the weight of the exchanger. The initial concentration of the electrolyte solution is adjusted such that it does not exceed 3% of the total exchange capacity of the exchanger taken.[9] K_d is an important factor for determining the analytical potential of an ion exchanger. This, along with the separation factor, α, point to the possibility of a particular separation on a column of the exchanger.

Varshney and co-workers[10] made a systematic pH titration and distribution studies of alkali metal ions on amorphous tin(IV) and chromium(III) arsenophosphate.[11] Tin(IV) arsenophosphate was prepared by mixing 0.1 *M* solutions of Sn(IV) chloride, sodium arsenate, and trisodium orthophosphate in the volume ratio 1:1:1. The pH of the mixture was kept at zero with HNO_3. The gel thus prepared was kept as such at room temperature for 24 h and then filtered off. It was washed several times with distilled water and dried at 40°C. The dried material, when placed in distilled water, cracked into small granules. They were then put in successive batches of 1 *M* HNO_3 to convert them into the H^+ form. The material was finally washed with DMW to remove excess acid and dried at 40°C. Granules of the required size were obtained by sieving. Chromium(III) arsenophosphate was prepared by mixing 0.1 *M* solutions of Cr(III) nitrate, sodium arsenate, and trisodium orthophosphate in the volume ratio 2:1:1. The rest of the procedure was as described above. The sodium ion-exchange capacities of Sn(IV) and Cr(III) arsenophosphates determined by column operation were found to be 1.75 and 0.74 meq/dry g, respectively. Potentiometric titrations were performed as follows.

a. Tin(IV) Arsenophosphate

Five hundred mg of the exchanger was mixed with 100 ml of a 0.015 *M* salt solution (NaCl, KCl, LiCl, or NH_4Cl). The mixture was kept for 1 h and titrated against 0.015 *M* solutions of the respective alkali, recording the pH of the solution after each 0.4 ml addition of the titrant until the pH became constant. The back titration was then carried out by adding the same fractions of 0.015 *M* HNO_3 to the solution.

b. Chromium(III) Arsenophosphate

The method followed was as above, except that the concentration of the salt and alkali solutions was taken as 0.008 *M*. The back titration was performed by adding 0.008 *M* HNO_3. The studies reveal that the uptake of the cations by the host lattice depends on the nature of the counter ions, and that there are inversions in the selectivity order. For instance, in the region between pH 3 and 6.5, tin(IV) arsenophosphate shows the selectivity sequence $NH_4^+ > Na^+ > K^+ > Li^+$, but on increasing the pH (6.5 to 7.5), it changes to $K^+ > Na^+ > NH_4^+ > Li^+$. A reversal in selectivity is again observed beyond pH 7.5, i.e., $NH_4^+ > Na^+ > K^+ > Li^+$. The behavior is quite common in amorphous and poorly crystalline ion exchangers. The amorphous materials have a range of different-sized cavities and, thus, steric effects can explain the change in the selectivity. Surprisingly, Cr(III) arsenophosphate shows the selectivity order $K^+ > NH_4^+ > Na^+ > Li^+$ at all pH values. This observation, together with a very low uptake of Li^+ ion by both materials, indicates that the ion exchange occurs by the diffusion of the hydrated counter ion into the ion exchanger (hydrated radii $Li^+ > Na^+ > K^+$). This also explains the high uptake of K^+ ion shown both in the pH titration curves and in the determination of distribution coefficients in various acidic media.

The pH titration curves obtained by the equilibrium process show a negligible hysteresis loop (Figures 6 and 7), while those obtained by the nonequilibrium process show a quite prominent one. This indicates that in the former case the ion exchange process is reversible, while in the latter case it is not.

5. Correlation Between the K_d Values and the Basic Properties of the Ion Exchangers

Since K_d values depend on the basic properties of the ion exchanger, it is useful to obtain a correlation of the basic properties, such as ion-exchange capacity, ionic radius, atomic number, and ionic charge, with the K_d values. An effort was made by Sunandamma[12] in this direction by selecting some amorphous metal antimonates as ion exchangers. The antimonates selected were chromium(III) antimonate, titanium(IV) antimonate, tin(IV) antimonate, zirconium(IV) antimonate, niobium(V) antimonate, cerium(IV) antimonate, iron(III) antimonate, lead(II) antimonate, and nickel(II) antimonate. On the basis of such a study, some general characteristics can be noted as follows:

1. Na^+ has the maximum average K_d for the antimonate exchangers, compared to the other metal ions. Thus, these ion exchangers can be used for the separation of Na^+ from K^+, Li^+, Rb^+, and Cs^+. This has been done by Abe et al.[13,14]
2. Antimonates have high K_d values for Ag^+, Hg^{2+}, and Pb^{2+}. Very low K_d values are reported for Mg^{2+}. A comparison shows that among the four alkaline earth ions, Mg^{2+} is the least adsorbed, while Ba^{2+} is the most adsorbed. Crystalline material shows a regular trend in this respect, but the amorphous one shows an erratic behavior.

Similar trends are noticed from the plots of log K_d vs. ionic radii. Thus, as the ionic radius increases, there is a regular increase in K_d values. This is because the ions are exchanged as hydrated ions, and the ions with the lowest ionic radii have the largest hydrated ionic radii. The trend on niobium, chromium, titanium, iron, and lead antimonates are similar.

The ion-exchange capacity depends upon two factors: (1) hydrated ionic radii and (2) selectivity. As the hydrated ionic radius increases, the ion-exchange capacity decreases because the exchange now becomes more difficult. Thus, for the alkali metals, the trend is $Cs^+ > Rb^+ > K^+ > Na^+ > Li^+$. This is observed on cerium, nickel, niobium, titanium, and, to some extent, lead antimonates.

The second factor is the specific selectivity of antimonic acid. Antimonic acid is known to be highly selective for Na^+ ions. This effect is apparent in stannic antimonate, zirconium antimonate, and antimonic acid. For titanium antimonate,[15] the authors state that except for Li^+ and Ba^{2+}, the ion-exchange capacity is negligibly affected by the size and charge of the exchanging ion. This does not appear to be true. Further, the authors state that "As the mixing ratio of antimony to titanium increases, the exchange capacity also increases. This becomes maximum at antimony to titanium ratio of 4:1." This is understandable, since antimonic acid loses its ion-exchange capacity for cations on partial replacement by metal ions. This replacement is minimum at the high Sb/Ti ratio of 4.0. A plot of ion-exchange

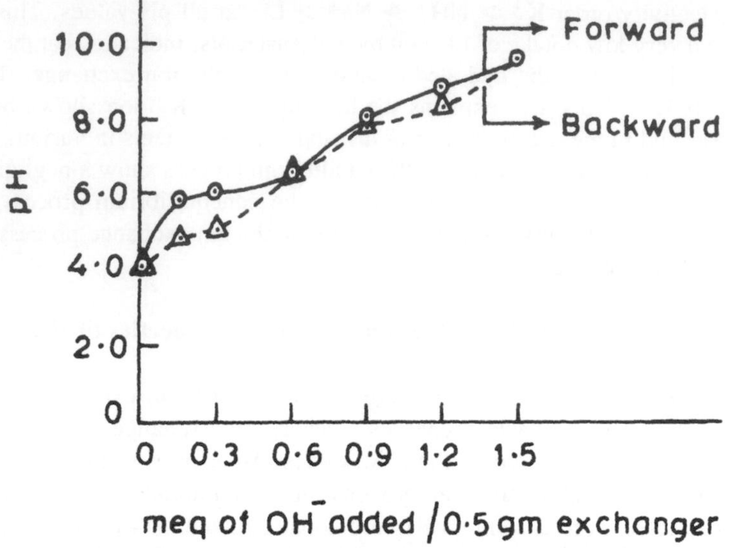

FIGURE 6. Forward and reverse pH titration curves on Sn(IV) arsenophosphate for Na^+ ions under equilibrium condition. (From Varshey, K. G. and Khan, A. A., *J. Inorg. Nucl. Chem.*, 41, 241, 1979. With permission.)

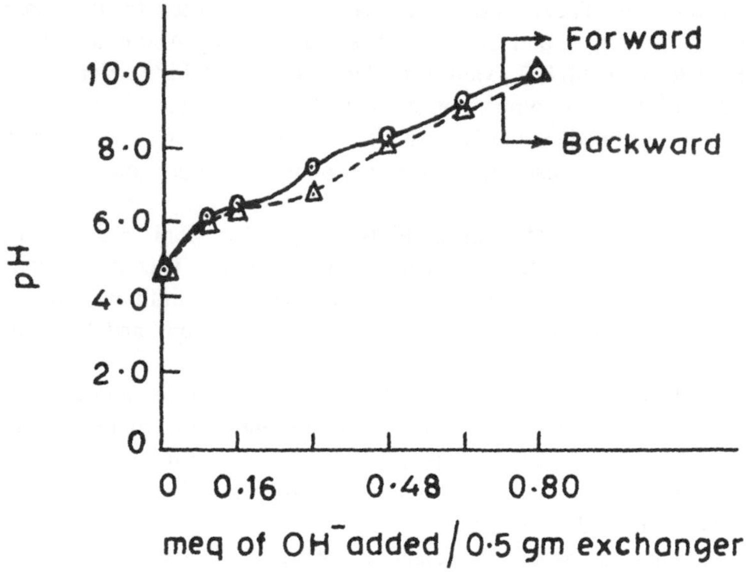

FIGURE 7. Forward and reverse pH titration curves on Cr(III) arsenophosphate for Na^+ ions under equilibrium condition. (From Varshey, K. G. and Khan, A. A., *J. Inorg. Nucl. Chem.*, 41, 241, 1979. With permission.)

capacity vs. hydrated ionic radii shows that the ion-exchange capacity decreases as the hydrated ionic radius increases. This is a normal behavior.

Zirconium antimonate synthesized by Mathew and Tandon[16] shows a higher exchange capacity for Na^+ ions than for K^+ and Li^+ ions, as in the case for antimonic acid. This is probably because the Sb/Zr ratio is 1.03. Abe and Ito[17] have observed a normal selectivity for alkali metals on the zirconium antimonate which they had synthesized. They have further showed that if the Sb/M ratio is greater than unity, then the unreacted antimonic acid is present in the antimonate. This is in agreement with the data of Mathew and Tandon[16] on zirconium antimonate.

On stannic antimonate, the ion-exchange capacity is maximum for Na^+ ions, and, in fact, the order of selectivity is the same as for antimonic acid. This supports the observations of Abe and Ito.[17] The Sb/Sn ratio for stannic antimonate is 5.0 and, therefore, it contains a large amount of antimonic acid. The exchange of Sn^{4+} with H^+ ions on antimonic acid may be represented by the equation:

$$2H_2Sb_2O_5\,(OH)_2 + Sn^{4+} \rightleftharpoons SnSb_2O_5\,(OH)_2 + 4H^+$$

These observations are applicable to cerium, nickel, and niobium antimonates also.

This explanation, however, does not apply to amorphous and crystalline lead antimonates. The exchange reaction in this case will be

$$(H_3O)_2Sb_2O_6 \cdot 2H_2O + Pb^{2+} \rightleftharpoons Pb \cdot Sb_2O_6 \cdot 2H_2O + 2[H_3O^+]$$

On a complete replacement, the Sb/Pb ratio should be 2. Thus, if the Sb/Pb ratio is greater than 2, then some antimonic acid should be left unreacted and the material should show a selectivity for Na^+ ions. However, it appears that the crystal structure of antimonic acid is so rigid that after most of the protons have been blocked out by Pb^{2+}, the material no longer shows the selectivity associated with antimonic acid.

The charge on the exchanging ions also affects significantly the K_d values. Chromium antimonate shows the largest K_d values for the di-, tri-, and tetravalent metals, while cerium(III) antimonate shows the lowest. These curves for crystalline and amorphous lead antimonates are identical. As a rule, the K_d values should increase with the charge on the ion. This trend is shown by zirconium and titanium antimonates. However, in most cases, the average K_d value first decreases and then increases. This may be due to the two contradictory factors, i.e., (1) an increase in charge increases the attraction for the cation and (2) an increase in the hydrated radii decreases the initial attraction of the ion exchanger for the cations.

6. Selectivity for Metal Ions

Inorganic ion exchangers are selective for certain ions. High adsorption is observed in the case of antimonates. Titanium[18] and stannic[19] antimonates are selective for alkaline earths, while zirconium antimonate[20] is selective for alkali metals. The order of adsorption on zirconium phosphate and organic cation exchange resins for some uni-, bi-, and trivalent metal ions follows Hofmeister's lyotropic row.

1. $Li < H < Na < NH_4 < K < Rb < Cs < Ag < Tl$
2. $Mn < Mg < Zn < Co < Cu < Cd < Ni < Ca < Sr < Pb < Ba$
3. $Al < Sc < Y < Eu < Sm < Nd < Pr < Ce < La$

The adsorption sequence for some metal ions on tantalum antimonate[21] has been found to be as follows:

At pH 5.4

$Cd > Zn > Cu > Ni > Co > Ba > Sr > Mg > Mn > Ca$; $Nd > Sm > Y >$
$La > Al > Ce > Pr$

At pH 3.4
Cd > Zn > Cu > Ca > Ni > Co > Sr > Ba > Mn > Mg; Nd > Sm > Y > La > Al > Ce > Pr
At pH 1.2
Ni > Co > Cu > Cd > Sr > Zn > Mn = Ba = Ca = Mg; Nd > Y > Al > La > Ce > Pr > Sm
At pH 0.2
Ba > Co > Cd > Ni = Cu = Zn = Mn = Sr = Ca = Mg; Pr > La > Sm > Nd > Y > Al = Ce

In most cases, there is only a light change in the molar distribution coefficients, λ, as the pH is altered from 5.5 to 2.5. At pH 2.5, the concentration of hydrogen ions is very small and the cations compete successfully for the exchange sites. When the pH is changed from 2.5 to 0.2, there is an increase in the hydrogen-ion concentration and a decrease in ionized exchange sites; hence, the hydrogen ions compete successfully and there is a sudden decrease in the λ value.

Abe and Hayashi[22] prepared tin(IV) antimonates (SnSbA) under various conditions and studied the changes in their selectivities for alkali metals. They observed that the ion-exchange reaction of the alkali metal ions, except for lithium ions, is relatively fast, about 10 h being required for the attainment of equilibrium. The reaction in the case of lithium ions is very slow, about 10 d being required. The behavior does not depend on the samples prepared under different conditions. The equilibrium K_d values of alkali metal ions in 0.01 *M* HCl are plotted as a function of the Sb/Sn ratio in the sample shown in Figure 8. The selectivity sequence showed Na < K < Rb < Cs << Li for 10^{-4} *M* of alkali metal ions over the entire range studied. The K_d values of alkali metal ions, except lithium ions, increase with increasing Sb/Sn ratio, while the K_d values of lithium ions show a maximum value at about 1.7 of the Sb/Sn ratio. The separation factor α^{Li}_{Na} calculated from the K_d values of lithium and sodium ions shows a maximum value at an Sb/Sn ratio of 1.4. The K_d values are also plotted (Figure 9) as a function of the hydrolysis temperature of the SnSbA. The K_d values of alkali metal ions, except lithium ions, are almost independent of the temperature of hydrolysis in the range studied, while the K_d values of lithium ions increase with the temperature. The highest K_d value of Li^+ ions in 0.1 *M* HCl was obtained on SnSbA hydrolyzed at 100°C and with an Sb/Sn ratio of 1.6. The equilibrium K_d values of alkali metal ions on a SnSbA are shown in Figure 10 as log-log plots of K_d vs. [HCl]. The slopes, d log K_d/d log [HCl], are −1 for all alkali metal ions, as expected for the ideal 1:1 ion-exchange reaction. This study shows an extremely high separation factor between lithium ions and other alkali metal ions on tin antimonate, compared to organic ion-exchange resins and other inorganic ion exchangers.

7. Stoichiometry of Exchange

The primary condition of an ion-exchange process is the stoichiometry. In organic resins, it is an established fact.[23] However, inorganic ion exchangers are not very good in this regard. Because of the inorganic nature of the matrix, they show an adsorption phenomenon in addition to the normal ion-exchange process on their surface. Thus, the uptake of ions is generally quite complicated on these materials. Gill and Tandon[24] made an attempt to study this aspect by selecting five ion exchangers: stannic ferrocyanide, zirconium ferrocyanide, ceric antimonate, titanic antimonate, and ceric tungstate. Cs^+, Hg^{2+}, and Tl^{3+} ions were chosen as the representatives of the different valence states. The exchangers were prepared by the reported methods,[25-29] and the samples selected were those having the highest exchange capacity and best mechanical strength.

^{134}Cs, ^{203}Hg, and ^{204}Tl radioisotopes were used to determine the distribution coefficients (Kd) by the following method. About 200 mg of the exchanger was loaded with 20 ml of ^{134}Cs, ^{203}Hg, or ^{204}Tl solutions. The initial pH of the tracer solution was adjusted with

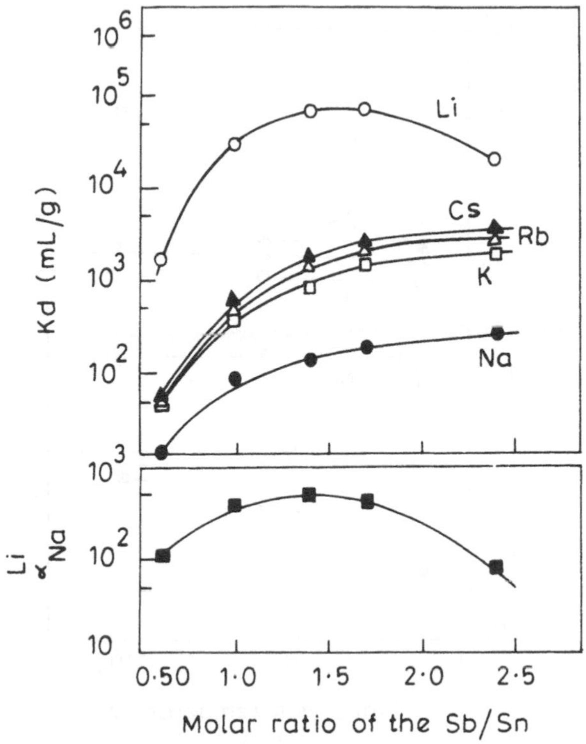

FIGURE 8. The K_d values of alkali metal ions and separation factors on tin antimonate with different Sb/Sn ratios. Initial concentrations = 10^{-4} M alkali metal ions in 0.1 *M* HCl. Tin antimonate hydrolyzed at 60°C = 0.10 g; total volume of solution = 10.0 ml. (From Abe, M. and Hayashi, K., *Solv. Extr. Ion Exchange*, 1, 97, 1983. With permission.)

perchloric acid. This was then shaken on a mechanical shaker to attain complete equilibrium. The K_d values were calculated using the formula:

$$K_d = \frac{I - F}{F} \times \frac{V}{A} \text{ (ml/g)}$$

where I is the initial activity of the cation, F is the final activity of the cation after equilibrium, A is the weight of the exchanger (g), and V is the volume of the cation solution (ml).

For the two ions, M^{n+} and H^+, competing for the exchanger, the equilibrium can be written as:

$$M^{n+} + m\overline{H}X \rightleftharpoons mH^+ + \overline{M}^{n+} (X^-)_n$$

In sufficiently dilute solutions where activity coefficients may be neglected, the expression may be written as

$$K_{HM} = \frac{[\overline{M}^{n+}]\ [H^+]^n}{[M^{m+}]\ [\overline{H}^+]^n}$$

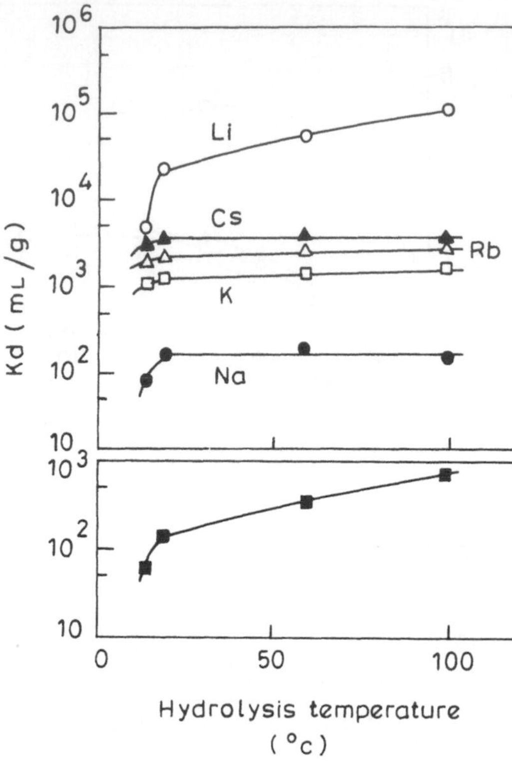

FIGURE 9. The K_d values of alkali metal ions and separation factors on tin antimonate with Sb/Sn ratio of 1.6 to 1.7. Initial concentration = 10^{-4} *M* alkali metal ions in 0.1 *M* HCl, exchanger = 0.10 g, total volume = 10.0 ml. (From Abe, M. and Hayashi, K., *Solv. Extr. Ion Exchange*, 1, 97, 1983. With permission.)

where $\overline{H}^+$, $\overline{M}^{n+}$ are the cations on the exchanger and H^+, M^{n+} are the cations in the solution.

If $[M^{n+}] \ll [H^+]$, as in tracer solutions of the ion M^{n+}, the variation in $[H^+]$ may be neglected, and we may write the distribution coefficient for M^{n+} as

$$K_{dM} = \frac{[\overline{M}^{n+}]}{[M^{n+}]} = K_{HM} \frac{[\overline{H}^+]^n}{[H^+]^n} \approx \frac{K}{[H^+]^n}$$

If the law of mass action is obeyed and the concentration of the metal ions is very small compared to H^+, a plot of log $[K_{dM}]$ versus $-\log [H^+]$ should be a straight line with a slope equal to n.

The plots of log K_d vs. pH for stannic ferrocyanide and zirconium ferrocyanide are linear, but the slopes do not really correspond to the valency of the ion exchanged. However, the slopes for Cs^+ approximate to unity. This deviation in behavior for the uptake of bivalent and tervalent ions may be due to the prominence of a mechanism other than ion exchange, like precipitation, surface adsorption, or simultaneous adsorption of anions. In the case of ceric antimonate, titanic antimonate, and ceric tungstate, the slopes are very close to 1, 2, and 3 for the sorption of Cs^+, Hg^{2+}, and Tl^{3+}, respectively. This suggests that for all practical purposes, the exchange on these three ion exchangers is stoichiometric in nature and obeys the law of mass action.

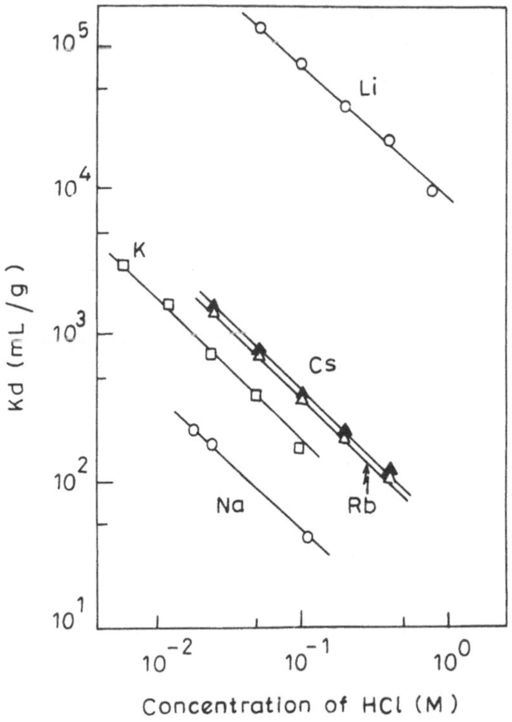

FIGURE 10. Plots of K_d of alkali metal ions against [HCl]. SnSbA (Sb/Sn = 1.6) hydrolyzed at 60°C = 0.10 g; initial concentration of alkali metal ions = 10^{-4} *M*. Total volume = 10.0 ml, equilibrated at 30°C. (From Abe, M. and Hayashi, K., *Solv. Extr. Ion Exchange,* 1, 97, 1983. With permission.)

8. Chemical Stability

Chemical stability of synthetic inorganic ion exchangers plays an important role in their analytical applications. Exchangers which have a high solubility in water, as well as in acidic media, may not be very useful for separation studies. It is, therefore, advisable to have a rough guide of the solubility of an ion exchanger. Qureshi and Thakur[30] made an effort in this direction and proposed a series for the solubility of synthetic inorganic ion exchangers. They plotted the solubilities of various inorganic ion exchangers in different solvents. The solvents selected were water and acid solutions of different concentrations because they are mostly used as eluants in column operations. On the basis of this analysis, the following sequence of solubility has been found: molybdates > arsenates > selenites > antimonates > phosphates ≥ tungstates. Only 3 of the 16 pairs of exchangers tested did not follow this trend. Thus, the proposed stability sequence is in 80% agreement with the literature data available. The abovc series suggests that, in general, tungstates are the most stable materials, while molybdates are the least stable, of the inorganic ion exchangers tested.

9. Radiation Stability

It is generally believed that inorganic ion exchangers are resistant to radiation. However, very few reports have appeared in the literature confirming this belief. Varshney et al.[31] synthesized some ion-exchange materials based on Si, P, and As which show promising ion-exchange behavior and high thermal and chemical stability. Their radiation stability was studied by varying the total dose of γ radiation from 10^8 to 3×10^8 rad obtained from a Co^{60} source and observing the effect on their ion-exchange properties. Zirconium(IV) arsenosilicate,[32] zirconium(IV) arsenophosphate (a heated phase),[33] tin(IV) arsenosilicate,[34] thorium(IV) phosphosilicate,[35] and antimony(V) silicate[36] were selected for such a study. The studies revealed that the materials are highly resistant to radiation as far as their ion-exchange capacity and elution behavior are concerned, except thorium(IV) phosphosilicate, which shows a slight variation in these properties with an increase in the absorbed dose. The organic counterpart undergoes about a 15% loss in its ion-exchange capacity (IEC) on such treatment. Also, no appreciable change has been observed in its infrared spectra, indicating no significant structural changes. pH titration curves, however, indicate a change on irradiation. The increase in pH is sharper for the irradiated samples containing the arsenate group (zirconium arsenosilicate, zirconium arsenophosphate and tin(IV) arsenophosphate) than for the normal samples. Thorium(IV) phosphosilicate and antimony(V) silicate, however, do not show any significant change in this behavior.

An interesting feature is observed in the distribution behavior of these materials on irradiation. The K_d values obtained in different media for four alkaline earths generally increase with the total dose absorbed. Also, a complete reversal in K_d values is observed in zirconium arsenosilicate. It shows a total adsorption of alkaline earths in DMW after irradiation, while on a normal sample, these metals have low K_d values.

Tandon et al.[37,38] made such a study on stannic ferrocyanide, zirconium ferrocyanide, ceric antimonate, titanic antimonate, ceric tungstate, and zirconium arsenophosphate. They observed that the exchange capacity of all these materials remains practically the same up to a γ dose of 10^9 rad. A comparison shows that Dowex-50Wx8 and Amberlite-IR-120 lose 10% of their IEC at a γ dose of 4×10^7 rad and about 25% at 10^9 rad. The solubility of these inorganic ion exchangers also remains unaffected, while organic resins develop some leaching tendency in a neutral salt solution. Similarly, the K_d values of inorganic materials are unaffected, while there is an appreciable decrease in the case of organic resins.

Zirconium(IV), tin(IV), and chromium(III) antimonates appear to have a different behavior under γ radiation. Mathew et al.[39] made a systematic study on these antimonates. Irradiation was performed with gamma rays for 147 h in the Hungarian reactor after shutdown. The absorbed dose was about 9×10^8 rad. This irradiation has a marked influence on the exchange characteristics of these antimonates. Zirconium antimonate loses 36.6% of its IEC, while tin and chromium antimonates lose 16.8 and 31.2%, respectively. Distribution coefficients decrease on irradiated samples, compared to the unirradiated ones. However, chemical stability still holds good for irradiated exchangers. The increased solubility of the anionic part of the exchanger, compared to the cationic part, implies a possible damage of Sb-OH groups on the surface. The damage of these exchangers at a high γ dose of $\simeq 10^9$ rad, however, does not rule out the possibility of their use for processing lower levels of radioactivity. The type of gamma radiation may have an influence on the damage imparted to these materials. Irradiation in a nuclear pile is a mixed exposure of γ radiation and neutrons, while a Co^{60} source is pure γ radiation. Dowex-50 irradiated in a nuclear pile is fully decomposed, while it loses only 15% of its initial IEC when irradiated in a Co^{60} source.

The kinetics of exchange on irradiated samples with 0.02 and 0.1 *N* RbCl solutions at 30°C shows the inability of irradiated zirconium antimonate to exchange for Rb^+, which indicates extensive superficial and internal damage. However, on irradiated tin(IV) antimonate, the rate of exchange at a 0.02 *N* solution is faster than that on an unirradiated one.

It can be concluded, therefore, that a generalization about the good resistance of inorganic exchangers against γ rays and/or charged particles cannot be made, as emphasized by Zsinka et al.[40]

C. THERMAL, REDOX, AND CATALYTIC CHARACTERIZATION

1. Thermal Behavior

Because of their unique structural and chemical properties, synthetic inorganic ion exchangers of the class of acid salts of multivalent metals may have practical end uses not only as ion exchangers or intercalating lattices, but also as catalysts. Knowledge of the thermal behavior of the wide number of known phases is, therefore, of basic importance for understanding their behavior either as ion exchangers or as catalysts. Hence, thermogravimetry must play an important role in characterizing these materials. UV, visible, IR, and EXAFS spectroscopies are also prerequisites for completing such a study.

Generally, heating causes a change in the crystalline structure, stereochemistry, and redox behavior and/or in the acid strength of inorganic ion exchangers. Therefore, there may be an immense change in other properties, such as the catalytic activity of these materials, with a change in the pretreatment temperature.

The thermal behavior of the hydrogen phases of these ion exchangers can be explained schematically as follows:

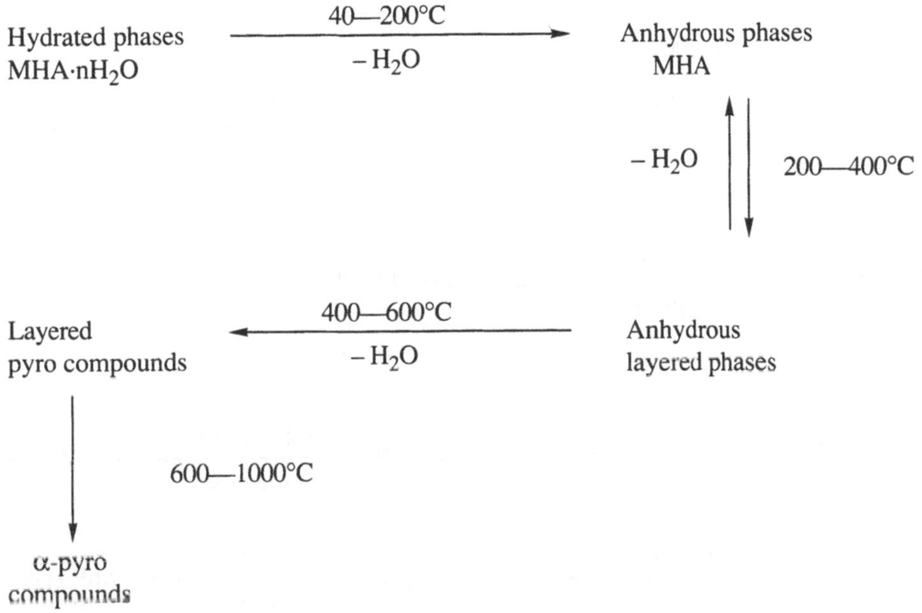

Since the interlayer distance of these compounds depends on the water content (cavity water) and the ionic radius of the exchanged ion, the dehydration process involves a decrease in the interlayer distance value. As an example, the results[41] of the interlayer distance variations on heating several α phases of Ti, Zr, and Sn acid phosphates or arsenates are given below:

$$\alpha\text{-Tip}\cdot H_2O\ (7.58) \xrightarrow[-H_2O]{170\text{—}250^\circ C} \alpha'\text{-TiP}\ (7.33) \underset{290^\circ C}{\overset{260\text{—}310^\circ C}{\rightleftharpoons}} \alpha''\text{-TiP}\ (7.10)$$

$$\alpha\text{-TiAs}\cdot H_2O\ (7.77) \xrightarrow[-H_2O]{60\text{—}200^\circ C} \alpha'\text{-TiAs}\ (7.59) \underset{230\text{—}200^\circ C}{\overset{200\text{—}230^\circ C}{\rightleftharpoons}} \alpha''\text{-TiAs}\ (7.21)$$

$$\alpha\text{-ZrP}\cdot H_2O\ 10/100\ (7.59) \underset{140^\circ C}{\overset{135\text{—}150^\circ C}{\rightleftharpoons}} \alpha'\text{-ZrP}\cdot H_2O\ (7.43)$$

(10/100) 100—200°C

$$\alpha\text{-ZrP}\cdot H_2O\ (7.59) \xrightarrow[-H_2O]{(dp)\ 120^\circ C} \alpha'\text{-ZrP}\ (7.43) \underset{200^\circ C}{\overset{200^\circ C}{\rightleftharpoons}} \alpha'''\text{-ZrP}\ (6.8)$$

$$\alpha\text{-ZrAs}\cdot H_2O\ (7.72) \underset{135^\circ C}{\overset{120\text{—}135^\circ C}{\rightleftharpoons}} \alpha'\text{-ZrAs}\cdot H_2O\ (7.49)$$

$$\alpha\text{-ZrAs}\cdot H_2O\ (7.75) \xrightarrow[-H_2O]{70\text{—}150^\circ C} \alpha''\text{-ZrAs}\ (7.49) \xrightarrow[\text{irreverse}]{160\text{—}190^\circ C} \alpha'''\text{-ZrAs}\ (6.2)$$

$$\alpha\text{-SnP}\cdot H_2O\ (7.75) \xrightarrow[-H_2O]{65\text{—}170^\circ C} \alpha'\text{-SnP}\ (7.47) \xrightarrow[-H_2O]{60\text{—}25^\circ C} \alpha''\text{-SnP}\cdot H_2O\ (7.7)$$

$$\alpha\text{-SnAs}\cdot H_2O\ (7.9) \xrightarrow[-H_2O]{80\text{—}150^\circ C} \alpha'\text{-SnAs}\ (7.65) \underset{300\text{—}180^\circ C}{\overset{200\text{—}300^\circ C}{\rightleftharpoons}} \alpha''\text{-SnAs}\ (7.33)$$

A phase transition of the anhydrous form, which generally involves a decrease of the interlayer distance, has been observed for Ge, Ti, and Zr acid phosphates and for Ti, Zr, and Sn arsenates. The lower the electronegativity of the tetravalent metal, the lower the temperature at which this phase transition occurs, e.g., it occurs (reversibly) at 370, 290, and 200°C for Ge, Ti, and Zr acid phosphates, respectively, and a similar trend is observed for Ti and Sn arsenates.

Given the relatively marked decrease in the interlayer distance caused by these phase transitions and the slow kinetics of the dehydration process, if the increasing temperature rate is not appropriate (2 to 5°C/min or higher), the dehydration process often is not complete when the anhydrous transition phase occurs, and the water still present in the inner part of the exchanger becomes trapped and can be evolved only at higher temperatures (340 to 450°C). Care must therefore be taken when a completely anhydrous phase is required, in order to prepare a compound with appropriate acid properties.

Temperatures as high as 800 to 900°C (or 1200°C for the highly crystalline materials) are required for the formation of cubic α-AP_2O_7, as indicated by the exothermic peaks in the DTA curves.

The thermal behavior of HNa[42] or Ag phases[43] shows that after the dehydration processes occurring below 200°C (cavity water) and between 450 and 600°C (condensation water), both indicated by endothermal effects, at 600 to 700°C the amorphous phases always give $NaZr_2(PO_4)_3$[42] or $AgZr_2(PO_4)_3$[43] independently of Na or Ag content. Glassy Ag or Na phosphates (meta- or pyro- are also formed, as shown by the cooling DTA curves obtained for the AgZrP system.[43]

Qureshi and co-workers[44] conducted dehydration studies of a new phase of amorphous niobium(V) phosphate and gave a theoretical interpretation of the heating effect. They observed that at 100°C, six water molecules attached to HPO_4 groups of the niobium(V) phosphate are lost. Up to 200°C, 1 to 12 more water molecules are irreversibly removed from the cavity, resulting in cavity shrinkage and a negligible effect on the IEC. Up to 350°C, the remaining absorbed (probably coordinated to niobium) water molecules are lost. The condensation of the phosphate groups starts above 400°C, and up to 550°C, two water molecules are lost as a result of the intermolecular condensation of four phosphate groups:

$$\left(HPO_4^{2-}\right)_2\left(HPO_4^{2-}\right)_4 \xrightarrow[550^\circ C]{-2H_2O} \left(HPO_4^{2-}\right)_2\left(P_2O_7^{4-}\right)_2$$

At 650°C, dimer formation starts due to intramolecular condensation of the monomers, and 0.5 H_2O molecules per formula weight is lost from one of the HPO_4^{2-} groups. Thus, the IEC is reduced to half:

$$2\Big[\underset{\text{Monomers}}{(HPO_4^{2-})_2(P_2O_7^{4-})}\Big] \xrightarrow[650^\circ C]{-H_2O} \underset{\text{Dimer}}{(HPO_4^{2-})(P_2O_7^{4-})_5}$$

Above 700°C, two dimers combine to give a tetramer. The condensation is complete up to 800°C, and at this stage, all the exchange sites are exhausted:

$$2\Big[\underset{\text{Dimers}}{(HPO_4^{2-})_2(P_2O_7^{4-})_5}\Big] \xrightarrow[700^\circ C]{-2H_2O} \underset{\text{Tetramer}}{(P_2O_7^{4-})_{12}}$$

The condensation process in this way agrees with IR, X-ray, TGA and IEC data, and the niobium-to-phosphorous mole ratio remains constant at all the temperatures studied.

2. Redox Behavior

These studies have been carried out quite extensively[41] for crystalline materials. The transition metal ion (TMI) phases of the compounds used for such a study are very important if these materials are to be employed as catalysts for oxidation or reduction processes.

The reduction of TMI in zirconium phosphate has been observed only in the case of Cu^{2+}, Ag^+, and VO^{2+} phases, while for the Mn^{2+}, Co^{2+}, Ni^{2+}, and Zn^{2+} forms, no reduction has been detected. Only slight changes to unknown phases have been found at 800°C with H_2 for the Mn and Co forms.

If the layered structure is maintained during reduction ($T \leq 400°C$), the oxidation again gives the starting oxidized phases; if the TMI is reduced in an unlayered phase (sample preheated at $T > 600°C$), the process is irreversible.

Even for the layered phases, care must be taken to avoid increases in metal crystallite formation, which is favored by high temperatures or by the duration of the thermal treatments. For example, the larger the crystallites of Cu or Ag, the less reversible is the reoxidation process to the starting phases. A similar behavior has been observed in the case of Cu and Ag zeolites.[45,46] The overall processes and their steps are schematized as follows:

Layered phases

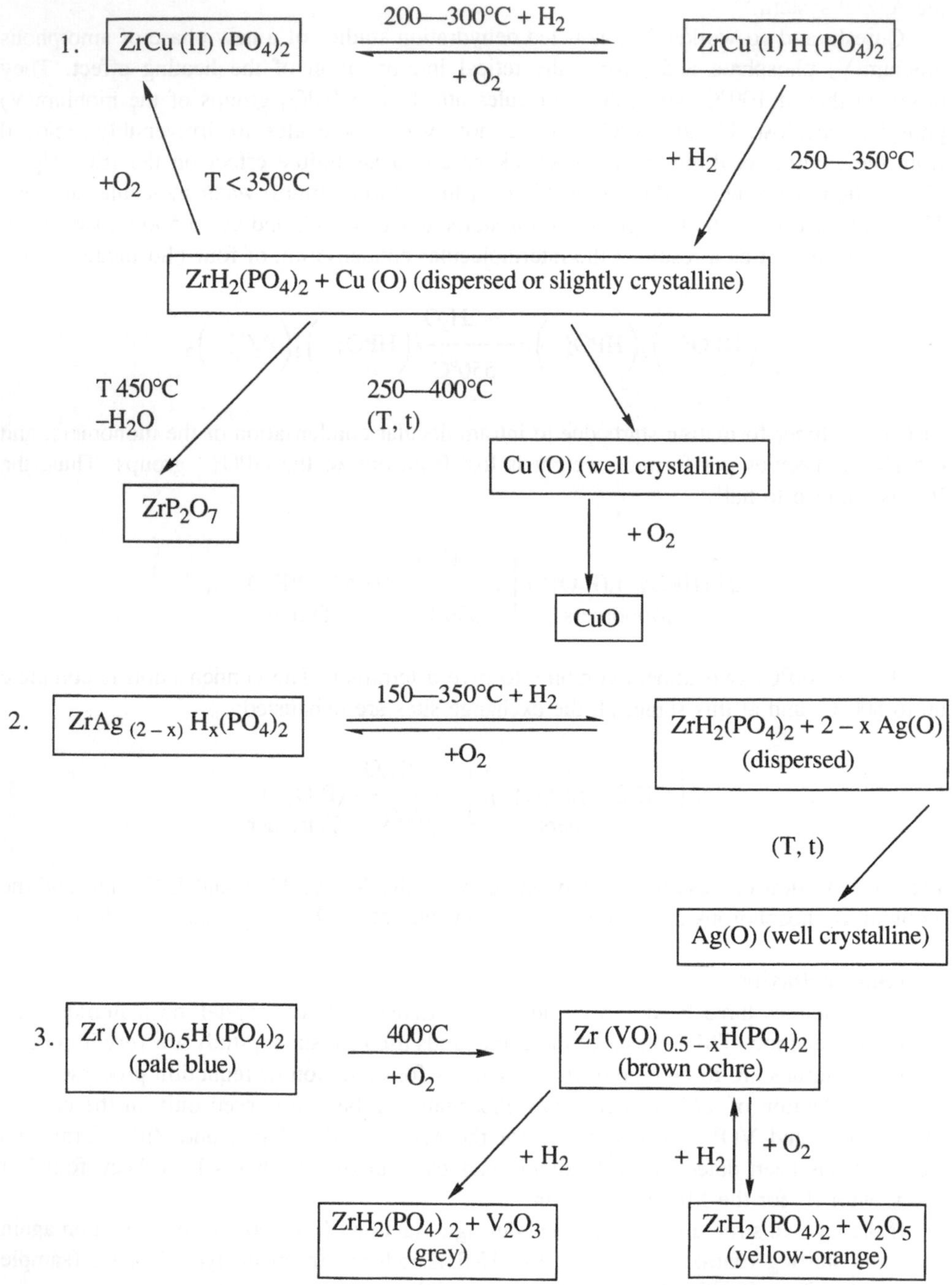

Nonlayered phases

$$1.\quad \boxed{CuZr(PO_4)_2} \xrightarrow{+H_2} \boxed{ZrP_2O_7 + Cu\ (O)} \xrightarrow{+O_2} \boxed{ZrP_2O_7 + CuO}$$

$$2.\quad \boxed{Ag_2Zr(PO_4)_2} \xrightarrow{+H_2} \boxed{ZrP_2O_7 + 2Ag\ (O)} \xrightarrow{+O_2} \boxed{ZrP_2O_7 + 2Ag_xO}$$

(amorphous)

3. Catalytic Behavior

Various similarities in structure and redox behavior between these exchangers and the zeolites have prompted their use as catalysts.

As for zeolites, which are now widely used as catalysts in industrial processes, the layered ion exchangers of the zirconium phosphate type have been tested as catalysts in some particular reactions; Cu or other TMI zirconium phosphates have been employed in oxidation reactions, such as $CO \rightarrow CO_2$ and propene → aldehydes.[47,48] Similarly, hydrogen phases have been used in alcohol dehydrations,[49,50] and hydrogen-palladium zirconium phosphate in the synthesis of isobutylmethyl ketone from acetone,[51] in a single step instead of the three normally necessary without the catalyst.

The crystallinity of the samples influences the catalytic activity of a substance because the latter is related to the surface area. Generally, amorphous or the less crystalline are initially more active, but condensation of the acidic sites occurs at a lower temperature and, hence, their catalytic activity rapidly decreases.

The presence of Bronsted acid sites permits these materials to activate some organic reactions catalyzed by acidic centers. The dehydration of alcohols, namely, cyclohexanol,[49] 2-propanol,[50,52] isobutanol[57], and methanol,[52] has been achieved by several authors.

On zirconium phosphate, Clearfield[9] obtained conversions of cyclohexanol up to 60% at 400°C, but the formation of cyclohexene occurs even at 300°C.

Apart from alcohol dehydrations, several zirconium phosphate acidic phases can also catalyze isomerization reactions, such as double-bond isomerization, 1-butene ⇌ 2-butene (cis or trans), or skeleton isomerization (1- or 2-butene ⇌ isobutene).

Oxidation reactions have been performed by using TMI forms of these ion exchangers. Kalman et al.[47] studied the oxidation of $CO + {}^1/_2O_2 \rightarrow CO_2$, and Iwamoto et al.[48] studied the oxidation of C_3H_6 on Mn, Co, Ni, and Cu zirconium phosphates at 400°C. Their results show that the presence of a TMI acts as a site activating the oxygen species, and that the Cu phase has the highest catalytic activity among the other TMI forms considered. At 400°C in a flow reactor, 10% of propene conversion was reached on the Cu form, obtaining CO, CO_2 together with CH_3CHO and CH_2–CH·CHO (5.6 and 7.0% of the total products, respectively).

The work reported so far on the catalytic activity of inorganic ion exchangers has been mainly on the crystalline materials. Amorphous ion exchangers, therefore, have great importance in this field. Because of their larger surface area, they have a greater promise.

III. ANALYTICAL APPLICATIONS

For an ion exchanger to be of much analytical use, the most important property it must show should be in regard to its selectivity for certain substances. Only then may it be put into practical use most judiciously. Although a large number of amorphous inorganic ion exchangers selective for certain metal ions[53-57] have come into existence, their main disadvantage has been that they have not been rigorously tested for real applications. In most of the works,[58-82] only simple binary and ternary separations have been tried and achieved at the laboratory level, which is mainly of academic interest. However, for wider use of these materials, their utility for some separations from real samples must be explored.

Recently, some attempts have been made in this direction in order to explore the utility of inorganic ion exchangers fully in the field of separation science. Zirconium(IV) and titanium(IV) arsenophosphates[83] and zirconium(IV) arsenosilicate[84,85] have been used for the analysis of alloys and rocks. Such materials have also been utilized for the separation and determination of metal ions from some antacid drugs[86] and multivitamin-multimineral formulations.[87] New chelating ion exchangers have been developed recently[88-93] and their analytical applications explored. A PAN sorbed zinc silicate[94] has been used for the recovery of precious metal ions such as Pt^{4+} and Au^{3+}. Also, a method has been developed for the recovery of silver from laboratory wastes.[95]

Detection and determination of metal ions is another novel aspect of analytical chemistry. Inorganic ion exchangers have found applications in this field, too. For instance, a complexo ion-exchange method has been developed for the specific detection of iron on zirconium(IV) arsenophosphate beads.[96] Similarly, a selective detection method for Fe(III) and Mo(VI) has been developed on antimony(V) silicate beads.[97]

Adsorption behavior of inorganic ion exchangers for pesticides is an important aspect of study which has so far been lacking. The adsorption of pesticides on soils is affected by the presence of metal ions in soils, as they have an important role in modifying its nutritional status. Inorganic materials possessing ion-exchange properties are known to be selective for various metal ions and, hence, their presence in soil may have some far-reaching consequences for its more judicious use in field crops. An attempt has been made in this direction by studying the adsorption of carbofuran on antimony(V) silicate.[98] This study has indicated a highly enhanced adsorption of carbofuran on antimony(V) silicate, compared to soil. It may be true for other materials of this class, too. But before a definite conclusion is drawn, a thorough and detailed study is warranted, which may be quite promising for soil chemists.

As is clear from the above, amorphous inorganic ion exchangers have multiphase applications. Apart from being useful as adsorbents in columns, they have a great utility as impregnants on papers and as thin layers on plates. The following pages summarize their applications in both column and planar chromatographic techniques.

A. COLUMN CHROMATOGRAPHY

The majority of applications of amorphous inorganic ion exchangers have been in columns. Table 1 summarizes some salient features of a few selected materials of this class and the separations achieved on their columns.

B. PLANAR CHROMATOGRAPHY

This type of chromatography includes paper and thin-layer chromatography.

1. Paper Chromatography

Paper chromatography is a very simple form that is used widely for both qualitative and quantitative analyses. A sample is spotted onto a strip of filter paper with a micropipet, and the chromatogram is developed by placing the bottom of the paper (but not the sample spot) in a suitable solvent, as shown in Figure 11. The solvent rises on the paper by capillary action, and the sample components move up the paper at different rates, depending on their solubility and their degree of retention by the paper. After development, the individual solute

TABLE 1
Some Salient Features of a Few Selected Amorphous Ion Exchangers and Their Column Applications

1. Antimony Silicate

Nature	Cation exchanger, poorly crystalline, monofunctional
Composition	Si(IV)/Sb(V) = 3; tentative formula: $[Sb_2O_5(H_2SiO_3)_6] \cdot nH_2O$
IEC (meq/g)	For different metal ions 1.05 (Li^+), 1.60 (Na^+), 1.49 (K^+), 1.53 (Mg^{2+}), 1.59 (Sr^{2+}), 1.10 (Ca^{2+}), 1.60 (Ba^{2+}), 0.8 (NH_4^+) At different temperatures 1.60 (45°C), 1.52 (100°C), 1.42 (300°C), 1.25 (400°C), 1.20 (600°C), 0.72 (800°C)
Main features	Highly stable and reproducible material, specifically high affinity for Rb^+

Separations achieved (M1—M2)	Amount taken (μg)		Amount found (μg)		Error (%)		Eluant and its volume used for different metals
	M1 (μg)	M2 (μg)	M1 (μg)	M2 (μg)	M1 (μg)	M2 (μg)	
Zn-Cd	200.0	304.0	204.0	309.0	0.0	+1.6	Zn: 0.1 *M* $HClO_4$, 50 ml Cd: 1 *M* $HClO_4$, 30 ml
Mn-Cd	71.0	187.0	67.0	187.0	−5.6	0.0	Mn: 0.1 *M* $HClO_4$, 80 ml Cd: 1 *M* $HClO_4$, 60 ml
Hg-Pb	240.0	517.0	240.0	497.0	0.0	−3.8	Hg: 10% HCl + 5% DMSO, 40 ml Pb: 1 *M* $HClO_4$, 40 ml
Mg-Ca	34.0	64.0	35.0	64.0	+2.6	0.0	Mg: 0.1 *M* $HClO_4$, 70 ml Ca: 1 *M* $HClO_4$, 50 ml
Mg-Ba	34.0	112.0	34.0	112.0	0.0	0.0	Mg: 0.1 *M* $HClO_4$, 50 ml Ba: 1 *M* $HClO_4$, 60 ml
Mg-Sr	48.0	123.0	49.0	123.0	+2.0	0.0	Mg: 0.1 *M* $HClO_4$, 80 ml Sr: 1 *M* $HClO_4$, 20 ml

Column conditions	Column diameter	0.6 cm
	Sample	2 g
	Particle size	50—100 mesh
	Flow rate	0.5 ml/min

Reference: Varshney, K. G., Sharma, U., Rani, S., and Premadas, A., Cation exchange study on a crystalline and thermally stable phase of antimony silicate. Effect of irradiation on ion exchange behavior and separation of Mg^{2+} from Ba^{2+}, Ca^{2+}, and Sr^{2+}, *Sep. Sci. Technol.*, 17, 1527, 1982—1983.

2. Ceric Antimonate

Nature	Cation exchanger
Composition	Sb(V)/Ce(IV) = 0.32
IEC (meq/g)	For different metals 0.91 (Li^+), 1.22 (Na^+), 1.25 (NH_4^+), 1.27 (K^+), 0.95 (Ba^{2+}), 1.20 (Ca^{2+}) At different temperatures 1.22 (14°C ± 3°), 1.19 (50°C), 1.18 (100°C), 0.86 (200°C), 0.49 (300°C), 0.36 (400°C)
Main features	Specific for Hg^{2+} ions
Separations achieved	Cd^{2+}-Hg^{2+}, Zn^{2+}-Hg^{2+}, Tl^+-Hg^{2+}, Pb^{2+}-Hg^{2+}, Th^{4+}-Zr^{4+}, Al^{3+}-Fe^{3+}, Mn^{2+}-Cu^{2+}

Conditions:	Column size = 5.0 cm × 0.19 cm^2 Flow rate = 0.15 ml/min
Eluants:	Water (Cd, Zn, Tl, Pb) 0.5 *M* NH_4Cl + 0.1 *M* HCl (Hg^{2+})

Reference: Adapted from Gill, J. S. and Tandon, S. N., Synthesis and ion exchange properties of ceric antimonate, *Talanta,* 19, 1335, 1972.

TABLE 1 (continued)
Some Salient Features of a Few Selected Amorphous Ion Exchangers and Their Column Applications

3. Ceric Tungstate

Nature	Cation exchanger
Composition	$Ce^{4+}/WO_4^{2-} = 0.49$
IEC (meq/g)	0.9 for Na^+
Main features	Negligible effect of pH on IEC
Separations achieved	Pb^{2+}-Hg^{2+}, Cd^{2+}-Hg^{2+}, Zn^{2+}-Hg^{2+}, Cr^{3+}-Al^{3+}

Conditions: Column size = 9.0 cm × 0.19 cm^2
Flow rate = 0.15 ml/min

Eluants: Water (Pb^{2+}, Cd^{2+}, Zn^{2+}, Cr^{3+})
0.4 *M* NH_4Cl + 0.05 *M* HCl (Hg^{2+})
0.2 *M* NH_4Cl + 0.02 *M* HCl (Al^{3+})

Reference: Adapted from Tandon, S. N. and Gill, J. S., Synthesis and ion exchange properties of ceric tungstate, *Talanta,* 20, 585, 1973.

4. Chromium(III) Antimonate

Nature: Cation exchanger, monofunctional

Composition: Sb(V)/Cr(III) = 5.06

IEC (meq/g): For different metal ions
1.37 (Li^+), 1.54 (Na^+), 1.45 (K^+), 1.12 (NH_4^+), 1.71 (Mg^{2+}), 2.67 (Ca^{2+}), 2.03 (Ba^{2+})
At different temperatures for Na^+
1.54 (32°C ± 3°), 0.84 (100°C ± 10°), 0.77 (200°C ± 10°), 0.75 (300°C ± 10°), 0.64 (400°C ± 10°), 0.26 (500°C ± 10°), 0.25 (600°C ± 10°)

Main features: High selectivity for Ag^+, Tl^+, Co^{2+}, Hg^{2+}, Pb^{2+}, Ca^{2+}, Sr^{2+}, Fe^{3+}, Bi^{3+}, Al^{3+}, La^{3+}, and Zr^{4+}

Separations achieved: Column details for metal ion separations on chromium antimonate: 2.0 g columns; feed solutions, 5 mol dm^{-3} of each cation; flow rate of feed and eluant, 10—12 cm^3/hr

Separation	Loading (meq of each cation per g)	Yields (%)
Cu^a-Pb^c	0.010; 0.022	Cu, 86; Pb, 85
Mg^a-Pb^c	0.010; 0.022	Mg, 87; Pb, 85
Zn^b-Pb^c	0.008; 0.022	Zn, 82; Pb, 85
Mn^a-Fe^c	0.010; 0.025	Mn, 100; Fe, 90
Ni^a-Fe^c	0.010; 0.025	Ni, 90; Fe, 90
Cu^b-Fe^c	0.010; 0.025	Cu, 85; Fe, 90
Ni^a-Co^c	0.010; 0.025	Ni, 90; Co, 89
Mn^a-Co^c	0.010; 0.020	Mn, 99; Co, 89
Mg^a-Ca^c	0.008; 0.040	Mg, 92; Ca, 98
Zn^a-Hg^c	0.008; 0.020	Zn, 82; Hg, 90
Cd^d-Hg^c	0.010; 0.020	Cd, 85; Hg, 90
In^e-Al^d	0.018; 0.100	In, 90; Al, 82
Ce^b-La^f	0.012; 0.120	Ce, 98; La, 94
Th^c-Zr^g	0.018; 0.100	Th, 92; Zr, 99
Th^c-La^f	0.018; 0.100	Th, 92; La, 95

Note: Separations were carried out in the given order. Superscripts denote eluants used.

TABLE 1 (continued)
Some Salient Features of a Few Selected Amorphous Ion Exchangers and Their Column Applications

[a] 0.1 *M* NH_4Cl + 0.01 *M* HCl.
[b] 0.1 *M* NH_4Cl + 0.05 *M* HCl.
[c] 0.5 *M* NH_4Cl + 0.10 *M* HCl.
[d] 0.5 *M* NH_4Cl + 0.5 *M* HCl.
[e] Water (pH 2—3).
[f] 2.0 *M* NH_4Cl + 1.0 *M* HCl.
[g] 0.1 *M* citric acid.

Reference: Adapted from Mathew, J. and Tandon, S. N., Chromium antimonate as an ion exchanger, *Chromatographia,* 9, 235, 1976.

5. Niobium(V) Phosphate

Nature: Cation exchanger, monofunctional
Composition: Nb(V)/P(V) = 0.67
IEC (meq/g): For different metal ions
0.70 (Li^+), 1.06 (Na^+), 1.18 (K^+), 0.96 (Mg^{2+}), 1.00 (Sr^{2+}), 1.56 (Ba^{2+})
At different temperatures
1.15 (45°C), 1.1 (100°C), 1.05 (200°C), 1.0 (300°C), 0.90 (400°C), 0.70 (500°C), 0.50 (600°C), 0.20 (700°C), 0.0 (800°C), 0.0 (1000°C)
Main features: High chemical and thermal stability, reproducibility in ion-exchange behavior
Separations achieved:

Sl. no.	Mixture loaded	Cation eluted	Eluant	Total volume of effluent collected (ml)	Amount loaded (μg)	Amount found (μg)	Error (%)
1.	Mg^{2+}-Ca^{2+}	Mg^{2+}	0.2 *M* HNO_3	100	216	205	−5.09
		Ca^{2+}	0.2% $H_2C_2O_4$	25	160	168	+5.00
2.	Mg^{2+}-Ba^{2+}	Mg^{2+}	0.2 *M* HNO_3	110	216	216	0.00
		Ba^{2+}	0.2% $H_2C_2O_4$	25	448	450	+0.45
3.	Zn^{2+}-Cd^{2+}	Zn^{2+}	0.2 *M* HNO_3 + 0.25% NH_4NO_3	60	293	299	+2.05
		Cd^{2+}	1.0 *M* HNO_3 + 0.5% NH_4NO_3	60	244	250	+2.46
4.	Ba^{2+}-Zn^{2+}	Ba^{2+}	2% Na_2CO_3	30	214	219	+2.34
		Zn^{2+}	0.2 *M* HNO_3 + 0.25% NH_4NO_3	70	290	294	+1.38

References: Qureshi, M., Ahmad, A., Shakeel, N. A., and Gupta, A. P., Synthesis, dehydration studies, and cation exchange behavior of a new phase of niobium(V) phosphate, *Bull. Chem. Soc. Jpn.,* 59, 3247, 1986.

6. Thorium Tetracyclohexylamine

Nature: Anion exchanger, ligand exchanger, monofunctional
Composition: Cyclohexylamine/Th = 4
IEC (meq/g): For various anions at different temperatures

TABLE 1 (continued)
Some Salient Features of a Few Selected Amorphous Ion Exchangers and Their Column Applications

Anion	Salt	Capacity 40°C	60°C	80°C
CrO_4^{2-}	K_2CrO_4	1.10	1.14	1.14
SCN^-	KSCN	0.66	0.79	0.79
Cl^-	KCl	0.82	0.87	0.87
I^-	KI	0.72	0.74	0.74
MnO_4^-	$KMnO_4$	1.02	1.06	1.06
Cu^{2+}	$Cu(NO_3)_2$	0.52[a]	—	—
Ni^{2+}	$Ni(NO_3)_2$	0.69[a]	—	—
SO_4^{2-}	K_2SO_4	1.08	1.10	1.10

[a] sorption capacity.

Main features: Influence of temperature studied on the exchange equilibria of SCN^--NO_3^-, MnO_4^--NO_3^-, Cl^--NO_3^-, and CrO_4^{2-}-NO_3^-.

Reference: Adapted from Rawat, J. P. and Kamoonpuri, S. I. M., Anion exchange and ligand exchange studies on thoriumtetracyclohexylamine, *React. Polym.*, 8, 153, 1988.

7. Tantalum Antimonate

Nature: Cation exchanger
Composition: Sb(V)/Ta(V) = 1.3
IEC: 0.99 meq/g for Na^+
Main features: Reproducible and chemically stable
Separations achieved: VO^{2+}-Fe^{3+}-Ti^{4+}, VO^{2+}-Al^{3+}-Ti^{4+}, UO_2^{2+}-Ti^{4+}

Conditions:
Column size = 6.9 mm (i.d.)
Sample = 1—2 g
Flow rate = 0.5 ml/min

Eluants:
0.1 *M* HNO_3 (VO^{2+})
1.0 *M* HNO_3 (Fe^{3+}, UO_2^{2+})
2 *N* H_2SO_4 + 3% H_2O_2 (9:1) (Ti^{4+})
1.0 *M* HCl (Al^{3+})

Reference: Qureshi, M., Gupta, J. P., and Sharma, V., Synthesis of a reproducible and chemically stable tantalum antimonate, *Anal. Chem.*, 45, 1901, 1973.

8. Tin(IV) Antimonate

Nature: Cation exchanger, monofunctional
Composition: Sn/Sb = 1.0; tentative structure

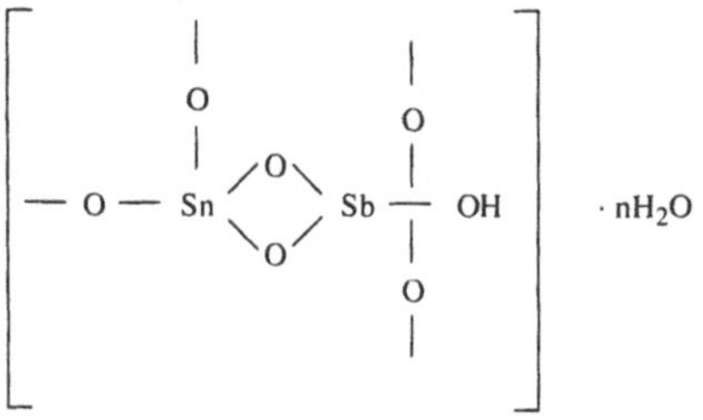

TABLE 1 (continued)
Some Salient Features of a Few Selected Amorphous Ion Exchangers and Their Column Applications

IEC	1.90 meq/g for Na^+
Main features	High thermal and chemical stability
Separations achieved	Ca^{2+}-La^{3+}, Mg^{2+}-La^{3+}, Mg^{2+}-Al^{3+}, Cd^{2+}-Cu^{2+}, Mg^{2+}-Cd^{2+}, Mg^{2+}-Sr^{2+}

Conditions: Column size = 3 mm (i.d.)
Sample = 2 g
Flow rate = 4—5 drops per/min

Eluants: 0.40 *M* NH_4NO_3 + 0.10 *M* HNO_3 (Ca^{2+})
1.50 *M* HNO_3 (La^{3+})
0.4 *M* NH_4NO_3 (Mg^{2+})
0.4 *M* NH_4NO_3 + 0.2 *M* HNO_3 (Al^{3+})
1.0 *M* HNO_3 (Cu^{2+})
0.1 *M* HNO_3 (Cd^{2+})
3% NH_4Cl + 0.1 *M* HNO_3 (Sr^{2+})

Reference: Adapted from Qureshi, M., Kumar, V., and Zehra, N., Synthesis of tin(IV) antimonate of high thermal and chemical stability, *J. Chromatogr.*, 67, 351, 1972.

9. Tin(IV) Antimonate

Nature	Cation exchanger
Composition	Sb(V)/Sn(IV) = 5.1
IEC (meq/g)	For different metal ions 0.88 (Li^+), 1.60 (Na^+), 1.09 (K^+), 1.94 (NH_4^+), 0.72 (Rb^+), 0.56 (Cs^+), 1.18 (Mg^{2+}), 1.55 (Ca^{2+}), 1.76 (Ba^{2+})
Main features	High selectivity for Ag^+, Hg^{2+}, and Pb^{2+}
Separations achieved:	Zn^{2+} (0.1 *M* NH_4Cl + 0.02 *M* HCl)-Cd^{2+} (0.5 *M* NH_4Cl + 0.1 *M* HCl) Hg^{2+} (0.5 *M* NH_4Cl + 0.1 *M* HCl)-Bi^{3+} (1% KI + 0.036 *M* H_2SO_4) Cs^+ (0.1 *M* NH_4Cl + 0.02 *M* HCl)-UO_2^{2+} (2 *M* NH_4Cl + 1 *M* HCl) UO_2^{2+} (2 *M* NH_4Cl + 1 *M* HCl)-ZrO^{2+} (0.1 *M* citric acid) Th^{4+} (2 *M* NH_4Cl + 1 *M* HCl)-La^{3+} (4 *M* NH_4NO_3 + 4 *M* HNO_3) Th^{4+} (2 *M* NH_4Cl + 1 *M* HCl)-ZrO^{2+} (0.1 *M* citric acid) Mn^{2+} (0.1 *M* NH_4Cl + 0.01 *M* HCl)-Fe^{3+} (0.5 *M* NH_4Cl + 0.1 *M* HCl) Mg^{2+} (0.5 *M* NH_4Cl + 0.1 *M* HCl)-Ca^{2+} (4 *M* NH_4Cl + 4 *M* HCl) Zn^{2+} (0.1 *M* NH_4Cl + 0.02 *M* HCl)-Hg^{2+} (2% KI + 0.036 *M* H_2SO_4)-Cd^{2+} (0.5 *M* NH_4Cl + 0.1 HCl) Al^{3+} (0.1 *M* NH_4Cl + 0.02 *M* HCl)-Fe^{3+} (0.5 *M* NH_4Cl + 0.1 *M* HCl)-UO_2^{2+} (2 *M* NH_4Cl + 1 *M* HCl)

Reference: Adapted from Mathew, J. and Tandon, S. N., Ion exchange behavior of stannic antimonate, *Acta Chim. Acad. Sci. Hung.*, 92, 1, 1977.

10. Tin(IV) Aresenate

Nature	Cation exchanger, monofunctional
Composition	Sn/As = 1.35
IEC (meq/g)	For different metal ions 0.81 (Na^+), 0.88 (K^+), 0.87 (Mg^{2+}), 1.13 (Ba^{2+})
Main feature	High thermal stability, loss in IEC at 500°C = 10%
Separations achieved	Cu^{2+}-Pb^{2+}, Cu^{2+}-Fe^{3+}, Fe^{2+}-Fe^{3+}, Fe^{2+}-Cu^{2+}, Fe^{2+}-Pb^{2+}

Conditions: Column size = 3.9 mm (i.d.)
Sample = 1 g
Flow rate = 7—9 drops per min

Eluants: 0.4% HNO_3 (Cu^{2+}, Fe^{2+})
3% HNO_3 + 1 *M* NH_4NO_3 (Pb^{2+}, Fe^{3+}, Cu^{2+})

Reference: Qureshi, M., Kumar, R., and Rathore, H. S., Synthesis and properties of stannic arsenate, *J. Chem. Soc. A*, 272, 1970.

TABLE 1 (continued)
Some Salient Features of a Few Selected Amorphous Ion Exchangers and Their Column Applications

11. Tin(II) Ferrocyanide

Nature	Cation exchanger, bifunctional	
Composition	Sn/Fe = 1, tentative formula: $[SnO \cdot H_4Fe(CN)_6] \cdot 2.5\ H_2O$	
IEC	2.03 meq/g for Na^+	
Main features:	High affinity for almost all ions except Mg^{2+} and In^{3+}	
Separations achieved	Mg^{2+}-Ba^{2+}, Mg^{2+}-Ca^{2+}, Mn^{2+}-Co^{2+}, Mn^{2+}-Ni^{2+}, Th^{4+}-Y^{3+}	
	Conditions:	Column size = 3.9 mm (i.d.) Sample = 1—3 g Flow rate = 1 ml/3 min
	Eluants:	0.005 *M* HNO_3 (Mg^{2+}) 0.1 *M* HNO_3 (Ba^{3+}, Ca^{2+}) 2 *M* HNO_3 (Co^{2+}, Ni^{2+}, Th^{4+}) 1 *M* $NaNO_3$ + 0.1 *M* HNO_3 (1:1) (Mn^{2+}, Y^{3+})

Reference: Qureshi, M., Varshney, K. G., and Khan, F., Synthesis, ion exchange behavior and analytical applications of stannous ferrocyanide, *J. Chromatogr.*, 65, 547, 1972.

12. Tin(IV) Molybdate

Nature	Cation exchanger, monofunctional	
Composition	Sn(IV)/Mo(VI) = 2.0	
IEC (meq/g)	0.54—0.73 for Na^+	
Main features	Many samples prepared to obtain conditions to synthesize a chemically stable material	
Separations achieved	Fe^{3+}-Al^{3+}, Mn^{2+}-Ce^{3+}, Ni^{2+}-Fe^{3+}, Ni^{2+}-Cu^{2+}, Al^{3+}-Cu^{2+}, Zn^{2+}-Cu^{2+}, Pr^{3+}-Ce^{3+}, Nd^{3+}-Ce^{3+}, Mg^{2+}-Al^{3+}, Fe^{3+}-Al^{3+}	
	Conditions:	Column size = 3.9 min (i.d.) Sample = 1—3 g Flow rate = 1 ml/3 min
	Eluants	0.1 *M* NH_4Cl (Fe^{3+}, Mn^{2+}, Ni^{2+}, Al^{3+}) 1.0 *M* HNO_3 (Al^{3+} from Fe^{3+}, Fe^{3+} from Mn^{2+}, Ni^{2+}, and Cu^{2+}) 0.1 *M* NH_4NO_3 (Zn^{2+}) 0.005 *M* HNO_3 (Pr^{3+}, Nd^{3+}) 2% NH_4Cl + 0.5% HNO_3 (Ce^{3+}) 1% HNO_3 (Al^{3+} from Mg^{2+} and Fe^{3+})

Reference: Qureshi, M., Husain, K., and Gupta, J. P., Synthesis and ion exchange characteristics of stannic molybdates, *J. Chem. Soc. A*, 30, 1971.

13. Tin(IV) Molybdosilicate

Nature	Cation exchanger, monofunctional
Composition	Sn:Mo:Si = 2:0.5:1 Tentative formula: $[(SnO_2)_2(MoO_3)_{0.5}SiO_2] \cdot nH_2O$
IEC (meq/g)	For different metal ions 0.45 NH_4^+), 0.40 (Li^+), 0.66 (Na^+), 0.70 (K^+), 0.44 (Hg^{2+}), 0.50 (Ca^{2+}), 0.55 (Sr^{2+}), 0.52 (Ba^{2+})
Main features	Thorium selective
Separations achieved:	

TABLE 1 (continued)
Some Salient Features of a Few Selected Amorphous Ion Exchangers and Their Column Applications

Sl. no.	Mixture loaded[a] (1.5 + 2.5) (ml)	Cation eluted	Eluant[b]	Amount loaded (μg)	Amount recovered (μg)	Error (%)
1	Hg^{2+} + Zn^{2+}	Hg^{2+}	A	301.0	284.0	−5.65
		Zn^{2+}	B	163.0	157.0	−3.68
		Hg^{2+}	A	301.0	286.0	−4.98
2	Hg^{2+} + Cd^{2+}	Cd^{2+}	B	281.0	293.0	+4.27
		Mn^{2+}	A	82.4	85.0	+3.16
3	Mn^{2+} + Al^{3+}	Al^{3+}	B	67.5	65.0	−3.70
		Mn^{2+}	A	82.4	87.0	+5.58
4	Mn^{2+} + Ni^{2+}	Ni^{2+}	B	146.8	134.3	−8.51
		Mn^{2+}	A	82.4	76.0	−7.77
5	Mn^{2+} + Zn^{2+}	Zn^{2+}	B	163.0	155.0	−4.91
		Al^{3+}	B	83.8	79.5	−5.13
6	Al^{3+} + Fe^{3+}	Fe^{3+}	C	67.5	64.0	−5.18
		Zn^{2+}	B	83.8	76.5	−8.71
7	Zn^{2+} + Fe^{3+}	Fe^{3+}	C	163.0	157.0	−3.68
		Cd^{2+}	B	83.8	80.0	−4.53
8	Cd^{2+} + Fe^{3+}	Fe^{3+}	C	281.0	290.0	+3.20

[a] All the cationic solutions used were 0.001 *M*.
[b] Eluants: A = demineralized water, B = 0.001 *M* HNO_3, C = 0.01 *M* HNO_3.

Reference: Adapted from Qureshi, M., Gupta, A. P., Rizvi, S. N. A., and Shakeel, N. A., Synthesis and ion exchange properties of thermally stable thorium(IV)-selective tin(IV) molybdosilicate. Comparison with other tin(IV)-based ion exchangers, *React. Polym.*, 3, 23, 1984.

14. Tin(IV) Tungstoarsenate

Nature	Cation exchanger, monofunctional
Composition	Sn:W:As = 12:5:2
Main features	High thermal and chemical stability
IEC (meq/g)	At different temperatures for Na^+
	1.18 (50°C), 1.24 (150°C), 0.51 (300°C), 0.31 (500°C), 0.04 (800°C)
Separations achieved	Mg^{2+}-Ba^{2+}, Mg^{2+}-Cu^{2+}, Ni^{2+}-Cu^{2+}

Conditions: Column size = 3.9 mm (i.d.)
Sample = 2 g
Particle size = 100—150 mesh

Eluants: 0.001 *M* HNO_3 (Mg^{2+}, Ni^{2+})
1% NH_4NO_3 in 1 *M* HNO_3 (Ba^{2+}, Cu^{2+})

Reference: Qureshi, M., Kumar, R., Sharma, V., and Khan, T., Synthesis and ion exchange properties of tin(IV) tungstoarsenate, *J. Chromatogr.*, 118, 175, 1976.

TABLE 1 (continued)
Some Salient Features of a Few Selected Amorphous Ion Exchangers and Their Column Applications

15. Titanium(IV) Antimonate

Nature	Cation exchanger	
Composition	Sb(V)/Ti(IV) = 1.10	
IEC (meq/g)	For different metals	
	0.43 (Li^+), 1.70 (Na^+), 1.77 (NH_4^+), 1.80 (K^+), 1.68 (Ca^{2+}), 2.03 (Ba^{2+})	
	At different temperatures	
	1.70 (RT), 1.71 (100°C), 1.70 (200°C), 1.72 (300°C), 0.98 (400°C), 0.25 (600°C)	
Main features	High chemical and radiation stability	
Separations achieved	Th^{4+}-UO_2^{2+}, Zn^{2+}-Cd^{2+}, Zn^{2+}-Hg^{2+}, Mn^{2+}-Co^{2+}, Ni^{2+}-Co^{2+}, Cr^{3+}-Al^{3+}, Fe^{3+}-Al^{3+}	
	Conditions:	Column size = 5.0 cm × 0.19 cm^2
		Flow rate = 0.15 ml/min
	Eluants:	0.1 *M* NH_4Cl + 0.01 *M* HCl (Th^{4+}, Zn^{2+}, Mn^{2+}, Ni^{2+})
		0.4 *M* NH_4Cl + 0.1 *M* HCl (UO_2^{2+}, Hg^{2+}, Co^{2+}, Al^{3+})
		H_2O (Cr^{3+}, Fe^{3+})

Reference: Gill, J. S. and Tandon, S. N., Ion exchange properties of titanic antimonate, *J. Radioanal. Chem.*, 20, 5, 1974.

16. Titanium(IV) Arsenate

Nature	Cation exchanger, monofunctional	
Composition	As(V)/Ti(IV) = 2.0; tentative formula: $[Ti(HAsO_4)_2] \cdot 2.5\ H_2O$	
IEC (meq/g)	For different metal ions	
	0.95 (Na^+), 1.05 (K^+), 1.04 (Ba^{2+}), 0.80 (Ca^{2+})	
Main features	Selective for Pb^{2+} ions	
Separations achieved	Cu^{2+}-Pb^{2+}, Zn^{2+}-Pb^{2+}, Mg^{2+}-Pb^{2+}, Ga^{3+}-Pb^{2+}	
	Conditions:	Column size = 3.9 (i.d.)
		Sample = 1.5 g
		Mesh size = 50—100
		Flow rate = 0.4—0.5 ml/min
	Eluants:	H_2O (Cu^{2+}, Zn^{2+}, Mg^{2+}, Ga^{3+})
		1.0 *M* NH_4NO_3 + O.1 *M* HNO_3 (Pb^{2+})

Reference: Qureshi, M. and Nabi, S. A., Preparation and properties of titanium arsenates, *J. Inorg. Nucl. Chem.*, 32, 2059, 1970.

17. Titanium(IV) Molybdate

Nature	Cation exchanger, monofunctional	
Composition	Mo(VI)/Ti(IV) = 2.0	
IEC	1.60 meq/g for Na^+	
Main features	Quantitative ternary separations achieved on column	
Separations achieved	Zn^{2+}-Pb^{2+}-Tl^+, Bi^{3+}-Pb^{2+}-Tl^+	
	Conditions:	Column size = 10 cm × 0.39 cm (i.d.)
		Sample = 2 g
		Rate of flow = 7—9 drops per min
	Eluants:	1% HNO_3 (Zn^{2+}, Bi^{3+})
		1% HNO_3 + 1 *M* NH_4NO_3 (Pb^{2+})
		1% HCl + 2 *M* NH_4Cl (Tl^+)

Reference: Qureshi, M. and Rathore, H. S., Synthesis and ion exchange properties of titanium molybdates, *J. Chem. Soc. A*, 2515, 1969.

TABLE 1 (continued)
Some Salient Features of a Few Selected Amorphous Ion Exchangers and Their Column Applications

18. Titanium(IV) Tungstate

Nature	Cation exchanger, monofunctional
Composition	W(VI)/Ti(IV) = 0.5; tentative structure:

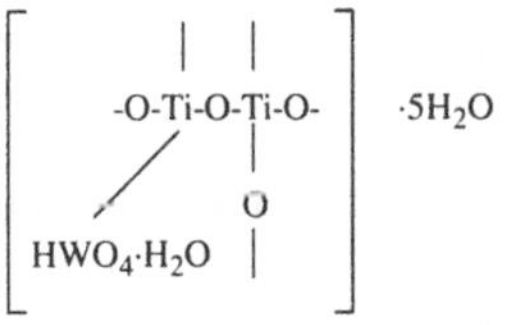

IEC (meq/g)	0.42 (Na^+), 0.42 (K^+), 0.77 (Ba^{2+}), 0.55 (Sr^{2+}), 0.46 (Ca^{2+}), 0.52 (Mg^{2+})
Main features	Reversal in selectivity for alkaline earth metals
Separations achieved	Sr^{2+}-Ca^{2+}, Mg^{2+}-Ca^{2+}

Conditions:	Column = 3.9 mm (i.d.) Sample = 2 g Mesh size = 100—150 Flow rate = 4 drops per/min
Eluants:	0.05 *M* HNO_3 (Sr^{2+}, Mg^{2+}) 0.5 *M* NH_4NO_3 + 0.05 *M* HNO_3 (Ca^{2+})

Reference: Qureshi, M. and Gupta, J. P., Preparation and properties of titanium(IV) tungstates, *J. Chem. Soc. A*, 1755, 1969.

19. Titanium(IV) Vanadate

Nature	Cation exchanger
Composition	Ti^{4+}/V^{5+} = 0.25; tentative formula: $[Ti_3(V_3O_9, 1.5\ H_2O)_4] \cdot 13H_2O$
IEC	0.70 meq/g for Na^+
Main features	High thermal stability and specificity for Sr^{2+}
Separations achieved	Ba^{2+}-Sr^{2+}, Mg^{2+}-Sr^{2+}, Ca^{2+}-Sr^{2+}

Conditions:	Column size = 3 mm (i.d.) Sample = 2 g Flow rate = 4—5 drops per/min
Eluants:	0.01 *M* HNO_3 (Ba^{2+}; Mg^{2+}, Ca^{2+}) 0.1 *M* HNO_3 (Sr^{2+})

Reference: Qureshi, M., Varshney, K. G., and Kabiruddin, S. K., Synthesis and ion exchange properties of thermally stable titanium(IV) vanadate: separation of Sr^{2+} from Ca^{2+}, Ba^{2+}, and Mg^{2+}, *Can. J. Chem.*, 50, 2071, 1972.

20. Titanium(IV) Arsenosilicate

Nature	Cation exchanger
Composition	Ti(IV)/As(V)/Si(IV) = 2:1:3; tentative structure: $[(TiO_2)_2 \cdot (H_3AsO_4)(H_2SiO_2)_3] \cdot nH_2O$
IEC	1.26 meq/g for Na^+ At different temperatures 1.26 (80°C), 0.90 (200°C), 0.85 (300°C), 0.72 (400°C), 0.05 (600°C), 0.0 (800°C)
Main features	Highly selective for Pb^{2+}
Separations achieved:	

TABLE 1 (continued)
Some Salient Features of a Few Selected Amorphous Ion Exchangers and Their Column Applications

Sample no.	Separation achieved	Amount loaded (μg)	Amount recovered (μg)	Error (%)	Eluant used	Volume of eluant used (ml)
1	Pb(II)-Mg(II)	730 Mg	742 Mg	+1.6	0.01 *M* HNO_3	40
		4140 Pb	4122 Pb	−0.43	1 *M* NH_4No_3	90
2	Pb(II)-Ca(II)	1200 Ca	1200 Ca	0	0.01 *M* HNO_3	60
		4140 Pb	4140 Pb	0	1 *M* NH_4NO_3	80
3	Pb(II)-Sr(II)	1750 Sr	1731 Sr	−1.6	0.01 *M* HNO_3	70
		4140 Pb	4140 Pb	0	1 *M* NH_4NO_3	80
4	Pb(II)-Ba(II)	750 Ba	750 Ba	0	0.01 *M* HNO_3	90
		4140 Pb	4111 Pb	−0.70	1 *M* NH_4No_3	70
5	Pb(II)-Mn(II)	2900 Mn	2850 Mn	−1.72	0.01 *M* HNO_3	100
		4140 Pb	4100 Pb	−0.96	1 *M* NH_4NO_3	80
6	Pb(II)-Ni(II)	2935 Ni	2900 Ni	−1.2	0.01 *M* HNO_3	110
		4140 Pb	4100 Pb	−0.96	1 *M* NH_4No_3	90
7	Pb(II)-Zn(II)	1960 Zn	1942 Zn	−0.92	0.01 *M* HNO_3	110
		4140 Pb	4188 Pb	+1.2	1 *M* NH_4NO_3	70
8	Pb(II)-Cd(II)	2250 Cd	2281 Cd	+1.4	0.01 *M* HNO_3	100
		4140 Pb	4092 Pb	−1.2	1 *M* NH_4NO_3	80
9	Pb(II)-Hg(II)	10030 Hg	9850 Hg	−1.8	0.01 *M* HNO_3	100
		4140 Pb	4140 Pb	0	1 *M* NH_4NO_3	90
10	Pb(II)-Sn(II)	5935 Sn	5900 Sn	−0.59	0.01 *M* HNO_3	110
		4140 Pb	4012 Pb	−1.2	1 *M* NH_4No_3	70
11	Pb(II)-Bi(III)	10449 Bi	10299 Bi	−2.1	0.01 *M* HNO_3	100
		4140 Pb	4188 Pb	+1.2	1 *M* NH_4No_3	80
12	Pb(II)-Fe(III)	2800 Fe	2890 Fe	+3.21	0.01 *M* $HClO_2$	100
		4140 Pb	4092 Pb	−1.2	1 *M* NH_4NO_3	80
13	Pb(II)-Al(III)	1350 Al	1400 Al	+3.73	0.01 *M* $HClO_4$	110
		4140 Pb	4092 Pb	−1.2	1 *M* NH_4No_3	90

Quantitative separation of lead from some synthetic lead alloys on TAS columns

Synthetic alloy and its composition (μg) per ml of stock solution	Amount of Pb obtained in effluant (μg)	Error (%)
Solder (Pb, 500; Sn, 500)	496.8	−0.64
-do- (Pb, 430; Sn, 570)	434.7	+1.1
Pewter (Pb, 250; Sn, 750)	253.6	+1.4
Fusible alloys		
Wood's metal (Pb, 250; Sn, 125; Bi, 500; Cd, 125)	243.2	−2.7
Lipowitz alloy (Pb, 270; Sn, 130; Bi, 500; Cd, 100)	274.3	+1.6
Rose's metal (Pb, 280; Sn, 220; Bi, 500)	284.6	+1.6
Newton's metal (Pb, 310; Sn, 190; Bi, 500)	315.7	+1.8

Reference: Varshney, K. G., Agrawal, K., Agrawal, S., Saxena, V., and Khan, A. R., Synthetic, kinetic and analytical studies on titanium(IV) arsenosilicate ion exchanger: separation of lead from its synthetic alloys, *Colloid Surf.*, 29, 175, 1988.

21. PAN-Sorbed Zinc Silicate

Nature	Chelating ion exchanger, cation exchanger
Composition	Si/Zn = 1.25
IEC (meq/g)	For different metals 2.38 (Ca^{2+}), 2.56 (Cu^{2+}), 2.36 (Zn^{2+}), 2.33 (Mg^{2+}), 2.29 (Fe^{3+})
Main feature	Useful in the recovery of precious metal ions such as Pt^{4+} and Au^{3+} from solutions containing Fe^{3+}, Co^{2+}, Cu^{2+}, and Ni^{2+}

TABLE 1 (continued)
Some Salient Features of a Few Selected Amorphous Ion Exchangers and Their Column Applications

Separations achieved	Pt^{4+}-Fe^{3+}, Au^{3+}-Fe^{3+}, Ag^{+}-Cu^{2+}, Au^{3+}-Cu^{2+}
	Conditions: Column size = 10 × 0.39 cm^2; Flow rate = 1.2 ml/min
	Eluants: DMW (Pt^{4+}, Au^{3+}); 0.1 *M* NH_4OH (Ag^{+}); 6 *M* NH_4OH (Fe^{3+}, Cu^{2+})

Reference: Rawat, J. P. and Iqbal, M., Separation and recovery of some metal ions using PAN sorbed zinc silicate as chelating ion exchanger, *J. Liq. Chromatogr.*, 3, 591, 1980.

22. Zirconium(IV) Antimonate

Nature	Cation exchanger
Composition	Sb(V)/Zr(IV) = 1.03
IEC (meq/g)	For different metal ions: 0.74 (Li^{+}), 1.20 (Na^{+}), 1.09 (K^{+}), 0.73 (NH_4^{+}), 1.21 (Mg^{2+}), 1.24 (Ca^{2+}), 1.85 (Ba^{2+})
	At different temperatures: 1.20 (RT), 0.99 (50°C), 0.90 (80°C), 0.80 (100°C), 0.76 (200°C), 0.71 (350°C), 0.67 (450°C), 0.36 (600°C)
Main features	High affinity for Ag^{+}, Hg^{2+}, Bi^{3+}, In^{3+}, and Tl^{3+}
Separations achieved	Pb^{2+}-Hg^{2+}, Cd^{2+}-Hg^{2+}, Zn^{2+}-Hg^{2+}, Hg^{2+}-Bi^{3+}, Pb^{2+}-Bi^{3+}, Tl^{+}-Ag^{+}, La^{3+}-In^{3+}, Fe^{3+}-In^{3+}, Ga^{2+}-In^{3+}, Sr^{2+}-Ba^{2+}, La^{3+}-Ba^{2+}, Mg^{2+}-Ca^{2+}, Sr^{2+}-Rb^{+}, Sr^{2+}-Cs^{+}
	Conditions: Columns containing 50—100 mesh particles
	Eluants: H_2O (pH 2—3) (Pb^{2+}, La^{3+}, Fe^{3+}, Sr^{2+}, Mg^{2+}); 0.5 *M* NH_4Cl + 0.1 *M* HCl (Hg^{2+}); 3 *M* NH_4NO_3 + 3 *M* HNO_3 (Bi^{3+}, Ag^{+}); 4 *M* NH_4Cl + 0.4 *M* HCl (In^{3+}); 0.1 *M* NH_4Cl + 0.01 *M* HCl (Ba^{2+}, Ca^{2+}, Rb^{+}, Cs^{+})

Reference: Mathew, J. and Tandon, S. N., Synthesis and ion exchange properties of zirconium antimonate, *J. Radioanal. Chem.*, 27, 315, 1975.

23. Zirconium Ferrocyanide

Nature	Cation exchanger
Composition	Fe^{3+}/Zr^{4+} = 0.63
IEC (meq/g)	For different metals: 2.78 (Li^{+}), 0.96 (Na^{+}), 0.82 (K^{+}), 0.89 (NH_4^{+}), 0.26 (Ca^{2+}), 0.41 (Ba^{2+})
	At different temperatures: 0.96 (25°C ± 3°), 0.92 (100°C), 0.43 (200°C), 0.12 (300°C), 0.08 (400°C)
Main features	Very efficient elution in column; only 20 ml eluant needed to leach out almost all H^{+} ions
Separations achieved	Cd^{2+}-Zn^{2+}, Co^{2+}-Zn^{2+}, Mn^{2+}-Zn^{2+}, Ca^{2+}-Ba^{2+}, Th^{4+}-Zr^{4+}, Th^{4+}-UO_2^{2+}
	Conditions: Column size = 5.0 cm × 0.19 cm^2; Flow rate = 0.15 ml/min
	Eluants: H_2O (Cd^{2+}, Co^{2+}, Mn^{2+}, Ca^{2+}, Th^{4+}); 0.5 *M* NH_4Cl + 0.1 *M* HCl (Zn^{2+}, Ba^{2+}, Zr^{4+}, UO_2^{2+})

Reference: Gill, J. S. and Tandon, S. N., Synthesis and ion exchange properties of zirconium ferrocyanide, *J. Radioanal. Chem.*, 13, 391, 1973.

TABLE 1 (continued)
Some Salient Features of a Few Selected Amorphous Ion Exchangers and Their Column Applications

24. Hydrous Zirconium Oxide (HZO)

Nature	Anion exchanger
Composition	OH/Zr = 2
IEC (meq/g)	At different temperatures for Cl^- ion 1.09 (32°C ± 3°), 0.91 (50°C), 0.75 (110°C), 0.74 (160°C), 0.74 (220°C), 0.72 (300°C), 0.44 (340°C), 0.28 (600°C)
Main features	Stable in alkalies, differential electrolyte solutions, alcohol, benzene, acetone, 0.1 *M* mineral acids; fairly resistant to heat treatment; retains ~70% capacity on heating up to 300°C
Separations achieved	Separation of anions on nitrate form of HZO

Ions separated	Elution system
Cl^-, CrO_4^{2-}	Cl^- with 8 ml of 1 *M* $NaNO_3$, 2 ml H_2O wash; CrO_4^{2-} with 8 ml of 3 *M* NH_3
Br^-, CrO_4^{2-}	Br^- with 8 ml of 1 *M* $NaNO_3$, 3 ml H_2O wash; CrO_4^{2-} with 8 ml of 3 *M* NH_3
I^-, CrO_4^{2-}	I^- with 9 ml of 1 *M* $NaNO_3$, 3 ml of H_2O wash
$S_2O_3^{2-}$, SO_4^{2-}	$S_2O_3^{2-}$ with 14 ml of H_2O; SO_4^{2-} with 12 ml of 3 *M* NH_3
$S_2O_3^{2-}$, CrO_4^{2-}	$S_2O_3^{2-}$ with 16 ml of H_2O; CrO_4^{2-} with 14 ml of 3 *M* NH_3
MoO_4^{2-}, CrO_4^{2-}	MoO_4^{2-} with 10 ml of 0.09 *M* NaCl/0.01 *M* HCl, 2 ml H_2O wash; CrO_4^{2-} with 8 ml of 3 *M* NH_3
MoO_4^{2-}, PO_4^{3-}	MoO_4^{2-} with 9 ml of 0.09 *M* NaCl/0.01 *M* HCl, 3 ml H_2O wash; PO_4^{3-} with 8 ml of 3 *M* NH_3
SCN^-, $Fe(CN)_6^{3-}$	SCN^- with 8 ml of 0.5 *M* NaCl/0.1 *M* HCl, 2 ml H_2O wash; $Fe(CN)_6^{4-}$ with 8 ml of 1 *M* NH_3/2 *M* NH_4Cl

Reference: Singh, N. J. and Tandon, S. N., Hydrous zirconium oxide as an anion exchanger, *Talanta*, 24, 459, 1977.

25. Zirconium(IV) Arsenophosphate (ZAP)

26. Titanium(IV) Arsenophosphate (TAP)

Nature	Cation exchangers
Composition	ZAP:Zr/As/P = 2:1:1; TAP: Ti/As/P = 3:1:1
IEC (meq/g)	For different metal ions ZAP: 0.94 (Na^+), 1.35 (K^+), 0.83 (Mg^{2+}), 0.91 (Ca^{2+}) TAP: 1.31 (Na^+), 1.82 (K^+), 1.17 (Mg^{2+}), 1.24 (Ca^{2+})
Main features	Analysis of alloys and rocks for metal ions
Separations achieved:	

A. Quantitative Separation of Iron on ZAP Column from Iron Base Alloys

	Composition of stock solution (μg/ml)					Stock solution found (μg)	
Sample	Fe	Cr	Ni	Mn	Si	Eluant 0.01 *M* HNO_3 (100 ml)	Eluant 1 *M* HNO_3 (100 ml)
AISI-303	222.8	56.5	26.7	4.7	1.6	Cr 55.1 Ni 26.7 Mn 4.8 Si 1.7	Fe 225
AISI-347	219.47	56.5	31.4	4.7	1.6	Cr 55.5 Ni 31.4 Mn 4.8 Si 1.7	Fe 221

TABLE 1 (continued)
Some Salient Features of a Few Selected Amorphous Ion Exchangers and Their Column Applications

B. Quantitative Separation of Various Constituents of Rocks on ZAP

Rock sample	Volume of stock solution loaded, (ml)	Elements present (μg) Al	Fe	Si	Ca	Mg	Mn	Elements eluted by eluant shown (μg) DMW	0.1 *M* $HClO_4$ (50 ml)	0.5 *M* HCl + 1 *M* NH_4Cl (50 ml)
G-2	0.5	7.70	1.34	34.6	0.98	0.37	0.01	Si 32.6 Ca 1.12 Mg 0.34	Al 8.4	Fe 1.4
G-2	1.0	15.4	2.69	69.2	1.96	0.75	0.03	Si 73.3 Ca 1.96 Mg 0.68	Al 15.8	Fe 2.7
AGV-1	0.5	8.59	3.89	29.8	2.47	0.76	0.05	Si 30.1 Ca 2.5 Mg 0.76 Mn 0.04	Al 8.6	Fe 3.8
AGV-1	1.0	17.2	6.78	59.6	4.94	1.52	0.10	Si 51.7 Ca 4.9 Mg 1.53 Mn 0.06	Al 17.2	Fe 6.7
BHVO-1	0.5	6.85	6.00	2.45	7.70	3.60	0.08	Si 2.45 Ca 7.5 Mg 3.6 Mn 0.06	Al 6.9	Fe 6.9
BHVO-1	1.0	13.7	12.0	4.90	11.4	7.20	0.17	Si 5.1 Ca 11.3 Mg 7.2 Mn 0.14	Al 14.0	Fe 11.6
BCR-1	0.5	6.86	6.70	27.26	3.48	1.74	0.09	Si 27.0 Ca 3.4 Mg 1.8 Mn 0.08	Al 7.1	Fe 6.8
BCR-1	1.0	13.7	13.4	54.5	6.97	3.48	0.18	Si 53.9 Ca 6.7 Mg 3.3 Mn 0.13	Al 14.1	Fe 13.6
PCC-1	0.5	0.36	4.14	21.1	0.27	21.7	0.06	Si 22.3 Ca 0.29 Mg 21.8 Mn 0.05	Al 0.36	Fe 4.1
PCC-1	1.0	0.73	8.28	42.1	0.55	43.5	0.12	Si 43.3 Ca 0.57 Mg 42.0 Mn 0.09	Al 0.72	Fe 8.2

References:
1. Varshney, K. G., Agrawal, S., Varshney, K., and Premadas, A., Analytical applications of Zr(IV) and Ti(IV) arsenophosphates as ion exchangers, *Talanta,* 30, 955, 1983.
2. Varshney, K. G. and Premadas, A., Synthesis, composition and ion exchange behavior of thermally stable Zr(IV) and Ti(IV) arsenophosphates: separation of metal ions, *Sep. Sci. Technol.,* 16, 793, 1981.

27. Zirconium(IV) Arsenophosphate (ZAP)

TABLE 1 (continued)
Some Salient Features of a Few Selected Amorphous Ion Exchangers and Their Column Applications

28. Zirconium(IV) Arsenosilicate (ZAS)

Nature	Cation exchangers
Composition	ZAP:Zr/As/P = 2:1:1 ZAS:Zr/As/Si = 1:2:1
IEC (meq/g)	For Na^+ ion at different temperatures 0.94 (45°C), 0.95 (100°C), 1.03 (200°C), 0.94 (400°C), 0.84 (600°C), on ZAP 1.30 (45°C), 1.30 (100°C), 1.25 (200°C), 0.46 (400°C), 0.20 (600°C) on ZAS
Separations achieved	Fe^{3+} and Al^{3+} from some iron alloys and USGS standard rocks. Results shown in Table 1, Parts A, B, and C.

TABLE 1 — PART A
Quantitative Separation of Iron/Aluminum from Some Synthetic Alloys Using α-ZAP and ZAS Columns

Synthetic alloy and its composition (μg) per ml of solution		Aluminum/iron obtained in the effluent[a] (μg)				Error (%)				SD (%)			
		α-ZAP		ZAS		α-ZAP		ZAS		α-ZAP		ZAS	
		Al	Fe	Al	Fe	Al	Fe	Al	Fe	Al	Fe	Al	Fe
Dow metal	(Al 40, Mn 15, Mg 945)	41.2	—	40.3	—	+3.00	—	+0.75	—	0.50	—	0.90	—
-do-	(Al 120, Mn 10, Mg 870)	118.3	—	122.4	—	−1.42	—	+2.00	—	0.80	—	0.95	—
Aluminum zinc binary alloy	(Al 750, Zn 250)	745	—	743	—	−0.67	—	−0.93	—	1.40	—	2.30	—
-do-	(Al 950, Zn 50)	939	—	941	—	−1.16	—	−0.95	—	2.10	—	1.60	—
Aluminum bronze	(Al 100, Cu 900)	99.5	—	101.5	—	−0.50	—	+1.50	—	0.46	—	1.20	—
-do-	(Al 120, Cu 875, Sn 5)	122.4	—	123.0	—	+2.00	—	+2.50	—	0.75	—	1.20	—
Magnalium	(Al 850, Mg 150)	843.6	—	847.6	—	+0.75	—	−0.28	—	1.80	—	2.80	—
-do-	(Al 900, Mg 100)	890.9	—	904.2	—	−1.01	—	+0.47	—	1.30	—	0.30	—
-do-	(Al 925, Mg 75)	932.4	—	918.9	—	+0.80	—	−0.66	—	0.65	—	0.34	—
Y-alloy	(Al 940, Ni 20, Mn 10, Cu 50)	937.3	—	946.1	—	−0.29	—	+0.65	—	0.70	—	0.46	—
Duralumin	(Al 950, Mn 5, Mg 6, Cu 40)	944.9	—	959.1	—	−0.54	—	+0.95	—	0.15	—	0.42	—
Manganese steel	(Fe 879, Mn 120)	—	887.2	—	884.2	—	+0.93	—	+0.59	—	1.03	—	0.94
-do-	(Fe 849, Mn 150)	—	841.6	—	853.6	—	−0.87	—	+0.54	—	1.20	—	0.58
Nickel steel	(Fe 944, Ni 36, Mn 18)	—	921.8	—	—	—	−2.35	—	—	—	0.66	—	—
Chrome steel	(Fe 965, Mn 19, Cr 15)	—	957.0	—	—	—	−0.83	—	—	—	0.85	—	—
-do-	(Fe 960, Mn 19, Cr 20)	—	949.0	—	—	—	−1.05	—	—	—	0.30	—	—

TABLE 1 — PART A (continued)
Quantitative Separation of Iron/Aluminum from Some Synthetic Alloys Using α-ZAP and ZAS Columns

Synthetic alloy and its composition (μg) per ml of solution		Aluminum/iron obtained in the effluent[a] (μg)				Error (%)				SD (%)			
		α-ZAP		ZAS		α-ZAP		ZAS		α-ZAP		ZAS	
		Al	Fe	Al	Fe	Al	Fe	Al	Fe	Al	Fe	Al	Fe
Invar	(Fe 620, Ni 360, Mn 19.5)	—	609.8	—	—	—	−1.65	—	—	—	1.40	—	—
Stainless steel	(Fe 734, Ni 80, Mn 5, Cr 180)	—	730.5	—	—	—	−0.48	—	—	—	1.98	—	—
Nichrome	(Fe 120, Ni 750, Mn 20, Cr 110)	—	119.2	—	—	—	−0.67	—	—	—	0.21	—	—

Note: ZAP heated at 200°C.

Average value of four replicates.

TABLE 1 — PART B
Quantitative Separation of Iron from Some Standard Steel Samples Using α-ZAP Columns

Steel analyzed	Volume of the stock solution loaded (ml)	Elements present as per the standard composition of the steel (μg)						Iron determined in the effluent[a] (μg)	Error (%)	SD (%)
		Fe	Cr	Ni	Cu	Mn	Si			
AISI-303[b]	1.0	222.80	56.52	29.69	—	4.71	1.57	230.46	+3.44	0.17
	2.0	445.60	113.04	53.38	—	9.42	3.14	454.40	+1.97	0.94
	2.5	557.0	141.30	66.725	—	11.775	3.925	566.40	+1.69	0.81
AISI-347[b]	1.0	219.40	56.52	31.40	—	4.71	1.57	225.13	+2.61	0.96
	2.0	438.80	113.04	62.80	—	9.42	3.14	446.26	+1.70	1.26
	2.5	548.50	141.30	78.50	—	11.775	3.925	554.43	+1.08	0.57
	3.0	658.20	169.56	94.20	--	14.13	4.71	626.42	−4.83	0.44
Inconel-600[c]	1.0	66.60	155.80	769.3	1.6	2.0	—	64.80	−2.70	0.75
Inconel-800[c]	1.0	451.4	209.90	319.8	2.9	8.1	—	450.20	−0.27	0.58

[a] Average value of five replicates.

[b] AISI Standard Steels.

[c] Huntington Alloy Products, Division of International Nickel Co., U.S.A.

TABLE 1 — PART C
Quantitative Separation of Aluminum and Iron from Various USGS Standard Rocks Using α-ZAP and ZAS Columns

Rock analyzed	Standard composition of the rock for Al and Fe per 5 ml of the stock solution (μg)		Al_2O_3 and Fe_2O_3 obtained[a] in the effluent (μg)				Error (%)				SD (%)			
			α-ZAP		ZAS		α-ZAP		ZAS		α-ZAP		ZAS	
	Al_2O_3	Fe_2O_3	Al_2O_3	Fe_2O_3	Al_2O_3	Fe_2O_3	Al_2O_3	Fe_2O_3	Al_2O_3	Fe_2O_3	Al_2O_3	Fe_2O_3	Al_2O_3	Fe_2O_3
G-2	77.00	13.45	78.00	13.80	79.00	18.80	+1.30	+2.60	+2.59	+2.60	1.94	9.65	1.64	1.06
AGV-1	85.95	38.90	85.40	39.20	87.00	38.50	−0.64	+0.77	+1.22	−1.02	1.38	1.97	0.94	1.10
BHVO-1	68.50	60.00	69.45	61.00	71.00	59.25	+1.39	+1.67	+3.64	−1.25	2.99	2.99	1.89	2.10
BCR-1	68.60	67.05	69.45	69.00	69.80	68.10	+1.24	+2.91	+1.74	+1.56	1.75	1.86	2.14	1.40
PCC-1	3.65	41.40	3.57	40.35	3.60	40.70	−2.19	−2.54	−1.36	−1.69	0.47	1.31	0.68	0.95

[a] Average values of five replicates.

References:

1. Varshney, K. G., Agrawal, S., and Varshney, K., Quantitative determination of iron and aluminum in some alloys and silicate rocks after a cation exchange separation on zirconium(IV) phospho and silicoarsentes, *J. Liq. Chromatogr.*, 6, 1535, 1983.
2. Varshney, K. G., Agrawal, S., and Varshney, K., Synthesis, exchange behavior and analytical applications of a new, crystalline, and stable zirconium(IV) arsenosilicate cation exchange: analysis of some silicate rocks, *Sep. Sci. Technol.*, 18, 59, 1983.

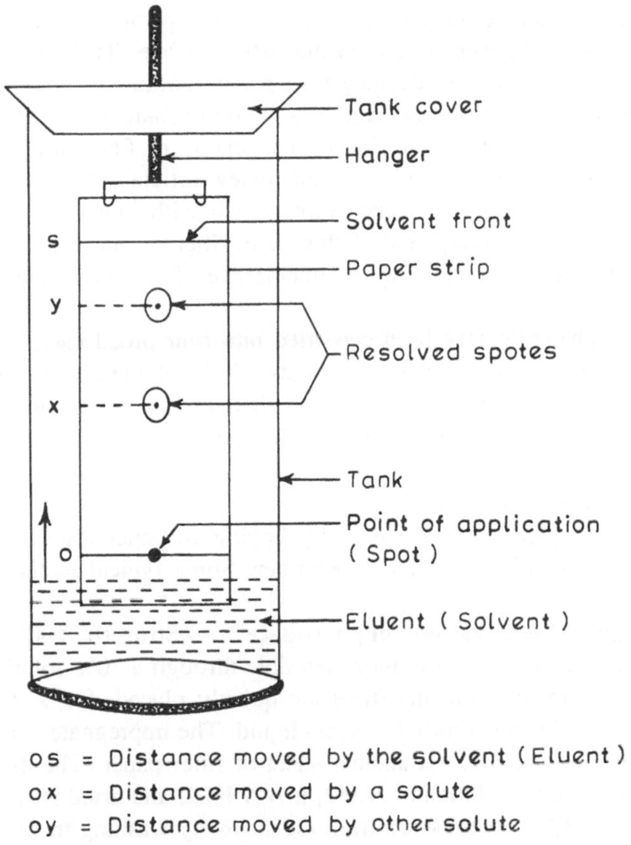

FIGURE 11. A typical paper chromatographic tank.

spots are noted or are made visible by treatment with a reagent that forms a colored derivative. The spots are characterized by the R_f value, defined as

$$R_f = \frac{\text{distance moved by the solute}}{\text{distance moved by the solvent}}$$

Distances are measured from the center of where the sample was spotted at the bottom of the paper. The solvent front is a line across the paper. The distance moved by the solute is measured at the center of the solute or at its maximum density if tailing occurs. The R_f value is, then, a characteristic for a given paper and solvent combination.

Because of the highly hydrophilic nature of the cellulose filter paper, it absorbs water from air which forms a thin coat. Distribution of the solute takes place, therefore, between the thin water-cellulose film on the paper and the developing solvent, thus forming a liquid-liquid partition mechanism. The developing solvent is usually a mixture of an organic solvent with water that may be buffered at a certain pH. The water causes the paper to swell, due to absorption. Other polar solvents like acids will also be absorbed by the paper. Sometimes water-immiscible solvents are used to develop the chromatogram.

The surface of the paper can be modified if it is soaked with a nonpolar solvent before its use in chromatography. The technique is then called ''reversed-phase chromatography''. It is advantageous in effecting separations for which a nonpolar adsorbent is required.

An ion exchanger, if present on the surface of the paper, will naturally impart its properties to it and the paper will have entirely different characteristics. Now, the paper can be used with greater advantage for the separation of ionic species because of the important role played by the ion exchanger present on the paper surface. Tuckerman,[99] Lederer and Kertes,[100] and Hale[101] studied the advantages of papers containing ion-exchange groups, compared with the usual chromatographic papers or ion-exchange columns.

Papers impregnated with inorganic ion exchangers were first introduced in 1949 by Flood[102] in the chromatography of ions. In their review articles, Alberti[103] and Qureshi[104] have highlighted the advantages of papers impregnated with inorganic ion exchangers at quite an early stage of the development of this field. Sherma and Fried[105] have reviewed the progress quite recently, emphasizing the importance of the technique and its versatile nature.

Inorganic ion exchangers have been classified into four broad categories: (1) hydrous oxides, (2) insoluble metal salts of polybasic acids, (3) insoluble metal ferrocyanides, and (4) heteropolyacid salts. All of them have found their use in paper chromatography, and some good and important separations haved been achieved.

a. Preparation of Impregnated Papers

Impregnation of the papers is done simply by dipping the paper strips into the component solutions, which get precipitated on the paper fiber. Some typical methods are described below.

Papers impregnated with zirconium phosphate[106] — Strips (6 × 40 cm) of Whatman No. 1 filter paper are drawn as uniformly as possible through a 30% solution of $ZrOCl_2$ in 4 *N* HCl. After this impregnation, the strips are quickly placed, for 5 min, on pieces of filter paper which immediately absorb the excess liquid. The impregnated paper is then dried at room temperature by placing it on another sheet of filter paper. The dry strips are then dipped for 2 min into a 60% solution of H_3PO_4 in 4 *N* HCl, and dried at room temperature. After 6 h, the excess H_3PO_4 is removed from the strips by washing them first in 2 *N* HCl for 10 min and then twice in H_2O. In order to increase the ZP exchange capacity, an additional treatment with the same 60% solution of H_3PO_4 in 4 *N* HCl is carried out, and afterwards the strips are put into an oven at 50°C for 75 min. As described above, the strips are subsequently washed with 2 *N* HCl and H_2O, and left to dry in air. By varying the concentration of the $ZrOCl_2$, it is possible to obtain ZP papers containing various quantities[107] of the ion exchanger per cm^2. It is also possible to obtain chromatographic strips with a ZP gradient by gradiented saturation[103] of the strips with $ZrOCl_2$. If glass fiber paper is used instead of cellulose paper, the resulting ion-exchange paper is completely inorganic, and is therefore highly resistant to heat and to eluants that attack cellulose.[108]

Papers impregnated with hydrous oxides — The paper is immersed in a solution of the desired metal ion, dried with filter paper to remove excess liquid, and then immersed in a 10% ammonia solution.[109]

Papers impregnated with ammonium molybdophosphate — The ammonium molybdophosphate is prepared by the method described by Smit et al.[110] and solubilized in ammonia solution (4 ml of 4 *N* ammonia solution per g of AMP). The chromatographic paper is immersed first in this solution and then in a solution consisting of 25 g of ammonium paramolybdate, 104 ml of 65% HNO_3, and 638 ml of distilled water, in order to precipitate the AMP in the paper.[111]

Papers impregnated with Sn(II) ferrocyanide[112] — Stannous ferrocyanide papers are prepared by dipping the paper strips in a 0.25 *M* solution of tin(II) chloride, drying, and then dipping in a 0.25 *M* solution of potassium ferrocyanide. After drying, the strips are washed thoroughly in distilled water to remove any excess tin(II) chloride or potassium ferrocyanide. They are finally dried before use.

Papers impregnated with a ligand-combined ion exchanger[113] — Papers impregnated with EDTA zinc silicate exchanger are prepared by first dipping the paper strips into a zinc nitrate solution for 3 to 5 s. The excess reagent is drained off. The strips are then dipped in sodium silicate solution for 15 s, the excess reagent is drained off, and the strips dried at room temperature. The zinc silicate-impregnated paper is made to the EDTA form by dipping in 10% EDTA solution for 10 to 15 s and then drying at room temperature.

b. Chromatography on Papers Impregnated with Hydrous Oxides

Divalent, tervalent, and tetravalent metal hydroxides have been used for both cation and anion separations with a judicious adjustment[114] of the pH of the developing solution to utilize the cation or anion exchange properties selectively of these amphoteric materials. Sakodinsky and Lederer[109] introduced zirconium(IV), titanium(IV), and iron(III) hydroxide papers and reported the separations of uranium, nickel, cobalt, copper, and iron in aqueous sodium chloride solvent. The efficiency of these materials in lithium chloride, ammonium sulfate, and sodium sulfate solutions has also been compared. These researchers have also performed[115,116] a novel and interesting study of the movement of 23 anions and four cobalt(III) complexes on seven hydroxide-treated papers, and the relation of the separation sequences with the electrophoretic mobility of components in lithium and potassium salt solutions has been established.

Tandon et al. have also used the papers of hydrous zirconium and titanium oxides to achieve a large number of binary and ternary separations for both metal ions[117] and anions.[118] Tables 2 and 3 summarize the separations achieved on these materials. Another important application of zirconium oxide paper has been reported by these workers for a rapid separation of amino acids,[119] as shown in Table 4. The same paper has also been used by them[120] for resolving six sets of quaternary mixtures of alkaloids.

Rawat et al.[121] have reported the separation of 19 phenols on ferric hydroxide papers. They have also suggested the possibility of detection during the chromatographic process. Movement of 34 organic acids on aluminum and cadmium hydroxide papers resulting in some important separations have been reported by Rathore et al.[122]

c. Chromatography on Papers Impregnated with Insoluble Metal Salts of Polybasic Acids

Among the numerous polybasic acid salts, zirconium phosphate, with considerably high ion-exchange capacity, marked heat resistance, and sound chemical form, has received the most attention.[123-128] These papers have been particularly useful for the chromatographic studies of anions, metal cations, complex molecules, metal nuclides,[129] and alkaloids.

Nunes Da Costa and Jeronimo[123] studied the behavior of the group IB metals on ZP columns and ZP paper. They showed that the behavior of Ag^{+}-Cu^{2+}-Au^{3+} on the ZP paper is identical to the one observed in column.

Sastri and Rao[125] used ZP paper for the separation of the different valence states of a number of elements, such as Fe^{2+}-Fe^{3+}, U^{4+}-U^{6+}, Ce^{3+}-Ce^{4+}, Mo^{5+}-Mo^{6+}, V^{4+}-V^{5+}, Cr^{3+}-Cr^{4+}, As^{3+}-As^{5+}, and Hg_2^{2+}-Hg^{2+}. The ZP paper has also been used as an electrophoretic support[130] for the separation of inorganic ions.

Later on, other materials of the zirconium phosphate type were extensively used by many research groups. Cabral[131] prepared zirconium molybdate papers and used them for studies of Ca^{2+}, Ba^{2+}, and Sr^{2+} in hydrochloric acid and ammonium chloride solvents. Systematic chromatographic studies were performed by Qureshi and co-workers[132,133] on stannic phosphate and stannic tungstate papers. As a result, specific separations were achieved[132] of Au^{3+}, Hg^{2+}, Pt^{4+}, Mg^{2+}, Mo^{6+}, and Se^{4+}. Stannic phosphate has also been used for the separation of Ni^{2+} and Tl^{+} at different temperatures.[134] Similarly, cerium phosphate paper was used by Alberti and co-workers[135] for the separation of metal ions.

A systematic chromatographic study of 48 metal ions on stannic arsenate papers by Qureshi and Sharma[136] has resulted in a prediction of the K_{sp} values of various metal arsenates

TABLE 2
R_f x 100 Values of Metal Ions on Plain and Hydrous Zirconium Oxide Paper in Different Solvent Systems

	0.01 *M* HCl		0.50 *M* KCl		0.01 *M* HCl + 0.50 *M* KCl		0.50 *M* NH_4Cl		0.01 *M* HCl + 0.5 *M* NH_4Cl	
Metal Ion	Impregnated paper	Untreated paper	Impregnated paper	Untreated paper	Impregnated paper	Untreated paper	Impregnated paper	Untreated paper	Impregnated paper	Untreated paper
Ag^+	18—0	36—0	0	0	0	0	0	0	0	0
Tl^+	79	69	67	88	74	80	72	85	68	100—65
Be^{2+}	2	55	0	45—0	0	40	1	57—0	0	37—0
Co^{2+}	85	90	a 90—57 b 13	87	73—49	94	95	92	86	95
Ni^{2+}	79	97—53	94—38	92—64	76—26	90	85	87	96—71	91
Cu^{2+}	4	61	6	88—37	06	86—30	8	76	4	94—52
Zn^{2+}	87	90	50	95	51	94	49	94	38	91
Pd^{2+}	30	68	51—15	a 82 b 45—0	12	98—67	78—14	85	74—04	86
Cd^{2+}	80	84—52	66—3	91	35	100—36	83	87	70—17	90
Sn^{2+}	4	91—49	0	37—14	0	97—38	0	100	0	98
Hg^{2+}	a 59 b 31	84—52	78	93	73	97—72	77	86	83	86
Pb^{2+}	72	58	69—0	85—02	66—0	83—31	58—0	80—27	45—0	96—0
Bi^{3+}	30—0	96—0	27—0	88—0	34—0	89	23—0	97—0	34—0	93—0
Al^{3+}	1	81—86	2	72—0	3	72—5	1	55—0	0	62—0
Fe^{3+}	2	54—14	2	53—0	2	56—0	1	52—0	1	48
As^{3+}	15—0	80	13	88	16	99—39	19—0	84	14—0	77
In^{3+}	0	64	0	72—0	0	60—30	0	75—0	0	55—0
Sb^{3+}	0	94	0	93	0	100	0	52	4	97
Au^{3+}	10	46	35	46	34	44	40	45	39	48
Ce^{3+}	80	90—47	8	100—51	6	100—0	8	90—0	45	100—0
Pt^{4+}	5	89	79	80	79	86	80	80	62	89
Ce^{4+}	60	100—69	6	80—0	5	83—87	4	85—0	40	94—0
Th^{4+}	3	55—0	0	30—0	0	40—0	0	6	0	5

V^{5+}	0	61	0	85	0	46	0	74—13	0	79—13
Se^{6+}	0	95	0	90	0	35—0	0	90	0	86
Mo^{6+}	5	95	0	98—10	0	97	0	91	0	90
Te^{6+}	0	92	0	95	0	98	0	92	0	89
UO_2^{2+}	8	61	0	10	0	30	0	7	0	7

Note: a and b denote the R_f values of the two spots observed.

TABLE 2A
Some Binary and Ternary Separations of Metal Ions on Hydrous Zirconium Oxide Paper

Sl. no.	Developer	Separations achieved
1	0.01 *M* HCl	Cu^{2+}-Cd^{2+}, Cu^{2+}-Ni^{2+}, Cu^{2+}-Co^{2+}, Zn^{2+}-Fe^{3+}, Cd^{2+}-Sn^{2+}, Cd^{2+}-As^{3+}, Co^{2+}-Fe^{3+}, Co^{2+}-Be^{2+}, Ni^{2+}-Pt^{4+}, Ni^{2+}-Au^{3+}, Pb^{2+}-Be^{2+}, Pb^{2+}-Pd^{2+}, Al^{3+}-Tl^{+}, In^{3+}-Tl^{+}, In^{3+}-Pb^{2+}, In^{3+}-Co^{2+}, In^{3+}-Ce^{4+}, In^{3+}-Cd^{2+}, Pt^{4+}-Pb^{2+}, Ce^{4+}-V^{5+}, Ce^{4+}-UO_2^{2+}, Ce^{4+}-Th^{4+}, Au^{3+}-Pd^{2+}, Ni^{2+}-Pd^{2+}-Pt^{4+}
2	0.5 *M* KCl	Ag^{+}-Zn^{2+}, Ag^{+}-Tl^{+}, Ag^{+}-Pt^{4+}, Ag^{+}-Hg^{2+}, Cu^{2+}-Au^{3+}, Cu^{2+}-Pt^{4+}, Zn^{2+}-Hg^{2+}, Au^{3+}-Pt^{4+}, Au^{3+}-Hg^{2+}
3	0.5 *M* NH_4Cl	Ag^{+}-Au^{3+}, Ag^{+}-Cd^{2+}, Ag^{+}-Ni^{2+}, Zn^{2+}-Ni^{2+}, Ni^{2+}-Sn^{2+}, Ni^{2+}-Fe^{3+}, Al^{3+}-Co^{2+}, Au^{3+}-UO_2^{2+}, Cu^{2+}-Zn^{2+}-Cd^{2+}, Cu^{2+}-Zn^{2+}-Ni^{2+}, Cu^{2+}-Zn^{2+}-Co^{2+}, Zn^{2+}-Co^{2+}-Fe^{3+}, Zn^{2+}-Ni^{2+}-Fe^{3+}, Zn^{2+}-Cd^{2+}-In^{3+}
4	0.01 *M* HCl + 0.5 *M* KCl	Cu^{2+}-Zn^{2+}, Cd^{2+}-Hg^{2+}, In^{3+}-Zn^{2+}, Ag^{+}-Pd^{2+}-Hg^{2+}, Ag^{+}-Au^{3+}-Pt^{4+}, Zn^{2+}-Cd^{2+}-Hg^{2+}, Cd^{2+}-Hg^{2+}-Pd^{2+}, Pd^{2+}-Au^{3+} Pt^{4+}
5	0.01 *M* HCl + 0.5 *M* NH_4Cl	Cu^{2+}-Zn^{2+}, Zn^{2+}-Co^{2+}

Reference: Adapted from Singh, N. J. and Tandon, S. N., Rapid separation of metal ions on hydrous zirconium oxide paper, *Chromatographia,* 10, 309, 1977.

on the basis of R_f values in butanol-HNO_3 systems on plain and impregnated papers. Butanol was selected because it gives clean and useful separations, and HNO_3 was chosen because it is a noncomplexing acid, and the understanding of the equilibria is simplified. The study resulted in the following relationship between R_f and K_{sp}:

$$R_f = 0.288 + 0.0066 \frac{10^3}{-\log K_{sp} Z^2 r^\circ}$$

where Z is the charge on the ion and r° is the base ion radii, Table 5 summarizes the separations achieved on stannic arsenate papers. Similar studies have been done by Varshney and Varshney[137] on stannous ferrocyanide papers.

Stannic tungstate papers have been employed for the recovery of microgram amounts of selenium from many other metal ions in ammonium tartarate and ammonia.[138] On the same papers, quantitative separation[139] of V^{5+} from Fe^{3+}, Ti^{4+}, Zr^{4+}, and Th^{4+} in acetone-water systems has also been reported. Further, a confirmation of the relation between R_f values of cations and the mole fraction of acetone in the solvents, $R_f = mx + c$ (where x = mole fraction of acetone), has been reported. Stannic selenite and stannic tungstate papers have been used for the chromatography of 47 cations in 15 solvents and 47 cations in 40 solvents.[140] Stannic tungstate papers have found application in electrophoretic studies[141] of metal ions. De and Chowdhury[142] studied the behavior of thorium phosphate and proposed the separation of Ag^{+}, Fe^{3+}, UO_2^{2+}, VO^{2+}, and ZrO^{2+} ions from several other cations. Rawat et al.[143,144] prepared zinc silicate papers and used them for the separation of amines and phenols. A useful separation of these groups of compounds has also been worked out by Singh and Darbari.[145] Qureshi et al.[146] studied the chromatographic behavior of 47 metal ions on tiatnium(IV) arsenate papers in various concentrations of nitric acid and attained four important separations. They introduced a new parameter, Ri, which is the difference of R_f values on untreated and treated papers. Studies on amino acids[147] and alkaloids[148] were also undertaken by Qureshi et al., and the quantitative separation of nicotine-related alkaloids in aqueous and mixed solvents was reported.

TABLE 3
Various Separations of Anions on Hydrous Titanium Oxide Paper

Anions separated	Solvent system used
CrO_4^{2-}-PO_4^{3-}	6, 8, 9, 12, 13
CrO_4^{2-}-VO_4^{3-}	6, 8, 9, 12, 13
CrO_4^{2-}-WO_4^{2-}	6, 8, 9, 12, 13
CrO_4^{2-}-AsO_4^{3-}	6, 8, 9, 12, 13
CrO_4^{2-}-MoO_4^{2-}	12, 13
MoO_4^{2-}-PO_4^{3-}	6, 8, 9
MoO_4^{2-}-VO_4^{3-}	6, 8
MoO_4^{2-}-WO_4^{2-}	6, 8, 9
MoO_4^{2-} – AsO_4^{2-}	6, 8, 9
I^--IO_3^-	1—16
ClO_3^--IO_3^-	3, 4, 6
BrO_3^--IO_3^-	3, 4, 6
ClO_3^--BrO_3^--IO_3^-	5, 10, 11, 15
SO_4^{2-}-$S_2O_3^{2-}$-$S_2O_8^{2-}$	9—16
$Fe(CN)_6^{4-}$-$Fe(CN)_6^{3-}$	1—4, 6—8, 10—16
CrO_4^{2-}-VO_4^{3-}-PO_4^{3-}	12
CrO_4^{2-}-MoO_4^{2-}-WO_4^{2-}	9
CrO_4^{2-}-MoO_4^{2-}-VO_4^{3-}	14
CrO_4^{2-}-MoO_4^{2-}-PO_4^{3-}	12, 13, 14
VO_4^{3-}-WO_4^{2-}-PO_4^{3-}	9
F^--Cl^--Br^--I^--CNS^-	10, 11, 15, 16
CrO_4^{2-} from a mixture of AsO_4^{3-}, WO_4^{2-}, PO_4^{3-}, MnO_4^-, and VO_4^{3-}	6, 8, 9, 12, 13
MoO_4^{2-} from a mixture of AsO_4^{3-}, WO_4^{2-}, PO_4^{3-}, MnO_4^-, and VO_4^{3-}	6, 8, 9, 12

Solvent System/Composition

1. 0.10 *M* $NaNO_3$
2. 0.10 *M* NaCl
3. 0.10 *M* NaAc
4. 0.10 *M* Na_2SO_4
5. 1.0 *M* Na_2SO_4
6. 0.10 *M* Na_2CO_3
7. 1.0 *M* Na_2CO_3
8. 0.10 *M* $NaHCO_3$
9. Acetone-3 *M* NH_4OH (1:1) v/v
10. Acetone-3 *M* NH_4OH (7:3) v/v
11. Acetone-*n*-butanol-3 *M* NH_4OH (4:3:3) v/v/v
12. Acetone-methanol-3 *M* HCl (4:3:3) v/v/v
13. Acetone-*n*-butanol-3 *M* HCl (4:3:3) v/v/v
14. Acetone-ethyleneglycol-3 *M* HCl (4:3:3) v/v/v
15. Acetone-ethanol-20% acetic acid (4:3:3) v/v/v
16. Acetone-*n*-butanol-20% acetic acid (4:3:3) v/v/v

Reference: Adapted from Surendranath, K. V. and Tandon, S. N., Chromatographic separation of anions on papers impregnated with hydrous titanium oxide, *J. Liq. Chromatogr.*, 11, 1433, 1988.

A systematic study of the chromatographic behavior of 48 metal ions on plain and stannic arsenate papers in a dimethyl sulfoxide-nitric acid system was reported by Qureshi and Sharma.[149] As a result, separation of six binary mixtures was achieved and the dependence of R_f values on the nitric acid concentration in the solvent was shown as $R_f = kc^1/_2$, where c is the acid concentration (Table 6). Complexation ion-exchange electrochromatography[150]

TABLE 4
Representative Separations of Amino Acids on Hydrous Zirconium Oxide Paper in 10^{-3} *M* HC1

Separation no.	Amino acids separated
1	L-Tyrosine (0.11), L-tryptophan (0.32), L-phenylalanine (0.61)
2	Cysteine (0.15), serine (0.33), L-leucine (0.50), DL-valine (0.76)
3	Cysteine (0.12), L-tyrosine (0.30), A-alanine (0.47), β-alanine (0.73)
4	Aspartic acid (0.0), aspargine (0.25), glutamine (0.42), L-histidine (0.60), arginine (0.70)
5	Glutamic acid (0.0), cysteine (0.11), L-tyrosine (0.27), L-methionine (0.40), L-leucine (0.53), L-phenylalanine (0.65)
6	Cysteine (0.13), L-threonine (0.27), L-methionine (0.42), L-leucine (0.53), L-histidine (0.62), arginine (0.71), L-lysine (0.82)
7	L-Tryptophan (0.11), L-threonine (0.26), L-methionine (0.54), L:-valine/L-leucine (0.70), L-phenylalanine/L-histidine (0.82), L-lysine (0.93)

Note: R_f values observed in the mixtures are given in parentheses.

Reference: Adapted from Singh, N. J., Rajeev, and Tandon, S. N., Rapid separation of amino acids on hydrous zirconium oxide papers, *Indian J. Chem.*, 15B, 581, 1977.

TABLE 5
Separations Achieved of Metal Ions on Stannic Arsenate Papers

Solvent system	Separations achieved, metal ion (R_T − R_L)	Time (h)
Butanol + 50% HNO_3 (7:3)	Cd^{2+} (0.11—0.26)-Sb^{3+} (0.65—0.81)	2
	Ag^+ (0.00—0.00)-Cu^{2+} (0.09—0.21)	
	Pt^{4+} (0.43—0.67)-Au^{3+} (0.90—1.00)	
	VO^{2+} (0.14—0.24)-UO_2^{2+} (0.43—0.63)	
	Ni^{2+} (0.06—0.18)-Pd^{2+} (0.42—0.60)	
Butanol + 50% HNO_3 (6:4)	Pb^{2+} (0.13—0.26)-UO_2^{2+} (0.48—0.64)	2
	Zr^{4+} (0.00—0.06)-Th^{4+} (0.14—0.28)	
	Ag^+ (0.00—0.00)-Cd^{2+} (0.16—0.29)	
	Hg^{2+} (0.69—0.85)-Ir^{3+} (0.05—0.20)	
	Pt^{4+} (0.40—0.62)-Au^{3+} (0.84—1.00)	
	Fe^{3+} (0.00—0.14)-VO^{2+} (0.21—0.30)	
	Mo^{6+} (0.00—0.20)-UO_2^{2+} (0.47—0.67)	
Butanol + 50% HNO_3 (1:1)	Th^{4+} (0.08—0.28)-UO_2^{2+} (0.40—0.68)	2
	Al^{3+} (0.24—0.38)-Be^{2+} (0.64—0.80)	
	Ag^+ (0.00—0.00)-Cd^{2+} (0.28—0.40)—Pd^{2+} (0.55—0.79)	
	Ag^+ (0.00—0.00)-Cu^{2+} (0.19—0.35)	
	Pd^{2+} (0.56—0.79)-Au^{3+} (0.94—1.00)	
	Ag^+ (0.00—0.00)-Tl^+ (0.24—0.37)	
	Fe^{3+} (0.00—0.22)-Cr^{3+} (0.28—0.42)	
	Ag^+ (0.00—0.00)-Pd^{2+} (0.19—0.35)	
Butanol + 50% HNO_3 (9:1)	Cr^{3+} (0.00—0.10)-Zn^{2+} (0.58—0.68)	2
	Mn^{2+} (0.00—0.10)-Zn^{2+} (0.58—0.70)	
	Al^{3+} (0.00—0.08)-Zn^{2+} (0.56—0.68)	
	Co^{2+} (0.00—0.10)-Zn^{2+} (0.55—0.70)	
	Ni^{2+} (0.00—0.10)-Zn^{2+} (0.55—0.68)	
	Ni^{2+} (0.00—0.80)-Pt^{4+} (0.30—0.42)	
	Cd^{2+} (0.80—0.18)-Zn^{2+} (0.58—0.68)	

Reference: Adapted from Qureshi, M. and Sharma, S. D., Chromatography of metal ions on stannic arsenate and plain papers in butanol-nitric acid media: prediction of K_{Sp} from R_f values, *Anal. Chem.*, 45, 1283, 1973.

TABLE 6
Separations of Metal Ions Achieved Experimentally on Stannic Arsenate Papers

Solvent	Separations achieved (R_T-R_L)	Time
DMSO	Ni^{2+} (0.00—0.08)-Pd^{2+} (0.72—0.92)	2 h
	Cd^{2+} (0.06—0.16)-Hg^{2+} (0.68—0.84)	
	Ni^{2+} (0.00—0.16)-Au^{3+} (0.66—0.84)	
	SeO_3^{2-} (0.00—0.06)-Au^{3+} (0.77—0.88)	
DMSO/1 mol dm^{-3} HNO_3 (9:1)	Ag^+ (0.00—0.00)-Cd^{2+} (0.20—0.36)-Hg^{2+} (0.64—0.90	2 h
	Ca^{2+} (0.00—0.00)-Mg^{2+} (0.48—0.70)	
DMSO/1 mol dm^{-3} HNO_3 (7:3)	Ga^{3+} (0.00—0.10)-Y^{3+} (0.60—0.80)	2 h
	Bi^{3+} (0.00—0.38)-Cu^{2+} (0.56—0.78)	
	Sr^{2+} (0.53—0.67)-Mg^{2+} (0.80—0.96)	
	Ag^+ (0.00—0.12)-Ni^{2+} (0.68—0.84)-Hg^{2+} (0.96—1.00)	
	Ag^+ (0.00—0.02)-Cu^{2+} (0.52—0.70)-Pt^{4+} (0.86—1.00)	
	Ag^+ (0.00—0.10)-Cd^{2+} (0.62—0.74)-Pd^{2+} (0.85—0.97)	
	Ag^+ (0.00—0.00)-Cu^{2+} (0.55—0.70)-Au^{3+} (0.90—1.00)	
	Bi^{3+} (0.00—0.33)-Tl^+ (0.44—0.53)-Hg^{2+} (0.97—1.00)	
	Ag^+ (0.00—0.06)-Cu^{2+} (0.69—0.83)-Hg^{2+} (0.95—1.00)	
DMSO/1 mol dm^{-3} HNO_3 (1:1)	Cr^{3+} (0.00—0.20)-UO_2^{2+} (0.70—0.98)	2 h
	Tl^+ (0.25—0.43)-Hg^{2+} (0.87—0.97)	
	Bi^{3+} (0.00—0.40)-Hg^{2+} (0.87—1.00)	
	Ag^+ (0.00—0.00)-Tl^+ (0.30—0.48)-Cd^{2+} (0.65—0.85)	
DMSO/1 mol dm^{-3} HNO_3 (3:7)	Cr^{3+} (0.00—0.00)-Mn^{2+} (0.69—0.87)	1 h
DMSO/1 mol dm^{-3} HNO_3 (1:9)	Fe^{3+} (0.00—0.32)-VO^{2+} (0.70—0.90)	30 min
	Ag^+ (0.00—0.04)-Pb^{2+} (0.64—0.84)	
	Fe^{3+} (0.00—0.33)-Al^{3+} (0.60—0.85)	
	Fe^{3+} (0.00—0.34)-Ni^{2+} (0.68—0.90)	
	Fe^{3+} (0.00—0.33)-Co^{2+} (0.66—0.90)	
	Bi^{3+} (0.00—0.30)-Pd^{2+} (0.64—0.90)	
DMSO/8 mol dm^{-3} HNO_3 (7:3)	In^{3+} (0.00—0.04)-Ga^{3+} (0.50—0.70)	2 h
	In^{3+} (0.00—0.05)-Al^{3+} (0.78—0.98)	
	Cr^{3+} (0.00—0.00)-Al^{3+} (0.77—1.00)	
	Th^{3+} (0.00—0.18)-VO^{2+} (0.72—0.94)	
	Th^{4+} (0.00—0.15)-UO_2^{2+} (0.78—0.98)	
	ZrO^{2+} (0.00—0.02)-UO_2^{2+} (0.80—1.00)	
	Cr^{3+} (0.00—0.00)-Zn^{2+} (0.80—1.00)	
	Cr^{3+} (0.00—0.00)-Mn^{2+} (0.87—1.00)	
	Cr^{3+} (0.00—0.00)-MoO_4^{2+} (0.85—1.00)	
	Th^{4+} (0.00—0.14)-MoO_4^{2+} (0.83—1.00)	
	Th^{4+} (0.00—0.12)-WO_4^{2-} (0.72—0.86)	

Reference: Adapted from Qureshi, M. and Sharma, S. D., A systematic study of the chromatographic behavior of 48 metal ions on plain and stannic arsenate impregnated papers in DMSO-HNO_3 systems, *Chromatographia,* 11, 153, 1978.

of 52 metal ions, leading to many interesting separations (Table 7), has also been performed with these papers. Similar studies have been reported on tin(IV) phosphate papers.[151] Separation of metal ions on titanium(IV) molybdate papers in sodium form using sodium nitrate solutions of various concentrations was performed by Rawat and Mujtaba.[152] Tandon and Mathew,[153] and Qureshi et al.[154] adopted stannic antimonate-treated papers for the separation (Table 8) and microdetermination of metal ions, the latter in DMSO-mixed solvents. Qureshi and co-workers[155] have also reported electrochromatography of metal ions on the same material in the same system. Tandon et al.[156,157] and Rajput et al.[158,159] achieved rapid metal ion separations on zirconium antimonate-treated papers. Many other separations have also been reported. Electrophoresis of various metal ions on zirconium tungstate paper and a

TABLE 7
Some Possible Separations of One Cation from Numerous Metal Ions by Electrochromatography on Stannic Arsenate Papers

Ion(s)	Background electrolyte	Interfering ion(s)
K^+ or Rb^+ or Cs^+ (+15.0) from 49 cations	0.5 *M* citric acid	None
Tl^+ or Co^{2+} (+7.0) from 48 cations	0.5 *M* citric acid	Sr^{2+}, Zn^{2+}
Sr^{2+} (+8.0) from 49 cations	0.5 *M* citric acid	Tl^+, Co^{2+}
Nb^{5+} (−5.4) from 50 cations	0.5 *M* tartaric acid	Mo^{6+}
VO^{2+} (+8.0) from 51 cations	0.5 *M* oxalic acid	None
Fe^{3+} or Tl^+ (+12.5) from 49 cations	0.5 *M* oxalic acid	K^+, Rb^+, Cs^+

Note: Values in parentheses are the migrations in cm of the metal ions under an electric field toward the cathode (+) and anode (−). Time = 6 h, V = 100 volts, I = 4.40 mA.

Reference: Adapted from Qureshi, M. and Sharma, S. D., Complexation ion exchange electrochromatography of 52 metal ions on stannic arsenate papers, *Sep. Sci. Technol.*, 13, 723, 1978.

specific separation of palladium by Qureshi et al.[160] is another highlight of the study. Gopal and Mishra[161] chose pyroantimonic acid papers for the separation of some selective radionuclides. Studies of phenols, including their identification, specific separation of Ga^{3+}, Au^{3+}, and Mg^{2+}, and the electrophoretic separation of metal ions, have involved the application of stannic molybdate papers, as reported by Rawat and co-workers.[162-164] Titanium(IV) tungstate papers have been utilized for the quantitative separation of magnesium and palladium, and of other metal ions by Qureshi et al.[166] and Husain,[165] respectively. De and Chakraborty[167] utilized titanium(IV) antimonate papers for the electrochromatography of metal ions, and Srivastava et al.[168] have reported studies on pyridinium tungstoarsenate papers. Tables 9 to 12 summarize the separations achieved on some selected impregnated papers of this class.

d. Chromatography on Papers Impregnated with Insoluble Metal Ferrocyanides

No significant contribution has been reported on the chromatographic study of metal ions on metal ferrocyanide papers. It may be due to the colored papers produced on impregnation and a comparatively less chemical stability of ferrocyanides. Only Varshney and co-workers[137,169,170] have reported studies of metal ions on papers impregnated with tin(II) and tin(IV) ferrocyanides. They have also achieved some metal ion separations on these papers, as shown in Table 13.

e. Chromatography on Papers Impreganted with Heteropolyacid Salts

Following the valuable contribution of Alberti,[111] who prepared ammonium molybdophosphate papers for the chromatography of alkali metals in acidic solutions, Shih and co-workers have used the papers for similar studies in ethanol-ammonium nitrate systems.[171] They have also prepared ammonium tungstophosphate papers[172] and developed them in methanol-ammonium nitrate-nitric acid systems to study alkali metal ions. Pe Hai Yin performed electrophoresis of potassium, rubidium, and cesium on ammonium molybdophosphate paper.[173] Marcu and Ghizdavu[174] separated Cs^+, Zr^{4+}, Ru^{3+}, UO_2^+, and Co^{2+} on papers impregnated with $H_6(SiW_{10}Nb_2O_{40})$ in 0.1 to 1 *M* HNO_3-NH_4NO_3 systems. They have also used[175] pyridine and 2-picolene-based materials for the study of various fission products. Sn(IV) arsenosilicate and arsenophosphate papers have been used by Varshney and co-workers[176] for the chromatographic study of some transition and alkaline earth metals. They have also separated amino acids on papers impregnated with tin(IV) and thorium(IV) phosphosilicate papers.[177]

TABLE 8
R_f Values of Metal Ions in Various Solvent Systems on Stannic Antimonate Paper

Metal ion	($R_f \times 100$) value									
	HCl		HNO_3		$HClO_4$		CH_3COOH		0.01 *N* HCl + 0.10 *M* NH_4Cl	Butanol: HCl (80:20)
	0.1 *N*	1.0 *N*	0.1 *N*	1.0 *N*	0.1 *N*	1.0 *N*	0.1 *N*	1.0 *N*		
Li^+	95	95	90	95	90	95	92	95	89	0
Na^+	92	95	92	95	92	98	95	98	91	0
K^+	90	95	95	90	91	92	89	98	90	0
Rb^+	95	98	95	92	90	95	92	96	88	0
Cs^+	95	95	95	98	89	95	93	93	92	0
Ag^+	0	0	0	0	0	0	0	0	0	0
Tl^+	0	0.18—0.0	0.20—0.0	0.30—0.0	0.20—0.0	0.24—0.0	0.0	0.12—0.0	0.0	0.0
Be^{2+}	100—74	95	100—63	89	100—54	85	93	88	93	43
Mg^{2+}	95	95	100—65	90	92	95	91	94	90	0
Ca^{2+}	68	88	79	88	75	95	78	82	77	0
Mn^{2+}	80	95	80	93	44	83	775	82	63	10
Co^{2+}	52—0	87	53	79	30	81	66	95	62	10
Cu^{2+}	24—0	82	30—0	80	30	68—0.32	34	37	34	45
Zn^{2+}	96—62	80	95—74	85	92—50	82	78	80	90	90
Sr^{2+}	90	88	88	88	82	95	92	84	25	0
Pd^{2+}	82	94	95	92	90	86	80	79	76	45
Cd^{2+}	84	90	76	85	85	85	86	85	67—40	78
Ba^{2+}	30	50	40	75—43	0	46	14	84	32	0
Hg^{2+}	85	90	82	82	84	85	72	70	81	30
Pb^{2+}	20	14	0	14	0	0	12	12	0	0
Al^{3+}	80	95	75—34	90	0—35	80	93	93	30	0
Cr^{3+}	90	91	88	93	80	92	77	90	93	0
Fe^{3+}	10	77	13	40	0	30	21	62	0	0
Ga^{3+}	76	92	62	98	90	92	86	88	88	32
In^{3+}	93	89	88	98	90	92	79	84	90	48
La^{3+}	85	88	57—15	90	66	84	22	43	88	0

TABLE 8 (continued)
R_f Values of Metal Ions in Various Solvent Systems on Stannic Antimonate Paper

Metal ion	$(R_f \times 100)$ value									
	HCl		HNO_3		$HClO_4$		CH_3COOH		0.01 *N* HCl + 0.10 *M* NH_4Cl	Butanol: HCl (80:20)
	0.1 *N*	1.0 *N*	0.1 *N*	1.0 *N*	0.1 *N*	1.0 *N*	0.1 *N*	1.0 *N*		
Au^{3+}	55	45	58	64	58	58	50	65	50	89
Bi^{3+}	0	09	0	0	0	0	0	0	0	24
Ti^{4+}	0	0	0	0	0	0	0	0	0	0
ZrO^{2+}	85	28	28	28	25	50	0	0	26	0
Ce^{4+}	30	90	0	23	0	0	0	0	76—55	0
Pt^{4+}	88	79	92	80	92	92	61	82	94	88—44
Th^{4+}	57	80	69	90	68	80—20	78	76	20	0
Mo^{6+}	0	0	0	0	0	0	0	0	0	0
VO^{2+}	68	91	81	86	50	84	17	88	50	18
UO_2^{2+}	88	82	80	78	53	80	50	84	28	42

Reference: Adapted from Mathew, J., Rajeev, and Tandon, S. N., Rapid separation of microgram amount of metal ions on stannic antimonate ion exchange paper, *Mikrochim. Acta*, 5, 1977.

TABLE 9
Separations of Metal Ions Achieved on Stannic Tungstate Papers

Sl. No.	Solvent System	Separations
1.	HNO_3:acetone:H_2O (1:1:1)	ZrO^{2+} (0.08—0.00)-La^{3+} (0.85—0.65)
		ZrO^{2+} (0.08—0.00)-Th^{4+} (0.65—0.37)
		Se^{4+} (0.12—0.00)-Tl^{+} (0.57—0.45)
		Se^{4+} (0.12—0.00)-Cu^{2+} (0.89—0.62)
		Tl^{+} (0.47—0.36)-Hg^{2+} (0.85—0.65)
		Sb^{3+} (0.27—0.00)-Hg^{2+} (0.95—0.70)
		Sb^{3+} (0.15—0.00)-Cd^{2+} (0.80—0.60)
		Hg_2^{2+} (0.10—0.00)-Tl^{+} (0.52—0.32)-UO_2^{2+} (0.89—0.65)
2.	HNO_3:acetone:H_2O (1:2:1)	ZrO^{2+} (0.07—0.00)-Ce^{4+} (0.62—0.47)-Hg^{2+} (1.0—0.80)
		ZrO^{2+} (0.05—0.00)-Pb^{2+} (0.42—0.20)-Hg^{2+} (1.0—0.82)
		ZrO^{2+} (0.08—0.00)-Rare earths (0.72—0.47)
		Ag^{+} (0.08—0.00)-Pb^{2+} (0.45—0.25)-Th^{4+} (0.78—0.62)
		As^{3+} (0.20—0.00)-Cd^{2+} (0.65—0.37)-Hg^{2+} (1.00—0.80)
3.	HNO_3:acetone:H_2O (1:3:1)	Ti^{4+} (0.15—0.00)-Th^{4+} (0.90—0.80)
		Se^{4+} (0.10—0.00)-Zn^{2+} (0.54—0.30)
4.	HNO_3:acetone:H_2O (1:4:1)	Cu^{2+} (0.45—0.18)-Bi^{3+} (0.82—0.62)
		Tl^{+} (0.12—0.00)-Be^{2+} (0.68—0.46)
		Fe^{3+} (0.12—0.00)-Be^{2+} (0.70—0.50)
		Ti^{4+} (0.13—0.00)-Cd^{2+} (0.50—0.28)
5.	HNO_3:acetone:H_2O (1:5:1)	Ag^{+} (0.16—0.00)-Bi^{3+} (0.83—0.48)
		Fe^{3+} (0.19—0.00)-UO_2^{2+} (0.90—0.68)
		UO_2^{2+} (0.93—0.70) from 28 cations (Bi, Th, Hg, and Sb interfere)
6.	HNO_3:acetone:H_2O (1:6:1)	Pb^{2+} (0.14—0.09)-Bi^{3+} (0.97—0.60)
		As^{3+} (0.28—0.00)-Sb^{3+} (0.80—0.60)
		Th^{4+} (0.80—0.77) from 28 cations (Bi, UO_2, Hg, and Sb interfere)
7.	HNO_3:acetone:H_2O (2:1:1)	Ag^{+} (0.05—0.00)-Cd^{2+} (0.90—0.60)
		As^{3+} (0.20—0.00)-Cd^{2+} (0.90—0.60)
		ZrO^{2+} (0.25—0.00)-Y^{3+} (0.95—0.60)
		Se^{4+} (0.16—0.00) from 26 cations (Ti, Fe, As, Ag, ZrO, and Hg_2^{2+} interfere)
		Ag^{+} (0.10—0.00) from 26 cations (Ti, Fe, As,Se, ZrO, and Hg_2^{2+} interfere)
8.	HNO_3:acetone:H_2O (3:1:1)	Ag^{+} (0.00)-Mn^{2+} (0.86—0.66)
		Ti^{4+} (0.18—0.00)-Rare earths (0.90—0.60)
9.	HNO_3:acetone:H_2O (1:1:2)	Se^{4+} (0.17—0.00) from 24 cations (Ti, Fe, ZrO, Ag, Sb, Bi, and Hg_2^{2+} interfere)
		ZrO^{2+} (0.10—0.00) from 24 cations (Ti, Fe, Se, Ag, Sb, Bi, and Hg_2^{2+} interfere)

Reference: Adapted from Qureshi, M., Varshney, K. G., Gupta, M. P., and Gupta, S. P., Cation chromatography on stannic tungstate papers: quantitative separation of vanadium from iron, titanium, zirconium and thorium, *Chromatographia,* 10, 29, 1977.

f. Separation of Radionuclides

Since separations on papers impregnated with inorganic ion exchangers are very fast, these papers have been used for the separation of radioactive elements with short half-lives. Adloff[129] has used various types of papers impregnated with various inorganic ion exchangers to separate radionuclides with short half-lives, the results of which are summarized in Table 14.

TABLE 10
R_f Values of Some Metal Ions on Zinc Silicate Papers in EDTA Solutions of Various Concentrations

		$R_f \times 100$				
		Molarity of EDTA				
Sl. No.	Metal ion	0.026	0.053	0.134	0.214	0.260
1	Hg_2^{2+}	21	32	46	56	69
2	Hg^{2+}	20	31	—	64	—
3	Ba^{2+}	80	91	86	74	75
4	ZrO^{2+}	0	15	26	38	87
5	Bi^{3+}	66	82	85	82	88
6	Cu^{2+}	82	83	86	87	96
7	Ni^{2+}	97	93	92	95	96
8	Co^{2+}	95	87	85	83	88
9	Al^{3+}	0	62	77	88	85
10	Be^{2+}	9	29	42	60	68
11	Rb^{+}	80	91	93	91	90
12	Fe^{3+}	0	32	54	62	67
13	K^{+}	91	91	93	95	97
14	Ag^{+}	0	0	0	0	0
15	Th^{4+}	0	38	54	73	83

Separations Achieved

Developers	Separations
1% EDTA	Hg_2^{2+}-Bi^{3+}-Be^{2+}, Cu^{2+}-Ag^{+}, Co^{2+}-Hg_2^{2+}, ZrO^{2+}-Ni^{2+}, ZrO^{2+}-Co^{2+}, Rb^{+}-Al^{3+}, ZrO^{2+}-Cu^{2+}, Ni^{2+}-Ag^{+}, Ni^{2+}-Hg^{2+}, Fe^{3+}-Hg_2^{2+}, Ag^{+}-Hg_2^{2+}
	Ba^{2+} from a mixture of Hg_2^{2+}, Hg^{2+}, ZrO^{2+}, Al^{3+}, Be^{2+}, Fe^{3+}, and Pb^{2+}
	Be^{2+}, Al^{3+}, Ag^{+}, and Th^{4+} individually from a mixture of Ba^{2+}, Bi^{3+}, Ni^{2+}, Co^{2+}, Rb^{+}, and K^{+}
	Ni^{4+}, Co^{2+}, Rb^{+}, and K^{+} individually from a mixture of Hg^{2+}, Hg_2^{2+}, ZrO^{2+}, Al^{3+}, Be^{2+}, Fe^{3+}, Pb^{2+}, and Ag^{+}
	Hg_2^{2+} or Hg^{2+} from a mixture of Ba^{2+}, Bi^{3+}, Cu^{2+}, Co^{2+}, Ni^{2+}, Rb^{2+}, and K^{+}
2% EDTA	Hg^{2+}-Ag^{+}-Pb^{2+}, Hg_2^{2+}-Ag^{+}-Bi^{3+}, Fe^{2+}-Pb^{2+}, ZrO^{2+}-Pb^{2+}, Pb^{2+}-Hg^{2+}, Be^{2+}-K^{+}, Be^{2+}-Rb^{+}, Bi^{3+}-Hg_2^{2+}, Be^{2+}-Bi^{3+}, Be^{2+}-Pb^{2+}
5% EDTA	ZrO^{2+}-Th^{4+}-Pb^{2+}, ZrO^{2+}-Th^{4+}-Cu^{2+}, Al^{3+}-Be^{2+}

Reference: Adapted from Rawat, J. P. and Chitra, Chromatography of certain metal ions on ligand combined ion exchange papers, *J. Indian Chem. Soc.*, 63, 857, 1986.

g. *Conclusion and Further Prospects*

As the preceding sections have emphasized, inorganic ion-exchange papers have a wide spectrum of applications. They give very quick and clean separations of various types of species, such as cations, anions, radionuclides, amino acids, and alkaloids. They are very suitable for specific separations of particular ions from complex mixtures. However, compared with the very large number of available inorganic ion-exchange materials and their column applications, paper chromatographic studies are not significant. This may be due to a rapid growth of technology, whereby newer and faster methods such as HPLC are available. However, paper chromatography, being an inexpensive and simple technique, is still useful and promising for the less well-equipped laboratories. It can produce some novel separations.

TABLE 11
Separations of Metal Ions Achieved on Ti(IV) Antimonate Papers

Developer	Separations achieved (R_f)	Time
DMSO (pure)	Cs^+ (0.02)-K^+ (0.31)	2 h
	Cu^{2+} (0.04)-Mn^{2+} (0.70)	
	Co^{2+} (0.00)-Mn^{2+} (0.62)	
	W^{6+} (0.00)-Pt^{4+} (0.73)	
	Be^{2+} (0.07)-Mg^{2+} (0.73)	
	Fe^{3+} (0.09)-Ru^{3+} (0.77)	
	Sm^{3+} (0.00)-Pt^{3+} (0.72)	
	In^{3+} (0.00)-Ga^{3+} (0.82)	
DMSO + 0.01 *M* HNO_3 (1:1)	Ba^{2+} (0.06)-K^+ (0.51)-Sr^{2+} (0.96)	2 h
	Ba^{2+} (0.04)-Cs^+ (0.41)-Mg^{2+} (0.94)	
	Co^{2+} (0.12)-Ni^{2+} (0.85)	
	Ca^{2+} (0.10)-Sr^{2+} (0.94)	
DMSO + 0.1 *M* HNO_3 (1:1)	Tl^+ (0.04)-K^+ (0.51)-VO^{2+} (0.95)	2 h
	Tl^+ (0.06)-Cs^+ (0.43)-UO_2^{2+} (0.94)	
	Fe^{3+} (0.06)-K^+ (0.51)-Mn^{2+} (0.96)	
	Cu^{2+} (0.07)-Cd^{2+} (0.97)	
	Co^{2+} (0.05)-Pd^{2+} (0.95)	
DMSO + 1 *M* HNO_3 (1:1)	Mo^{6+} (0.10)-Cr^{3+} (0.97)	2 h
	W^{6+} (0.00)-Cr^{3+} (0.98)	
	Mo^{6+} (0.14)-Pt^{4+} (0.94)	
DMSO + 3 *M* HNO_3 (1:1)	Tl^+ (0.12)-K^+ (0.54)-Al^{3+} (0.94)	2 h
	Mo^{6+} (0.08)-Ni^{2+} (0.86)	
	Tl^+ (0.12)-Bi^{3+} (0.95)	
	Pb^{2+} (0.08)-Cd^{2+} (0.94)	
	Tl^+ (0.11)-Hg^{2+} (0.94)	
DMSO + 6 *M* HNO_3 (1:1)	Pb^{2+} (0.23)-Zn^{2+} (0.87)	2 h
	Tl^+ (0.14)-In^{3+} (0.96)	
	Nb^{5+} (0.14)-VO^{2+} (0.95)	
	Hf^{4+} (0.00)-Th^{4+} (0.95)	
	Zr^{4+} (0.15)-Th^{4+} (0.95)	

Reference: Adapted from Rajput, R. P. S. and Seth, N. S., Chromatographic behavior of metal ions on tin(IV) and titanium(IV) antimonate papers, *Chromatographia*, 13, 219, 1980.

2. Thin-Layer Chromatography

Thin layer chromatography (TLC) is very similar to paper chromatography, except that the stationary phase is a thin layer of finely divided adsorbent supported on a glass or aluminium plate, plastic strip, etc. Any of the solids used in column chromatography can be used, provided a suitable binder is found for good adherence to the plate. Sometimes the adsorbents are spread over the plates without any binder. This technique has a distinct advantage over paper chromatography because of its convenience and rapidity, its greater sharpness of separation, and its high sensitivity.

In thin-layer chromatography, a variety of coating materials are available as stationary phase. Silica gel is used more often than other materials. The separation of cations on silica gel is not, however, always satisfactory, as many cations have similar R_f values and remain grouped together on this adsorbent. Cellulose powder is recommended as an adsorbent for the separation of cations by TLC even though separations may be slower than those obtained on silica gel. The use of cellulose may be regarded as a substitute for paper chromatography, and data obtained for inorganic paper chromatography are generally applicable to inorganic TLC on cellulose. Another popular adsorbent has been alumina. Both silica gel and alumina contain hydroxyl groups or oxygen atoms. Alumina is preferred for the separation of weakly

TABLE 12
Separations of Metal Ions Achieved on Titanium(IV) Arsenate Papers

Separation	Solvent
Fe^{3+} (0.0—0.28)-Cr^{3+} (0.65)	0.001 *M* HNO_3
Bi^{3+} (0.0)-Cu^{2+} (0.86)	0.01 *M* HNO_3
Bi^{3+} (0.0)-Hg^{2+} (0.46—0.75)	0.01 *M* HNO_3
Th^{4+} (0.1)-Y^{3+} (0.55)	0.01 *M* HNO_3
Pt^{4+} (0.85)-Ir^{3+} (0.00)	0.1 *M* HNO_3
Pt^{4+} (0.85)-Au^{3+} (0.55)	0.1 *M* HNO_3
Sb^{3+} (0.0)-Cu^{2+} (0.85)	0.1 *M* HNO_3
Sb^{3+} (0.0)-Cd^{2+} (0.64)	0.1 *M* HNO_3
Cd^{2+} (0.62)-Tl^{+} (0.0-0.3)	0.1 *M* HNO_3
Bi^{3+} (0.08-Cd^{2+} (0.65)	0.01 *M* HNO_3
Ag^{+} (0.0)-Cu^{2+} (0.69)	0.01 *M* HNO_3
Ag^{+} (0.0)-Au^{3+} (0.85)	0.01 *M* HNO_3
Th^{4+} (0.0—0.2)-La^{3+} (0.51)	0.01 *M* HNO_3
Ca^{2+} (0.89)-Al^{3+} (0.4)	0.1 *M* HNO_3
Nb^{3+} (0.0—0.28)-Zr^{4+} (0.85)	1.0 *M* HNO_3
Nb^{3+} (0.0—0.28)-UO_2^{2+} (0.70)	1.0 *M* HNO_3
Sb^{3+} (0.06)-Cu^{2+} (0.90)	2.0 *M* HNO_3
Sb^{3+} (0.06)-Cd^{2+} (0.92)	2.0 *M* HNO_3
Au^{3+} (0.55)-Sb^{3+} (0.08)	3.0 *M* HNO_3
Sb^{3+} (0.06)-Sn^{2+} (0.75)	3.0 *M* HNO_3
Sb^{3+} (0.15)-Numerous	4.0 *M* HNO_3
Ag^{+} (0.0)-Pt^{4+} (0.84)	1.0 *M* NH_4NO_3 + 0.1 *M* HNO_3
Pt^{4+} (0.76)-Ir^{3+} (0.10)	Water
Pt^{4+} (0.81)-Pb^{2+} (0.07)	Water
Pt^{4+} (0.81)-Bi^{3+} (0.05)	Water
Pt^{4+} (0.81)-Ni^{2+} (0.07)-Co^{2+} (0.0)	Water
Pt^{4+} (0.81)-Ag^{+} (0.0)	Water

Reference: Adapted from Qureshi, M., Rawat, J. P., and Sharma, V., Chromatographic behavior of 47 metal ions on titanium(IV) arsenate papers, *Talanta*, 20, 267, 1973.

polar compounds, but silica gel is preferred for polar compounds, such as amino acids and sugars. Magnesium silicate, calcium silicate, and activated charcoal are also used as adsorbents. Ion-exchange resins in particle sizes of 40 to 80 μm are also suitable for preparing thin-layer plates. Examples are Dowex-50W strong-acid cation exchange and Dowex-1 strong anion exchange resins, usually in the sodium or hydrogen or the chloride forms, respectively.

The presence of an ionogenic material on the plate is advantageous, particularly for the study of ionic species. Here, the ion exchange property of the adsorbent plays a more prominent role than its simply adsorption behavior. Apart from silica gel and alumina, which possess hydroxyl groups showing ion-exchange characteristics, other inorganic ion exchangers have also found use in TLC in the same manner as organic resins.

For the preparation of the stationary phase, an aqueous slurry of the material is prepared, usually with a binder such as plaster of paris, gypsum, or poly (vinyl alcohol) to help it adhere to the backing material. The slurry is spread on the plate in a thin film, typically 0.1 to 0.3 mm, using an applicator to assure uniform thickness.

Sherma and Fried,[178] in their recent review, have maintained the analytical capabilities of synthetic inorganic ion exchangers as adsorbents in TLC. For the sake of convenience, the studies made so far in this field are summarized below under the same heads used for the paper chromatographic studies described in the last section.

TABLE 13
Binary Separations of Metal Ions Achieved on Stannous Ferrocyanide Papers

Solvent system	Separations achieved
0.2 *M* NH_4NO_3 + 0.1 *M* HNO_3 (1:2)	Ca^{2+} (0.82—0.66)-Ba^{2+} (0.58—0.32)
	Mg^{2+} (0.98—0.84)-Sr^{2+} (0.40—0.18)
	Mg^{2+} (0.92—0.80)-Ba^{2+} (0.70—0.34)
	Ca^{2+} (0.00)-Sr^{2+} (0.93—0.81)
2 *M* NH_4NO_3 + 0.5 *M* HNO_3 (1:1)	Cr^{3+} (0.00)-Al^{3+} (0.96—0.87)
	Ga^{3+} (0.87—0.00)-Al^{3+} (1.0—0.93)
	In^{3+} (0.21—0.00)-Al^{3+} (1.0—0.87)
	Zr^{4+} (0.21—0.00)-Th^{4+} (0.93—0.68)
	Ca^{2+} (0.00)-Ba^{2+} (1.0—0.93)
	Sb^{3+} (0.15—0.00)-Pb^{2+} (1.0—0.75)
	Pt^{4+} (0.00)-Pb^{2+} (0.75—0.31)
2 *M* NH_4NO_3 + 2 *M* HNO_3 (1:1)	Au^{3+} (0.34—0.00)-Pb^{2+} (0.93—0.46)
	Cu^{2+} (0.25—0.00)-Pb^{2+} (0.93—0.75)
	Cd^{2+} (0.50—0.03)-Pb^{2+} (0.93—0.75)
	Ag^{+} (0.12—0.00)-Pb^{2+} (1.0—0.84)
	Ru^{3+} (0.00)-Pb^{2+} (0.97—0.75)
	Ir^{4+} (0.12—0.00)-Pb^{2+} (0.87—0.59)
	Hg^{2+} (0.25—0.00)-Pb^{2+} (0.86—0.62)
	As^{3+} (0.12—0.00)-Pb^{2+} (1.0—0.86)
	Ga^{3+} (0.40—0.00)-Al^{3+} (0.97—0.68)
0.8 *M* $NaNO_3$ + 4 *M* HNO_3 (1:1)	Ag^{+} (0.13—0.00)-Tl^{+} (0.92—0.34)
	Zr^{4+} (0.50—0.00)-La^{3+} (1.0—0.75)

Reference: Adapted from Qureshi, M., Varshney, K. G., and Khan, F., Chromatographic separation of metal ions on stannous ferrocyanide papers, *Sep. Sci.*, 8, 279, 1973.

a. Thin Layers of Hydrated Oxides

Zabin and Rollins[179] used, probably, for the first time zirconium oxide for the separation of Ni^{2+}, Co^{2+}, Pb^{2+}, Fe^{3+}, Ag^{+}, Hg^{2+}, Cd^{2+}, and Cu^{2+}. Berger et al.[180,181] also used it for the study of ferrocyanide, ferricyanide, and sulfocyanide in 3 *M* HCl, and of iodate, borate, and chlorate in 2 *M* HCl. Kirchner et al.[182] resolved terpenes on silicic acid and magnesium oxide layers.

The resolution of Bi^{3+} from 25 ternary and 12 quaternary mixtures was achieved by Sen and co-workers[183] in combinations of 0.1 *M* HCl, 0.1 *M* HNO_3, acetone, and dioxane (Table 15). They have separated up to 20 μg of bismuth by this method. The same authors also have chromatographed some anions and separated Cr(III) from several other metal ions on stannic oxide[184] plates eluted by various solvents. Rathore and co-workers[185] have isolated citric acid from ten carboxylic acids on calcium sulfate, ammonium molybdate, zinc oxide, and titanium oxide thin plates in ethyl acetate medium. Titanium oxide was also selected by Grace[186] for the chromatography of ortho-, para-, and meta-amino phenols. Cremer and Seid[187] prepared indium oxide plates and studied the movement of various anions in organic solvents.

b. Thin Layers of Insoluble Metal Salts of Polybasic Acids

As in the case of impregnated papers, the application of zirconium phosphate as a medium for thin-layer chromatography has found preference. Studies of 8 cations in hydrochloric acid, ammonium nitrate, and ammonium chloride systems by Zabin and Rollins,[179] 19 cations in 0.1 to 0.5 *M* hydrochloric acid by Keonig and Demiel,[188] and 8 cations in 1 *M* perchloric acid by Alberti et al.[189] are worth noting. Cerium(IV) phosphate sulfate layers were used by Keonig and Graf[190] for the chromatography of 10 cations in 0.2 *M* HCl. Cerium

TABLE 14
Some Separations of Radionuclides with Short Half Lives on Different Inorganic Ion-Exchange Papers

Radionuclides	Ion-exchange paper	Eluant	Time of development (min)	$R_f \times 100$
^{211}Pb (AcB, $t_{1/2}$ = 36 min) in equilibrium with ^{207}Th (AcC, $t_{1/2}$ = 4.76 min)	ZP	0.5 *M* HCl	5	Tl(10), Pb(100)
^{204}Tl-$^{204}Tl^{3+}$				
	ZO	0.1 *M* NH_4NO_3	12	Tl(60) Pb(0)
	ZP	0.5 *M* HCl	30	Tl^+(20), Tl^{3+}(90)
	ZO	0.1 *M* NH_4NO_3 0.1 *M* HCl	30	Tl^+(0), Tl^{3+}(80)
^{212}Pb (ThB, $t_{1/2}$ = 10.6 h) in equilibrium with ^{212}Bi (ThO, $t_{1/2}$ = 1 h)	ZO	0.05 *M* NH_4NO_3	30	Bi(0), Pb(40)
^{140}Ba-^{140}La	ZP	0.5 *M* NH_4Cl	20	La(0), Ba(90)
	ZO	0.1 *M* NH_4NO_3	30	La(0), Ba(90)
^{90}Sr-^{90}Y	ZP	0.5 *M* HCl	30	Y(30), Sr(100)
	ZO	0.1 *M* NH_4NO_3	30	Y(0), Sr(95)
^{228}Ra (MTH_1) in equilibrium with ^{220}Ac (MTh_2, $t_{1/2}$ = 6.13 h) ^{223}Ra (AcX)-^{223}Fr	ZP	0.5 *M* NH_4Cl	30	Ra(85), Ac(10)
	ZO	0.1 *M* NH_4NO_3	20	Fe(0), Ra(85)
(AcK, $t_{1/2}$ = 21 min) ^{137}Cs-^{223}Fr	ZW	0.1 *M* NH_4Cl	20	Ra(0), Fr(50)
	ZW	1 *M* NH_4Cl	20	Fr(50), Cs(60)

Note: ZP = zirconium phosphate, ZO = zirconium oxide, ZW = zirconium tungstate.

Reference: Adapted from Adloff, J. P., Separation of radioelements by chromatography on papers impregnated with ion exchange minerals, *J. Chromatogr.*, 5, 366, 1961.

molybdate was used by Srivastava et al.[191] to obtain data on 15 cations. They also compared the results on plain silica gel and mixed layers in benzene, ethylacetate, acetic acid, and hydrochloric acid systems. Separation of noble metals, and thereby the resolution of many binary and ternary mixtures on stannic phosphate in ammonia-hydrochloric acid and acetone-butanol-pyridine systems, and thin-layer electrophoresis of noble metals has been reported by Yin and associates.[193]

Qureshi et al.[194] prepared stannic tungstate thin layers for the study of 15 binary metal ion mixtures in three eluants (Table 17) and reported the separation of gold from numerous metal ions. Studies of 28 phenolic compounds have been performed by Nabi et al.[195] who have also explored the immunophysiological and bioanalytical aspects of the work and separated phenols from tea, cocoa,and grapes. Thin films of thorium tungstate for the quantitative separation of mercury(II) and gold(III) from various other metal ions in nitric acid-dioxane, butyl alcohol, and isobutyl methyl ketone solvents, followed by spectrophotometric determination of Hg(II), has been reported by De and Pal.[196]

Chromatography of 57 metal ions on stannic arsenate layers was performed by Husain and associates,[197,198] who separated many ions (Tables 18 and 19) and drew an interesting relationship between the pH of the solvent, activation time of the exchanger, and the thickness of the layer. Movement of platinum metals on some semicrystalline inorganic ion exhangers in six solvents has been reported,[199] as well as many separations predicted by the same workers. Qureshi et al.[200] reported 20 binary separations of metal ions on nonrefluxed stannic

TABLE 15
R_f Values of Separated Metal Ions on Hydrous Zirconium Oxide Layers

Adsorbent	Hydrous zirconium oxide
Developer	0.1 *M* HNO_3 + acetone (2:1) (v/v)
Conditions	Run: 11 cm; development times: 2 h; technique: ascending
Detection	Standard spot-test reagents
Main feature	Quantitative separation of Bi^{3+} from some ternary and quaternary mixtures of metal ions

Developer	(R_f × 100) values of separated metal ions
0.1 *M* HNO_3-acetone (2:1)	Bi^{3+}(0)-Au^{3+}(21), Mn^{2+}(91) or Co^{2+}(84)
	Bi^{3+}(0)-Au^{3+}(13)-Zn^{2+}(75) or Tl^{+}(65)
	Bi^{3+}(0)-Cu^{2+}(40)-Mn^{2+} or Co^{2+}(87)
	Bi^{3+}(0)-Cu^{2+}(30)-Cd^{2+} or Tl^{+}(67)
	Ag^{+}(0)-Au^{3+}(22)-Ni^{2+}(84) or Co^{2+}(90)
	Ag^{+}(0)-Au^{3+}(12)-Zn^{2+}(82) or Tl^{+}(62)
	Ag^{+}(0)-Cu^{2+}(42)-Mn^{2+}(90) or Co^{2+}(85)
	Pt^{4+}(0)-Au^{3+}(13)-Zn^{2+}(72) or Cd^{2+}(65)
	Pt^{4+}(0)-Au^{3+}(12)-Mn^{2+}(86) or Co^{2+}(80)
	Ag^{+}(0)-Hg^{2+}(16)-Pb^{2+}(27)-Mn^{2+}(88)
	Ag^{+}(0)-Hg^{2+}(16)-Pb^{2+}(27)-Tl^{+}(90)
	Ag^{+}(0)-Hg^{2+}(17)-Pb^{2+}(28)-Co^{2+} or Ni^{2+}(89)
	Bi^{3+}(0)-Hg^{2+}(19)-Pb^{2+}(29)-Mn^{2+} or Co^{2+}(90)
	Bi^{3+}(0)-Hg^{2+}(17)-Pb^{2+}(28)-Tl^{+}(70) or Cd^{2+}(89)

Reference: Sen, A. K., Das, S. B., and Ghosh, U. C., *J. Liq. Chromatogr.*, 8, 2999, 1985.

TABLE 16
R_F Values of Separated Anions on Stannic Oxide Layers

Adsorbent	Hydrated stannic oxide
Developers	D1: 0.05 *M* HCl + 0.09 *M* KCl, pH 2
	D2: 0.1 *M* CH_3COOH + 0.1 *M* CH_3COONa, pH 4
	D3: 0.1 *M* CH_3COOH + 0.1 *M* CH_3COONa, pH 6
Conditions	Run: 11 cm; technique: ascending
Detection	Standard spot-test reagents

Developer	(R_f × 100) values of separated ion
D1	$Fe(CN)_6^{3-}$(34)-W^{6+} or Mo^{6+}(05)
	$Fe(CN)_6^{3-}$(33)-As^{5+} or V^{5+}(03)
	$Fe(CN)_6^{4-}$(36)-V^{5+} or W^{6+}(00)
	$Fe(CN)_6^{4-}$(33)-As^{5+} or Mo^{6+}(06)
D2	BO_2^-(05)-Cr^{6+}(38)
	SCN^-(44)-W^{6+} or V^{5+}(04)
	SCN^-(41)-BO_2^-(07)
	$Fe(CN)_6^{4-}$(36)-P^{5+}(15)
D3	BO_2^-(05)-$Fe(CN)_6^{3-}$(33) or I^-(24)
	I^-(22)-V^{5+} or W^{6+}(03)

Reference: Sen, A. K. and Ghosh, U. C., *J. Liq. Chromatogr.*, 3, 71, 1980.

TABLE 17
R_f Values of Separated Metal Ions on Stannic Tungstate Layers

Adsorbent	Stannic tungstate in H+ form
Developers	D1: HNO_3 + dimethyl sulfoxide (6:4)
	D2: HCl + dimethyl sulfoxide (6:4)
	D3: HNO_3 + water (1:1)
Condition	Run: 10 cm; technique: ascending; development time: 6 h for D1 and D2 and 2 h for D3
Detection	Conventional spot-test reagents
Main feature	Quantitative separation of gold from numerous metal ions

Developer	(R_f × 100) range of separated metal ions
D1	Cu^{2+}(0)-Cd^{2+}(36—40)
	Fe^{3+}(0)-Al^{3+}(30—56)
	La^{3+}(0)-Pr^{3+}(80—100)
	La^{3+}(0)-Nd^{3+}(78—100)
	La^{3+}(0)-Sm^{3+}(76—100)
	Fe^{3+}(0)-Au^{3+}(90—100)
D2	Mn^{2+}(0)-Zn^{2+}(16—30)
	Cr^{3+}(0)-Al^{3+}(30—50)
	Pt^{4+}(30—0)-Au^{3+}(62—100)
	Pd^{2+}(30—0)-Au^{3+}(65—100)
	Hg^{2+}(25—0)-Au^{3+}(63—100)
D3	Cr^{3+}(0)-Al^{3+}(85—100)
	Tl^{+}(0)-In^{3+}(32—55)
	Fe^{3+}(0—35)-Mn^{2+}(70—100)
	Cr^{3+}(0)-Mn^{2+}(72—100)

Reference: Qureshi, M., Varshney, K. G., Gupta, S. P., and Gupta, M. P., Cation chromatography on stannic tungstate thin layers in DMSO-acid and aqueous DMSO systems: quantitative separation of gold from numerous metal ions, *Sep. Sci.*, 12, 649, 1977.

arsenate layers using various concentrations of HNO_3, HCl, HNO_3-$NaNO_3$, acetic acid, and tartaric acid (Table 20). Similar studies on refluxed material have yielded five separations (Table 21). They have also chromatographed 34 metal ions in six DMSO-HCl or DMSO-HNO_3 combinations on β stannic arsenate plates and separated Cr^{6+} from numerous metal ions.[201] Similarly, separations of Zr^{4+} and Mo^{6+} from various other cations were performed.[202] Sharma et al.[203] have recorded the movement of 47 cations on stannic arsenate, predicted K_{sp} (solubility product) values from R_f values, and resolved many binary and ternary mixtures of metal ions in 1 to 5 *M* HCl. A dimethyl formamide system on titanium(IV) antimonate and DMSO-based solvents on cerium((V) antimonate layers have resulted in the quantitative separation of Pd(II).[204,205] Qureshi et al.[206] have used DMSO-mixed eluants on stannic antimonate thin layers and achieved the separation of uranium from 48 metal ions. De and associates have isolated gold(III) from other metal ions with various solvents of hydrochloric acid-nitric acid-butanol combinations on thorium(IV) phosphate films.[207] Ammonium molybdate-impregnated silica gel has been utilized by Srivastava et al.[208] for the thin-layer chromatography of 32 synthetic dyes. The authors compared and discussed the data on plain and impregnated plates.

c. *Thin Layers of Metal Ferrocyanides*

This group of materials has yet to generate interest as a stationary phase in TLC. Zinc ferrocyanide layers were prepared by Fog and Wood,[209] who chromatographed 16 samples of sulfonamides in various concentrations of acetic acid, and the dependence of the R_f values on the degree of coprecipitation has been worked out. Kawamura and co-workers[210] have

TABLE 18
R_f Values of Separated Metal Ions on Stannic Arsenate Layers

Adsorbent	Stannic arsenate
Developers	D1: 1 *M* CH_3COONa + 1 *M* CH_3COOH (1:2)
	D2: 1 *M* citric acid + phenol (0.5 g in 600 ml) + NH_3 until a pH of 4.5 is reached
	D3: 1 *M* NH_4NO_3 + 0.5 *M* HNO_3 (1:1)
	D4: 1 *M* NH_4Cl + 0.1 *M* HCl (1:1)
	D5: 3 *M* citric acid + 1 *M* ammonium citrate (1:2)
	D6: ammonium molybdate + 0.2 *M* HNO_3 (1:1)
Condition	Run: 11 cm; technique: ascending
Detection	Conventional spot-test reagents

Developer	(R_f × 100) range of separated metal ions
D1	Pb^{2+}(0—23)-Cd^{2+}(91—100)
	Cr^{3+}(0—18)-Mn^{2+}(86—100)
	Ag^{+}(0)-Ce^{3+}(80—86)-Hg^{2+}(95—100)
D2	Se^{4+}(0—45)-Hg^{2+}(91—100)
	Pb^{2+}(14—64)-Hg^{2+}(91—100)
	Ag^{+}(0)-Bi^{3+}(36—95)
	As^{3+}(64—77)-Mo^{6+}(91—100)
D3	Sb^{3+}(0—4)-As^{3+}(64—73)-Cd^{2+}(82—95)
	Ag^{+}(0)-Pb^{2+}(4—27)-Hg^{2+}(91—100)
	Pb^{2+}(0)-Mn^{2+} (91—100)
D4	Ag^{+}(0)-Se^{44}(4—36)-Pd^{2+}(82—100)
	Pb^{2+}(0—36)-Pd^{2+}(82—100)
D5	Ag^{+}(0)-Pb^{2+}(4—50)-Cu^{2+}(82—95)
	Se^{4+}(0—32)-Mo^{6+}(87—91) or Au^{3+}(87—91)
D6	Cd^{2+}(77—85)-Zn^{2+}(88—95)
	Pb^{2+}(0—9)-Zn^{2+}(86—95)
	Pb^{2+}(0—9)-As^{3+}(73—76)-Hg^{2+}(86—95)

Reference: Husain, S. W. and Kazmi, S. K., Thin layer chromatography of metal ions on a new synthetic inorganic ion exchanger, *Experientia*, 28, 988, 1972.

analyzed various combinations of alkali metals in ammonium nitrate eluants and suggested many separations on the same material.

d. Thin Layers of Heteropolyacid Salts

Analytical applicability of the heteropolyacid salts (double salts) has been a recent trend. An excellent beginning in this direction was made by Lesigang and his group, who chromatographed alkali metals on thin plates of ammonium phosphododeca molybdate, arsenododeca molybdate, germanododeca molybdate, and oxinium and pyridinium germanododeca molybdates[211,212] in an ammonium nitrate system. Similar studies on ammonium germanododeca molybdate and pyridinium germanododeca molybdate in EDTA or ammonium acetate eluants has yielded successful separations of nuclides.[213] Lepri and Desideri[214] have presented exhaustive explorations of ammonium molybdophosphate and ammonium tungstophosphate layers. Thirty-five primary aromatic amines on AMP and ATP plates in nitric aid-ammonium nitrate systems have been studied and many separations suggested. In another study, 14 sulfanamides on calcium sulfate-AMP layers have been chromatographed, and the separation of 6 compounds was proposed.[215] The dependence of R_f on the thickness and temperature of plates, and the acidity, salt concentration, and nature of the organic solvent in the eluants has been worked out. The movement of amino acids, their derivatives, and peptides on ATP layers was tabulated in acetic aid and methanol mixtures of nitric acid and ammonium nitrate, and the trends in R_M values versus number of amino acid residues were established.[216] The

TABLE 19
R_f Values of Metal Ions in Some Mixed-Solvent Systems on Stannic Arsenate Layers

Adsorbent	Stannic arsenate
Developers	D1: ethyl malonate + methyl isobutyl ketone + acetone + 50% HNO_3 (1:2:1:0.5)
	D2: *n*-butyl acetate + acetic acid + acetone + HCl (3:1:2:1)
	D3: tri-*n*-butyl phosphate + acetone + 50% HNO_3 (2:4:1)
	D4: acetyl acetone + acetone + 10% HCl (5:3:1)
	D5: 2% diacetylmonooxime in ethanol + dioxane + 10% HNO_3 (1:2:2)
Condition	Run: 11 cm; techniques: ascending
Detection	Conventional spot-test reagents

	(R_f × 100) range				
Metal ion	**D1**	**D2**	**D3**	**D4**	**D5**
Cu^{2+}	—	—	49—59	—	63—95
Ag+	—	—	0	—	—
Cd^{2+}	0—4	—	45—59	—	77—90
Hg^{2+}	87—90	—	—	—	90—100
Hg_2^{2+}	0	—	—	—	0—18
Tl^+	0	0	0	—	45—69
Pb^{2+}	0	—	0	—	0
Bi^{3+}	9—54	—	—	—	—
Sb^{3+}	—	—	—	—	0—4
Se^{4+}	—	—	0—13	—	—
Mo^{6+}	60—81	—	95—100	—	—
W^{6+}	0—10	68—77	0	—	—
Al^{3+}	—	86—100	18—45	—	52—68
Ga^{3+}	—	—	0	—	—
Rh^{3+}	0—6	32—48	38—53	—	—
Pb^{2+}	31—54	—	72—88	—	—
In^{4+}	—	0.25	—	—	—
Pt^{4+}	—	—	0	—	—
Au^{3+}	86—90	78—88	95—100	—	—
Ti^{4+}	—	22—35	0	—	—
V^{5+}	—	—	18—63	—	—
Cr^{3+}	—	44—54	31—60	—	0—27
Mn^{2+}	—	68—95	—	—	—
Fe^{3+}	—	87—94	—	—	—
Co^{2+}	—	65—74	—	—	—
Ni^{2+}	—	45—54	32—44	—	—
Zn^{2+}	0	38—50	85—100	—	—
Zr^{4+}	—	—	0	—	—
Nb^{5+}	—	0	—	—	—
Be^{2+}	—	—	27—54	95—100	—
Mg^{2+}	—	—	85—100	—	81—100
Ba^{2+}	—	—	—	—	0
Y^{3+}	—	—	31—55	—	—
La^{3+}	—	—	95—100	—	—
Ce^{4+}	—	—	0—45	—	—
Pr^{3+}	—	—	95—100	—	—
Nd^{3+}	—	—	85—100	—	—
Sm^{3+}	—	—	0—36	—	—
Gd^{3+}	—	—	90—100	—	—
Tb^{3+}	—	—	0—22	—	—
Dy^{3+}	—	—	0—30	—	—
Ho^{3+}	—	—	90—100	—	—

TABLE 19 (continued)
R_f Values of Metal Ions in Some Mixed-Solvent Systems on Stannic Arsenate Layers

Er^{3+}	—	—	0	—	—
Tm^{3+}	—	—	88—100	—	—
UO_s^{2+}	81—86	—	95—100	—	—

Reference: Husain, S. W. and Eivazi, F., Thin layer chromatography of 57 metal ions on an inorganic ion exchanger in mixed solvent system, *Chromatographia,* 8, 277, 1975.

TABLE 20
R_f Values of Separated Metal Ions on Nonrefluxed Stannic Arsenate Layers

Adsorbent	Nonrefluxed stannic arsenate
Developers	D1: 0.5 *M* HNO_3
	D2: 1 *M* HNO_3
	D3: 0.5 *M* HCl
	D4: 1 *M* HCl
	D5: 1 *M* HNO_3 + 1 *M* $NaNO_3$ (1:1)
	D6: 0.1 *M* H_3AsO_4
	D7: 10% tartaric acid
Condition	Run: 10 cm; technique: ascending
Detection	Conventional spot-test reagents
Main feature	Quantitative separation of mercury from cadmium, zinc, and copper

Developer	(R_f × 100) values of separated metal ions
D1	Pd^{2+}(83)-Nb^{5+}(0)
	Au^{3+}(90)-Tl^{+}(0)
D2	Pt^{4+}(80)-Tl^{+}(0)
D3	Cd^{2+}(78)-Sb^{3+}(2)
	V^{5+}(100)-Nb^{5+}(10)
	V^{5+}(100)-Ta^{5+}(6)
D4	Co^{2+}(30)-Fe^{2+}(0)
	Ni^{2+}(24)-Fe^{2+}(0)
	Cd^{2+}(75)-Cu^{2+}(10)
D5	Pd^{2+}(90)-Mo^{6+}(0)
D6	Hg^{2+}(85)-Pb^{2+}(0)
	Hg^{2+}(85)-Ag^{+}(0)
	Hg^{2+}(85)-Cd^{2+}(0)
	Hg^{2+}(85)-Bi^{3+}(10)
	Hg^{2+}(85)-Cu^{2+}(15)
	Hg^{2+}(85)-Fe^{2+}(7)
	Hg^{2+}(85)-Ni^{2+}(15)
D7	Cd^{2+}(70)-Fe^{2+}(0)
	Cd^{2+}(70)-Hg^{+}(0)
	Cd^{2+}(70)-Ag^{+}(0)

Reference: Qureshi, M., Varshney, K. G., and Fatima, N., Thin layer chromatography of metal ions on stannic arsenate: quantitative separation of Hg(II) from Cd(II), Zn(II) and Cu(II), *Sep. Sci.*, 12, 321, 1977.

TABLE 21
R_f Values of Separated Metal Ions on Refluxed Stannic Arsenate Layers

Adsorbent	Refluxed stannic arsenate
Developers	D1: 1 *M* HNO_3
	D2: 0.5 *M* HCl
	D3: 1 *M* HNO_3 + 1 *M* $NaNO_3$ (1:1)

Condition and Reference as described in Table 20

Developer	(R_f × 100) values of separated metal ions
D1	Co^{2+}(40)-Cu^{2+}(0)
	Ni^{2+}(44)-Cu^{2+}(0)
	Pd^{2+}(99)-Tl^{+}(0)
D2	Hg^{2+}(100)-Hg^{+}(0)
D3	Tl^{+}(100)-W^{6+}(0)

material has also been used for the high-performance thin-layer chromatography of diastereomeric di- and tripeptides and dinitrophenyl amino aids and phenyl amines.[217-220] Silica gel-supported ammonium molybdophosphate has been adopted by Kaletca and Koneeny[221] for the separation of Cs.[137] Pyridinium tungstoarsenate thin films were introduced by Srivastava et al., and successful separations of amino acids and metal ions were obtained.[222,223] Andreiv[224] mixed silica gel with tungstophosphoric acid and reported the separation of carbohydrates. Varshney and associates[225,226] have performed thin-layer chromatography of alkaline earths, transition metals, and amino acids on tin(IV) arsenosilicate and arsenophosphate layers in buffered EDTA solutions.

e. Unusual Selectivities on Stannic Arsenate Layers

The selectivitity of the ion exchanger depends upon its nature as well as the medium of exchange. The conventional ion-exchange resins do not usually show striking selectivities. Efforts were made earlier to enhance their selectivities by using organic solvents. Before 1968, the organic solvents of low dielectric constants were used, e.g., acetone and butanol, which favored the formation of undissociated metal complexes. In 1968, a new solvent, dimethyl sulfoxide (DMSO), was introduced in ion exchange simultaneously by Janauer,[227] Birze et al.,[228] Fritz and Lehoczky,[229] and Phipps.[230] They noticed that DMSO is highly useful for altering the selectivities of different metal ions on ion-exchange resins and achieved some important separations of metal ions.

DMSO was introduced, probably for the first time, as a solvent in the chromatographic studies of metal ions on stannic antimonate papers by Qureshi and co-workers,[154] who obtained remarkably good results. Along the same lines, this solvent was used by them in the TLC of metal ions on stannic arsenate layers.[202] This study has given fantastic separation possibilities (Tables 22 and 23), such as Ag^+ from Pb^{2+}, Hg_2^{2+}, Cu^{2+}, and Au^{3+}; Mo^{6+} from W^{6+}; Zr^{4+} from Th^{4+}; and Se^{4+} from Te^{4+}. These interesting possibilities arise from three effects.

1. DMSO, being an aprotic dipolar solvent with hard oxygen and soft sulfur, is a strong solvating agent for cations. It also decreases the dielectric constant, leading to the formation of nondissociated complexes.
2. The presence of HCl is responsible for the formation of chloro complexes.
3. Stannic arsenate layers selectively adsorb certain cations.

TABLE 22
R_f Values of Separated Metal Ions in Some DMSO-HCl Systems on Stannic Arsenate Layers

Adsorbent	Stannic arsenate
Developers	D1: DMSO + 2 *M* HCl (1:5) D2: DMSO + 2 *M* HCl (2.4) D3: DMSO + 2 *M* HCl (3:3) D4: DMSO + 2 *M* HCl (4:2) D5: DMSO + 2 *M* HCl (5:1) D6: DMSO + 1 *M* HCl (1:1) D7: DMSO + 2 *M* HCl (1:1) D8: DMSO + 3 *M* HCl (1:1) D9: DMSO + 4 *M* HCl (1:1) D10: DMSO + 5 *M* HCl (1:1) D11: DMSO + 6 *M* HCl (1:1)
Condition	Run: 10 cm; technique: ascending
Detection	Conventional spot-test reagents

Developer	**(R_f × 100) values of separated metal ions**
D1	Tl^{+}(0—10)-Bi^{3+}(42—60)-Hg^{2+}(88—100)
	Tl^{+}(0—12)-Cu^{2+}(86—100)-Pb^{2+}(48—70)
	Ag^{+}(0—08)-Pb^{2+}(50—72)-Hg^{2+}(86—95)
	Te^{4+}(0—0)-Bi^{3+}(47—60)-Pd^{2+}(70—93)
	Mo^{6+}(0—10)-UO_2^{2+}(50—68)-VO^{2+}(82—100)
	Th^{4+}(0—30)-UO_2^{2+}(45—67)
	Th^{4+}(0—32)-VO^{2+}(84—96)
	Th^{4+}(0—27)-La^{3+}(65—90)
	Sb^{3+}(0—10)-Cd^{2+}(60—82)
	Fe^{3+}(0—37)-VO^{2+}(78—98)
	Ir^{3+}(55—69)-Pt^{4+}(78—100)
D2	Ag^{+}(0—08)-Bi^{3+}(40—52)-Cu^{2+}(82—100)
	Fe^{2+}(91—100)-Fe^{3+}(0—32)
	Cd^{2+}(90—100)-Bi^{3+}(35—50)
	Fe^{3+}(0—40)-Al^{3+}(77—98)
	Fe^{3+}(0—35)-Zn^{2+}(78—100)
	Fe^{3+}(0—36)-Co^{2+}(79—100)
	Fe^{3+}(0—38)-Cr^{3+}(85—100)
D3	K^{+}(18—32)-Ti^{4+}(93—100)
	Hg^{2+}(0—0)-Hg^{2+}(84)—100)
	Te^{4+}(0—04)-Au^{3+}(76—100)
	Fe^{2+}(75—100)-Mo^{6+}(0—15)
	Rb^{+}(15—28)-Sr^{2+}(75—90)
	Tl^{+}(0—04)-Ni^{2+}(87—100)
	Mo^{6+}(0—15)-Cr^{3+}(92—100)
	Se^{4+}(0—03)-Au^{3+}(78—97)
D4	Ag^{+}(0—05)-Pb^{2+}(55—73)-Cu^{2+}(98—100)
	Mo^{6+}(0—10)-Pb^{2+}(52—72)-UO_2^{2+}(85—100)
	Sn^{2+}(0—18)-Pb^{2+}(52—70)-UO_2^{2+}(83—98)
	Tl^{+}(0—10)-Ga^{3+}(87—100)
D5	Ag^{+}(0—04)-Cu^{2+}(45—68)-Ni^{2+}(75—97)
	Cu^{2+}(50—75)-Sn^{2+}(0—10)
	Cu^{2+}(48—70)-Cd^{2+}(86—100)
D6	Ag^{+}(0—03)-Cd^{2+}(85—100)
	Bi^{3+}(88—100)-Pb^{2+}(0—18)
	Bi^{3+}(84—98)-Sb^{3+}(0—04)
	Hg^{2+}(0—0)-Cd^{2+}(87—100)
	Tl^{+}(0—08)-Tl^{3+}(88—100)

TABLE 22 (continued)
R_f Values of Separated Metal Ions in Some DMSO-HCl Systems on Stannic Arsenate Layers

	La^{3+}(90—100)-Se^{4+}(0—08)
	Te^{4+}(0—05)-Au^{3+}(70—95)
D8	Tl^{+}(0—08)-Pb^{2+}(60—72)-Hg^{2+}(88—100)
	Ag^{+}(0—05)-Hg^{2+}(58—75)-Hg_2^{2+}(90—100)
	Se^{4+}(0—12)-Ni^{2+}(84—100)
	Tl^{+}(0—10)-Zn^{2+}(90—100)
	Ag^{+}(0—08)-Pd^{2+}(83—100)
	Fe^{3+}(65—95)-Mo^{6+}(0—10)
D9	Te^{4+}(0—08)-Ni^{2+}(80—97)
	Te^{4+}(0—12)-Bi^{3+}(78—95)
	Te^{4+}(0—11)-Cd^{2+}(82—100)
	Sb^{3+}(0—04)-Bi^{3+}(84—100)
D10	Zr^{4+}(0—03)-Mo^{6+}(26—41)-a mixture of Ca^{2+}, Y^{3+}, In^{3+}, Th^{4+}, Ti^{4+}, Nb^{5+}, Ta^{5+}, and rare earths (90—100)
	Y^{3+}(80—100)-Se^{4+}(0—10)
D11	Mo^{6+}(55—77)-W^{6+}(0—0)
	Se^{4+}(0—08)-Te^{4+}(36—75)
	Zr^{4+}(0—10)-UO_2^{2+}(88—100)

Reference: Qureshi, M., Varshney, K. G., and Sharma, S. D., Unusual selectivities of stannic arsenate layers in DMSO-HCl systems. Separation of zirconium from numerous metal ions, *Sep.Sci. Technol.*, 13, 917, 1978.

On the basis of this study, the following trends were noticed on the adsorption behavior of various metal ions with the variation of the DMSO mole fraction in the solvent system:

1. For most cations, there is no change in R_f values. Hg^{2+}, Pd^{2+}, Tl^{3+}, Zn^{2+}, Mn^{2+}, Cr^{3+}, Ba^{2+}, Sr^{2+}, Ni^{2+}, Cd^{2+}, Co^{2+}, Be^{2+}, Al^{3+}, Ga^{3+}, Au^{3+}, Pt^{4+}, Mg^{2+}, Ca^{2+}, Y^{3+}, La^{3+}, Fe^{2+}, Ir^{3+}, Nd^{3+}, Pr^{3+}, Sm^{3+}, Ce^{3+}, In^{3+}, Ce^{4+}, Ti^{4+}, Nb^{5+}, and Ta^{5+} have very high and constant R_f values. Zr^{4+}, Sn^{2+}, Se^{4+}, Te^{4+}, Mo^{6+}, W^{6+}, Sb^{3+}, Tl^{+}, and Ag^{+} have very low and constant R_f values.
2. The R_f values of K^{+}, Rb^{+}, Cs^{+}, and Pb^{2+} decrease as the DMSO mole fraction increases.
3. The ions which appear to be most affected by the DMSO mole fractions are Cu^{2+}, UO_2^{2+}, VO^{2+}, and Bi^{3+}.
4. Hg_2^{2+} is unique. Its R_f value increases with an increase in the mole fraction of DMSO.

The high but constant R_f values are due to the solvation of the cations or their chlorocomplexes. The low, constant R_f values are due to the precipitation of the chlorides or arsenates. The R_f values of the alkali metals decrease with an increase in the DMSO mole fraction because KCl, RbCl, and $CaCl_2$ are completely insoluble in DMSO. The behavior of Pb^{2+} may be explained along similar lines. An increase in the R_f value of Hg_2^{2+} takes place partly because of a decrease in the Cl^{-} concentration and partly because of the solvation of the chlorocomplex of the Hg_2^{2+} ion. The Cu^{2+}, UO_2^{2+}, VO^{2+}, and Bi^{3+} probably are strongly solvated by DMSO and show a marked change in R_f value with the DMSO concentration.

Most of the earlier studies on the TLC of inorganics has been done on silica gel and alumina layers, and reported in the *Handbook of Chromatography (Inorganics)*.[231] Tables 24 to 28 summarize the R_f values and separations of the various metal ions on the thin layers of some more materials.

TABLE 23
R_f Values of Metal Ions on DMSO-HCl/HNO_3 Systems on β-Stannic Arsenate Layers

Adsorbent	β-Stannic arsenate
Developers	D1: dimethyl sulfoxide + 4 *M* HNO_3 (1:1)
	D2: dimethyl sulfoxide + 1 *M* HCl (1:1)
	D3: dimethyl sulfoxide + 1 *M* HCl (1:4)
	D4: dimethyl sulfoxide + 1 *M* HCl (4:1)
	D5: dimethyl sulfoxide + 4 *M* HCl (1:1)
	D6: dimethyl sulfoxide + 6 *M* HCl (1:1)
Condition	Run: 10 cm; technique: ascending
Detection	Conventional spot-test reagents
Main feature	Quantitative separation of chromium from numerous metal ions

	(R_f × 100)					
Metal ions	**D1**	**D2**	**D3**	**D4**	**D5**	**D6**
Cu^{2+}	92	T	56	0	0,96	89
Ag^{+}	5	0	0	0	—	—
Cd^{2+}	81	80	80	0	91	90
Hg^{+}	5	80	T	T	—	—
Hg^{2+}	T	T	0.80	T	89	T
Tl^{+}	0	8	0	5	—	—
Pb^{2+}	17	T	5	T	—	—
Bi^{3+}	T	0	T	T	—	—
Sb^{3+}	0	15	5	0	T	T
Se^{4+}	95	96	0	0	—	—
Mo^{6+}	10	T	11	5	—	—
W^{6+}	12	10	0	0	—	—
Al^{3+}	95	60	0	0	82	87
Ga^{3+}	97	0	90	0	—	—
In^{3+}	95	—	91	0	—	—
Pd^{2+}	0.97	T	0.90	T		—
Pl^{4+}	T	95	90	0.90	0.95	96
Au^{3+}	T	0.95	0.85	T	93	0.85
Ti^{4+}	0	15	0	7	T	T
V^{5+}	0	0	0	0	7	0
Cr^{3+}	0	0	0	0	0	0
Mn^{2+}	90	75	0	0	80	82
Fe^{2+}	0	T	6	0	—	—
Fe^{3+}	0	T	T	0	0.95	T
Co^{2+}	95	T	80	0	0.97	95
Ni^{2+}	98	T	83	5,100	0.96	95
Zn^{2+}	90	77	0	0	80	82
Zr^{4+}	7	0	0	0	15	15
Nb^{5+}	92	87	77	0.96	90	90
To^{5+}	92	73	70	0	75	80
La^{3+}	92	96	77	7	—	83
Ce^{3+}	97	T	53	0.90	88	80
Th^{4+}	97	86	10	7	87	87
U^{6+}	95	T	T	T	T	T

Note: T, tailing.

Reference: Qureshi, M., Varshney, K. G., and Fatima, N., A novel quantitative separation of Cr(III) from numerous metal ions on β-stannic arsenate thin layers in DMSO-HCl systems, *Sep. Sci. Technol.*, 13, 321, 1978.

TABLE 24
R_f Values of Separated Metal Ions on Stannic Antimonate Layers

Adsorbent	Stannic antimonate
Developers	D1: dimethyl sulfoxide + 6 *M* HNO_3 (1:1) D2: dimethyl sulfoxide + 3 *M* HNO_3 (1:1) D3: 12 *M* HNO_3
Condition	Run: 11 cm; technique: ascending
Detection	Conventional spot-test reagents
Main feature	Quantitative separation of uranium from numerous metal ions

Developer	(R_f × 100) range of metal ions separated
D1	La^{3+}(0-20)-Pr^{3+}(74—100)
	La^{3+}(0-18)-Nd^{3+}(70—100)
	La^{3+}(0-20)-Sm^{3+}(76—100)
	Ce^{3+}(0)-Pr^{3+}(72—100)
	Ce^{3+}(0)-Nd^{3+}(80—100)
	Ce^{3+}(0)-Sm^{3+}(78—100)
	Th^{4+}(0—20)-UO_2^{2+}(85—100)
	Ti^{4+}(0)-VO^{2+}(75—100)
	Nb^{5+}(0)-VO^{2+}(75—100)
	Ta^{5+}(0)-VO^{2+}(75—100)
	In^{3+}(0)-Ga^{3+}(80—100)
	Tl^{+}(0—18)-Ga^{3+}(80—100)
	Y^{3+}(0)-Ga^{3+}(80—100)
D2	Fe^{3+}(0)-VO^{2+}(80—100)
	Fe^{3+}(0)-Be^{2+}(85—100)
	Mg^{2+}(0—20)-Al^{3+}(75—100)
D3	Cd^{2+}(0)-Hg^{2+}(85—95)
	Zn^{2+}(0)-Hg^{2+}(85—100)
	Al^{3+}(0)-Tl^{+}(22—32)

Reference: Qureshi, M. and Varshney, K. G., Thin layer chromatography of 49 metal ions on stannic antimonate in aqueous and mixed solvent systems containing dimethylsulfoxide: quantitative separation of uranium from numerous metal ions, *Sep.Sci.*, 11, 533, 1976.

Tin(IV) arsenosilicate has been found useful for the separation of amino acids on their thin layers, the results of which are summarized in Table 29.

IV. KINETICS AND THERMODYNAMICS

As has been emphasized in the previous sections, inorganic ion exchangers have various applications in analytical chemistry, owing to their resistance to heat and radiations. In addition to the ion-exchange procedures in which a high chemical and thermal stability or a high selectivity for a particular ion are required, new applications are in the areas of heterogeneous catalysis, solid electrolytes, inorganic ion-exchange membrances, ion-selective electrodes, and intercalation compounds. In most of these fields, information on ion-exchange kinetics and the mobility of counter ions in the lattice structure are needed. Similarly, ion-exchange equilibria are of great practical and theoretical importance. The theoretical aspects of these two fundamental approaches have been discussed very well in Chapter 2 by Dyer, on zeolite models which appropriately reproduce the results on synthetic inorganic ion exchangers too. Here, therefore, only some of the salient features regarding the results of such studies will be discussed on inorganic ion exchangers, particularly the amorphous ones.

TABLE 25
R_f Values of Separated Metal Ions on Thorium Antimonate Layers

Adsorbent	Thorium antimonate
Developers	D1: dioxane
	D2: 1.0 *M* HNO_3
	D3: 0.1 *M* HNO_3
	D4: 0.1 *M* CH_3COOH
	D5: 1.0 *M* CH_3COOH
	D6: DMSO + 0.1 *M* HNO_3 (4:6)
Conditions	Run: 11 cm (layer thickness), 0.1 mm air-dried plates used without activation; technique: ascending
Detection	Standard spot-test reagents
Main feature	The method is applied for quantitative separation of micro amounts of Hg^{2+} from several metal ions

Developer	(R_f × 100) values of separated ions
D1	Hg^{2+}(80)-Co^{2+} or Cu^{2+}(0)
D2	Ce^{4+}(0)-Pr^{3+} or Sm^{3+}(74)
	Ce^{4+}(0)-La^{3+}(73)
D3	Pb^{2+}(0)-Cd^{2+}(37)-Hg^{2+}(71)
	Ag^{+}(0)-Cd^{2+}(32)-Cu^{2+}(85)
	Bi^{3+}(0)-Cu^{2+}(81)-Cd^{2+}(37)
	Bi^{3+}(0)-Cd^{2+}(42)-Hg^{2+}(75)
	Zn^{2+}(65)-Cd^{2+}(35)
D4	Fe^{3+}(0)-Ni^{2+}(90) or Co^{2+}(87)
	Al^{3+}(0)-Mg^{2+}(93)
	Cr^{3+}(0)-Zn^{2+}(85)
D5	Bi^{3+} or Ag^{+}(0)-Pd^{2+}(36)-Hg^{2+}(75)
	Pb^{2+}(0)-Pd^{2+}(36)-Hg^{2+}(80)
D6	Fe^{3+} or Fe^{2+}(0)-UO_2^{2+} or VO^{2+}(95)
	Ag^{+}(0)-Hg^{2+}(35)-Zn^{2+}(80)
	Bi^{3+}(10)-Hg^{2+}(36)-Zn^{2+}(81)

Reference: De, A. K., Rajput, R. P. S., and Das, S. K., *Sep. Sci. Technol.*, 14, 735, 1979.

A. KINETICS

The theory of ion-exchange kinetics is not as far advanced as that of ion-exchange equilibria. The more recent quantitative approaches to ion-exchange equilibria include effects such as swelling pressure and specific interactions. The theory of ion-exchange kinetics, on the other hand, is still in the stage of first approximation, where such effects are not quantitatively considered.

Kinetic studies give clues regarding the following three aspects of ion exchange:

1. The mechanism of ion exchange
2. The rate-determining step
3. The rate laws obeyed by the system

Though many studies on the kinetics of ion exchange on organic resins have been reported, relatively less information exists on the kinetics of exchange on inorganic ion exchangers. Moreover, most of the studies[232-245] are based on the Bt criterion,[246] which is of limited use because of the different mobilities[247] of the competing ions. The criterion is useful only for the isotopic exchange, i.e., the ions having similar effective diffusion coefficients. The use of Nernst-Planck equations is suggested[248] for nonisotopic exchange to obtain more precise values of the various kinetic parameters. Varshney et al.[249-253] and Bohman et al.[254] have recently reported some studies based on this new criterion.

TABLE 26
R_f Values of Separated Metal Ions on Titanium(IV) Antimonate Layers

Adsorbent	Ti(IV) antimonate in H^+ form
Developers	D1: 1.0 *M* HNO_3
	D2: 0.1 *M* HNO_3
	D3: 0.01 *M* HNO_3
	D4: dimethyl formamide + 0.1 *M* HNO_3 (4:6)
	D5: dimethyl formamide + 0.1 *M* HNO_3 (6:4)
Conditions	Run: 11 cm (layer thickness) 0.1 mm plates were dried at room temperature for 3 h; development time: 6—10 h; technique: ascending
Detection	Standard spot-test reagents
Main features	32 metal ions were chromatographed. Pd^{2+} was determined quantitatively

Developer	(R_f × 100) values of separated ion
D1	Cu^{2+}(48)-Hg^{2+}(92)
	Zr^{4+}(10)-Y^{3+}(67)
D2	Cd^{2+}(72)-Ho^{3+}(18) or Nd^{3+}(0)
	La^{3+}(65)-Pr^{3+}(0)
	La^{3+}(70)-Ce^{4+}(15)
D3	Cu^{2+}(0)-Hg^{2+}(36)-Pd^{2+}(95)
	Au^{3+}(0)-Hg^{2+}(37)-Pt^{4+}(94)
D4	Pd^{2+}(91)-Co^{2+} or Ni^{2+}(0)
	Hg^{2+}(90)-Zn^{2+}(05)
D5	Pt^{4+}(74)-Fe^{3+}(07)
	Pt^{4+}(81)-Bi^{2+}(0)
	Pt^{4+}(73)-Cr^{3+}(10)

Reference: Seth, N. S. and Rajput, R. P. S., *Indian J. Chem. A*, 122, 1088, 1983.

1. Evaluation of the Exchange Mechanism

A kinetic approach can very well explain the exchange mechanism. On tin(IV) tungstate,[249] for example, it has been observed that the K_d values of alkaline earths are in the same order as their mobilities and activation energies, i.e., $Ba^{2+} > Sr^{2+} > Ca^{2+} > Mg^{2+}$.

Studies on antimonic acid and antimony(V) silicate cation exchangers[250] illustrate that the ion-exchange process on these materials is controlled by the particle diffusion phenomenon at and above a metal ion concentration of 2×10^{-2} *N*. Below this concentration, film diffusion is more prominent. The fractional attainment of equilibrium is faster on SbOH than on SbSi. Also, it is faster at a higher temperature, as for other materials of this class. In addition, other kinetic parameters, such as the diffusion coefficient (D_o), energy of activation (E_a), and entropy of activation (ΔS^*), have also been determined. It has been found that the energy and entropy of activation are maximal for the Co^{2+}-H^+ exchange and minimal for the Fe^{2+}-H^+ exchange among the transition metal ions, both on SbOH and SbSi exchangers. In the case of the alkaline earths, these values are maximal for the Mg^{2+}-H^+ exchange on both the exchangers. However, the two materials differ when we consider the minimum values of these quantities. They are minimal for the Ba^{2+}-H^+ on SbOH and for the Ca^{2+}-H^+ exchange on SbSi. The half times, $t^1/_2$, of exchange indicate that the exchange reactions are slow.

A kinetic study of heavy pollutants (Mn^{2+}, Fe^{3+}, Co^{2+}, Ni^{2+}, Cu^{2+}, Zn^{2+}, Cd^{2+}, Hg^{2+}, and Pb^{2+}) on titanium(IV) arsenosilicate indicates[251] that the ion-exchange phenomenon is a particle diffusion controlled at and above a 0.03 *M* concentration of the metal ions. The entropy (ΔS^*) values are highest for the H^+-Pb^{2+} exchange, perhaps due to the highest mobility (60.8) of the Pb^{2+} ion among the metal ions studied. This is supported by the high selectivity of the exchange for this metal ion.

TABLE 27
R_f Values of Metal Ions on Silica Gel-Ceric Molybdate-Impregnated Layers

Adsorbent	Ceric molybdate-silica gel
Developer	D1: benzene, ethyl acetate, acetic acid, HCl (60:30:10:0.5) D2: benzene, ethyl acetate, acetic acid, HCl (60:30:10:1)
Condition	Temp of development: 30 ± 1°C; technique: ascending
Detection	Conventional spot-test reagents

	(R_f × 100) values			
	Plain silica gel		Ceric molybdate-silica gel	
Metal ion	**D1**	**D2**	**D1**	**D2**
Cu^{2+}	5	12	80	83
Ag^{+}	0	0	51	53
Cd^{2+}	0	0	73	78
Hg^{+}	14	26	95	95
Pb^{2+}	0	0	60	64
As^{3+}	16T	26T	29	31
Sn^{4+}	0	0	62	64
Sb^{5+}	30T	32T	56	61
Se^{4+}	44T	45T	65	69
Al^{3+}	—	—	31	34
Cr^{3+}	0	0	18	23
Fe^{2+}	7T	8T	70	72
Co^{2+}	0	0	53	58
Ni^{2+}	0	0	49	52
Zn^{2+}	3	5	84	87

Note: T, tailing.

Reference: Srivastava, S. P., Dua, V. K., and Gupta, K., TLC separation of some inorganic ions on silica gel-ceric molybdate impregnated layers, *J. Indian Chem. Soc.*, 56, 529, 1979.

Similar studies have been performed on zirconium(IV) and thorium(IV) phosphosilicate cation exchangers.[252] The particle diffusion mechanisms on these systems are found to be applicable at and above 0.01 *M* solutions. The energy and entropy of activation corroborate with the actual ion exchange taking place on these materials.

The kinetics of exchange of transition metal ions on the tin(IV) arsenosilicate[253] cation exchanger, however, indicates a film-diffusion phenomenon to be predominent up to a metal ion concentration $< 0.10\ M$. The particle diffusion is observed at and above this concen-

TABLE 28
Binary Separation of Alkaline Earth and Transition Metal Ions on Tin(IV) Arsenosilicate and Arsenophosphate Thin Layers

Adsorbents	Tin(IV) arsenosilicate and tin(IV) arsenophosphate					
Developers	Buffered EDTA solutions obtained by mixing 10 ml of 0.1 *M* EDTA solution with 5 ml each of the various buffers prepared by adding the following amounts of 0.2 *N* NaOH into a 100 ml mixture of phosphoric, acetic, and boric acids (each 0.04 *M*)					
NaOH, 0.2 *N* (ml)	0.0	25.0	42.5	60.0	77.0	100.00
pH	1.81	4.10	6.09	7.96	9.90	1.98
Condition	Technique: ascending, layer thickness: 0.1 mm					
Detection	Conventional spot-test reagents					
Main features	Following separations achieved on the basis of R_f values: Ca^{2+}-Ba^{2+}, Ca^{2+}-Sr^{2+}, Zn^{2+}-Cd^{2+} Pb^{2+}-Hg^{2+}, Ni^{2+}-Mn^{2+}, Co^{2+}-Mn^{2+} Ni^{2+}-Cr^{3+}, Co^{2+}-Cr^{3+}, Fe^{3+}-Cr^{3+}					

Reference: Varshney, K. G., Khan, A. A., and Anwar, S., Chromatography of alkaline earths and transition metals on tin(V) arsenosilicate and arsenophosphates thin layers in buffered EDTA solutions, *J. Liq. Chromatogr.*, 8, 1347, 1985.

tration. The study further indicates that Fe^{2+} is the fastest diffusing ion, while Co^{2+} is the slowest. As for zeolites,[255,256] the energies of activation do not follow any simple relationship with the ion size.

The exchange kinetics on zirconium(IV) phospho- and silico-arsenates, applying the Nernst-Planck equations, correlates[257] the practically achieved separations of Cd^{2+} from other metal ions such as Mn^{2+}, Fe^{2+}, Co^{2+}, Ni^{2+}, and Zn^{2+}. A linear variation of the energy and entropy of activation with the ionic radii and ionic mobility of alkaline earth metal ions has also been established[258] on zirconium(IV) arsenophosphate, on the basis of such a study. Similar studies have been performed on the phosphosilicates of zirconium(IV) and thorium(IV), and the arsenosilicate of tin(IV), for the exchange of some divalent metal ions. Tables 30 and 31 summarize the values of the various parameters obtained in the kinetic studies of some inorganic ion exchangers for the exchange of alkaline earths and transition metals, respectively.

Varshney et al.[261] have also performed the forward and reverse ion-exchange kinetics for the Na^+-H^+ and K^+-H^+ exchanges on crystalline antimony(V) silicate. They have observed that the values of the self-diffusion coefficient (D_o), energy of activation (E_o), and entropy of activation (ΔS^*) decrease with the decrease in hydrated radii, as shown in Table 32. Further, the exchange is faster at a higher temperature (Table 33) and the exchanger is more selective for the K^+ ion than for the Na^+ ion. This is in accordance with an earlier observation for resins.[262]

2. Evaluation of the Dimensionless Time Parameter

In all the calculations involving ion-exchange kinetics using the Nernst-Planck equations, evaluation of the dimensionless time parameter (τ) is a prerequisite. A graphical method is generally used[249-253] for this purpose, which is tedious and gives only approximate (τ) values resulting in less accurate kinetic parameters. In view of this, Varshney and co-workers[263] have established a table for a direct computation of (τ) values for a particular fractional attainment of equilibrium, $U(\tau)$, for some metal ions, such as Mn^{2+}, Co^{2+}, Ni^{2+}, Fe^{2+}, Cr^{3+}, Mg^{2+}, Cu^{2+}, Cd^{2+}, Zn^{2+}, Sr^{2+}, Ca^{2+}, Ba^{2+}, and Pb^{2+} against H^+ ions, in both the

TABLE 29
R_f Values of Amino Acids on Tin(IV) Arsenosilicate Thin Layers

Adsorbent	Tin(IV) arsenosilicate
Developers	Universal buffer mixtures of varying pH obtained by mixing the following amounts (ml) of a 0.2 *M* solution of NaOH with a 100 ml mixture of phosphoric, acetic, and boric acids (0.04 *M*)
NaOH	0.0 25.0 42.5 60.0 77.0 100.0
pH	1.8 4.1 6.1 8.0 9.9 11.9
Condition	Technique: ascending; layer thickness: 0.1 mm
Detection	2% alcoholic solution of ninhydrin
Test solution	2% solutions of amino acids in demineralized water
Main feature	Some binary separations of amino acids at pH6 achieved as follows: DL-methionine (0.80) from L-cysteine HCl (0.0) L-cystine (0.0), or L-ornithine (0.0) DL-B-phenyl alanine (0.80), DL-tryptophan (0.60), and L-tyrosine (0.70) from L-histidine HCl (0.0), L-arginine HCl (0.0), or L-ornithene (0.0) DL-threinine (0.75) from L-histidine HCl (0.0), L-arginine HCl (0.0), L-ornithene (0.0), or L-hydroxyproline (0.0) DL-norleuane (0.80) from DL-isoleucine (0.0)

Amino acid	(R_f × 100) at different pH 2	4	6	8	10	12
DL-aspartic acid	30	35	25	25	20	15
L-glutamic acid	10	10	50	50	50	50
L-histidine HCl	0	0	0	0	0	0
L-lysine HCl	10	10	15	0	0	0
L-arginine HCl	0	0	0	0	0	0
L-ornithene HCl	0	0	0	0	0	0
L-cystine	0	0	0	0	0	0
L-cystein HCl	0	0	0	0	0	0
DL-methionine	80	80	80	80	80	80
L-tyrosine	100	90	80	70	70	60
DL-tryptophan	100	80	80	80	80	80
DL-B-phenylalanine	80	80	80	80	60	80
DL-dopa	80	90	90	90	90	80
Glycine	20	20	20	20	20	20
DL-alanine	40	40	50	30	30	30
DL-valine	10	10	20	15	30	30
L-proline	80	80	80	60	20	30
DL-serine	60	80	60	70	60	60
DL-isoleucine	60	60	30	20	20	0
DL-norleucine	90	90	80	80	80	80
DL-threonine	80	90	90	90	90	80
L-hydroxyproline	20	10	30	20	20	0
DL-2-aminobutyric acid	20	20	20	20	20	20
DL-leucine	80	70	40	40	40	40

Reference: Varshney, K. G., Anwar, S., and Khan, A. A., Thin layer chromatography of amino acids on tin(IV) arsenosilicate cation exchanger in some buffer solutions, *Anal. Lett.*, 19, 543, 1986.

TABLE 30
Self-Diffusion Coefficients, Energies of Activation, and Entropies of Activation of Alkaline Earths on Some Inorganic Ion Exchangers Applying Nernst-Planck Equations

	Parameters		
Migrating ion	D_o (m^2/s)	E_a (kJ/mol)	Δ S* (J deg/mol)
	1. Tin(IV) Tungstate[249]		
Mg^{2+}	6.60×10^{-10}	8.150	−72.12
Ca^{2+}	1.26×10^{-9}	10.03	−66.76
Sr^{2+}	1.82×10^{-9}	10.45	−63.70
Ba^{2+}	3.47×10^{-9}	11.92	−58.34
	2. Antimonic Acid[250]		
Mg^{2+}	3.16×10^{-9}	12.93	−59.11
Ca^{2+}	7.24×10^{-10}	9.72	−71.36
Sr^{2+}	9.33×10^{-10}	10.98	−69.25
Ba^{2+}	5.50×10^{-10}	9.17	−73.65
	3. Antimony(V) Silicate[250]		
Mg^{2+}	7.59×10^{-9}	15.63	−51.83
Ca^{2+}	5.13×10^{-10}	11.53	−74.23
Sr^{2+}	5.50×10^{-10}	11.19	−73.65
Ba^{2+}	8.32×10^{-10}	11.24	−70.21
	4. Zirconium(IV) Arsenophosphate[258]		
Mg^{2+}	2.0×10^{-8}	14.74	−43.79
Ca^{2+}	9.5×10^{-8}	18.40	−30.77
Sr^{2+}	1.7×10^{-7}	19.65	−25.79
Ba^{2+}	3.5×10^{-7}	21.11	−20.05
	5. Zirconium(IV) Phosphosilicate[259]		
Mg^{2+}	1.9×10^{-9}	14.53	−63.32
Ca^{2+}	1.91×10^{-8}	10.02	−44.17
Sr^{2+}	5.25×10^{-8}	20.90	−35.75
Ba^{2+}	1.82×10^{-6}	31.46	−6.27
	6. Thorium(IV) Phosphosilicate[259]		
Mg^{2+}	1.26×10^{-9}	16.20	−66.76
Ca^{2+}	2.00×10^{-9}	16.41	−62.93
Sr^{2+}	2.63×10^{-8}	22.05	−41.49
Ba^{2+}	5.25×10^{-9}	18.71	−54.89
	7. Tin(IV) Arsenosilicate[259]		
Mg^{2+}	2.51×10^{-8}	20.07	−41.88
Ca^{2+}	2.00×10^{-7}	28.54	−24.65
Sr^{2+}	1.66×10^{-7}	26.86	−26.18

TABLE 30 (continued)

Migrating ion	Parameters D_o (m^2/s)	E_a (kJ/mol)	ΔS^* (J deg/mol)
	8. Zirconium(IV) Arsenosilicate[260]		
Mg^{2+}	1.36×10^{-4}	34.53	−29.57
Ca^{2+}	2.93×10^{-4}	36.05	−35.98
Sr^{2+}	3.85×10^{-4}	37.55	−38.25
Ba^{2+}	4.47×10^{-4}	38.28	−39.49

forward and reverse exchanges, using a computer, thus simplifying the treatment and improving the accuracy. Tables 34 and 35 summarize these values.

B. THERMODYNAMICS

The equilibrium studies on synthetic inorganic ion exchangers have been of great interest because of their wide range of applications. Thermodynamics is the fundamental approach for describing equilibria and helps in evaluating the adsorption exchange processes occurring on the surface of the exchanger. The thermodynamics of ion exchange on zirconium phosphate has been studied by Larsen and Vissers,[264] Baetsle,[265] Nancollas and Tilak,[266] and Allulli and Costantino.[267] Recently, such studies have also been performed on similar materials,[268-275] which have given some interesting findings.

1. Selectivity for Alkali Metals

The maximum uptake and, hence, ion-exchange capacity of inorganic ion exchangers vary with the nature of the exchanging species, and the temperature and concentration of the ions in solution. For organic resins, however, it is fairly constant, due to their elastic structure. Qureshi and Ahmad[276] observed that on crystalline niobium phosphate (C-NbP), the column ion-exchange capacities for Li^+, Na^+, K^+, and Cs^+ are 0.70, 1.06, 1.37, 0.86, and 0.84 meq/g, respectively. The maximum uptake data of thermodynamic studies of these ions also show the same order as the IEC data obtained by the batch process, i.e.,

$$Li^+ < Cs^+ < Rb^+ < Na^+ < K^+$$

The C-NbP appears to have two types of selectivity behavior. The selectivities of Li^+, Na^+, and K^+ ions increase, while those of K^+, Rb^+, and Cs^+ decrease, with the increase of ionic radii. It appears that the selectivities of the former group are a function of ionic radii, while those of the latter are a function of hydrated ionic radii. In this way, K^+ is acting as a bridge ion for the two selectivity trends, due to its medium ionic size and medium hydration number.

Ion-exchange isotherms of C-NbP for the hydrogen ion/alkali metal ion exchange show an irregular or an S-shaped behavior without phase transitions and, hence, a selectivity reversal occurs over the entire range of composition. The Kielland plots, obtained by plotting $\ln K_H^M$ vs. $\bar{X}_M$, are linear at all temperatures and for all systems studied. The selectivity coefficients at $\bar{X}_M = 0$ and $\bar{X}_M = 1$ are calculated by extrapolation of the fitted data. The thermodynamic equilibrium constant, K, is evaluated by linear integration of the following equation:

$$\ln K = \ln K_H^M + 2.303C\,(1 - 2\,\bar{X}_M)$$

where C is the Kielland coefficient.

The behavior of C-NbP for the various alkali ions is unique for each ion, as has recently been reported by Kullberg and Clearfield[277,278] for zirconium phosphate. They have explained the uniqueness of the behavior on the basis of a new thermodynamic model which is assumed to have two types of cavities, viz., large and small. In this model, the selectivity for the

TABLE 31
Self-Diffusion Coefficients, Energies of Activation, and Entropies of Activation for the Exchange of Transition Metal Ions on Some Inorganic Ion Exchangers Applying Nernst-Planck Equations

	Parameters		
Migrating ion	D_o(m^2/s)	E_a (kJ/mol)	ΔS* (JK^{-1} mol^{-1})
	1. Antimonic Acid[238]		
Mn^{2+}	2.19×10^{-9}	12.55	−62.16
Fe^{2+}	4.57×10^{-10}	11.15	−75.19
Co^{2+}	3.98×10^{-9}	14.48	−57.19
Ni^{2+}	1.95×10^{-9}	12.33	−63.12
	2. Antimony(V) Silicate[250]		
Mn^{2+}	1.45×10^{-9}	12.34	−65.59
Fe^{2+}	2.88×10^{-10}	8.94	−79.03
Co^{2+}	8.71×10^{-9}	17.19	−50.68
Ni^{2+}	1.58×10^{-9}	13.48	−64.87
	3. Tin(IV) Arsenosilicate[253]		
Mn^{2+}	1.82×10^{-7}	20.17	−25.41
Fe^{2+}	4.79×10^{-3}	49.36	+59.20
Co^{2+}	7.24×10^{-8}	16.31	−33.07
Ni^{2+}	7.59×10^{-6}	1.98	+5.60
	4. Zirconium(IV) Phosphosilicate[252]		
Mn^{2+}	3.55×10^{-10}	13.07	−77.29
Fe^{2+}	8.41×10^{-9}	16.85	−31.83
Co^{2+}	1.88×10^{-8}	23.52	−44.27
Ni^{2+}	1.99×10^{-8}	23.25	−43.79
	5. Thorium(IV) Phosphosilicate[252]		
Mn^{2+}	2.51×10^{-8}	22.21	−47.88
Fe^{2+}	5.01×10^{-9}	19.86	−55.28
Co^{2+}	6.68×10^{-8}	22.60	−33.74
Ni^{2+}	1.00×10^{-8}	20.51	−49.53
	6. Zirconium(IV) Arsenophosphate[257]		
Mn^{2+}	7.08×10^{-9}	12.50	−52.40
Fe^{2+}	5.62×10^{-9}	12.02	−54.32
Co^{2+}	1.51×10^{-9}	8.48	−65.23
Ni^{2+}	3.54×10^{-9}	10.74	−58.15
Zn^{2+}	8.71×10^{-8}	20.96	−33.49
Cd^{2+}	2.09×10^{-7}	21.74	−24.26
	7. Zirconium(IV) Arsenosilicate[257]		
Mn^{2+}	3.55×10^{-8}	12.26	−58.15
Fe^{2+}	2.95×10^{-8}	16.62	−40.53
Co^{2+}	$1.74 \times$s 10^{-8}	15.78	−44.94
Ni^{2+}	7.76×10^{-8}	13.69	−51.64
Zn^{2+}	7.41×10^{-8}	19.52	−32.88
Cd^{2+}	2.04×10^{-7}	22.42	−24.45

TABLE 32
Self-Diffusion Coefficients, Energies of Activation, and Entropies of Activation for the Forward and Reverse Exchanges of Na^+-H^+ and K^+-H^+ on Antimony Silicate

Migrating ion	Ion present in the exchanger	D_o (m^2s)	E_a (kJ/mol)	ΔS^* ($JK^{-1}mol-^1$)
H^+	Na^+	8.32×10^{-7}	27.70	−12.78
Na^+	H^+	2.88×10^{-8}	19.96	−40.74
H^+	K^+	2.19×10^{-8}	18.03	−43.02
K^+	H^+	3.98×10^{-9}	14.63	−57.19

TABLE 33
Half Times of Exchanges for Forward and Reverse Exchanges of Na^+-H^+ and K^+-H^+ on Antimony(V) Silicate at Different Temperatures

Migrating ion	Ion present in the exchanger	Half-time exchange, $t^1/_2$ (S), at different temperatures			
		20°C	35°C	50°C	65°C
H^+	Na^+	42	21	18	12
Na^+	H^+	66	45	33	25
H^+	K^+	45	33	27	21
K^+	H^+	45	39	33	28

ingoing ion depends entirely upon the manner in which it can accommodate itself inside the cavity and, of course, the overall steric hindrance experienced by it. The behavior of C-NbP may also be explained on the basis of the theory developed by Barrer et al.[279,280] The theory postulates that the irregular behavior results from the fact that the occupancy of an exchange site by ion A or B affects the relative affinities of the adjacent sites for these ions.

With the progress of exchange in irregular systems, the steric hindrance for a large counter ion increases, resulting in a difficult and, hence, energetically less favorable occupancy of the two neighboring sites by two large ions, compared to those of one large and one small or two smaller ones. On the basis of statistical thermodynamics, Barrer gave an equation identical to that of Kielland's semiempirical relationship, and defined Kielland's empirical constant, C, by

$$C = -\frac{n_A + n_B}{\bar{n}_z} \times \frac{Ew}{kT}$$

where the n's are the mole fractions of species in the solid phase and Ew is the energy of interaction between the neighboring sites occupied by the B ions. In this way, C is a function of the interaction energy, Ew. For irregular systems in which the occupancy of the two neighboring sites by two B ions, is unfavorable, coefficient C is negative, and for $|C| > \ln K_H^M$, the isotherms are S-shaped and a selectivity reversal occurs. The same behavior is exhibited by C-NbP. For the H^+/Cs^+, H^+/Rb^+, and H^+/K^+ systems, $|C|$ increases in the order of $K^+ < Rb^+ < Cs^+$ and is proportional to their ionic sizes. The $|C|$ values for the H^+/Li^+, H^+/Na^+, and H^+/K^+ systems decrease with the increase in ionic size. The lithium ion however, has the highest $|C|$ value despite having the lowest ionic size. On the other hand, the lowest $|C|$ value is observed for the potassium ion, which does not possess the

TABLE 34
Dimensionless Time as a Function of Fractional Attainment of Equilibrium for Some Particle Diffusion-Controlled H(I)-Metal(II) Exchanges Involving Different Mobility Ratios

Fractional attainment of equilibrium, U(T)	Dimensionless time, T				
	$\alpha = \frac{\bar{D}_H}{\bar{D}_{MN}} = 11.39$	$\alpha = \frac{\bar{D}_H}{\bar{D}_{Co}} = 7.42$	$\alpha = \frac{\bar{D}_H}{\bar{D}_{Ni}} = 7.25$	$\alpha = \frac{\bar{D}_H}{\bar{D}_{Fe/Cr}} = 7.09$	$\alpha = \frac{\bar{D}_H}{\bar{D}_{Mg/Cu}} = 6.95$
0.01	2.501×10^{-5}	2.040×10^{-5}	2.019×10^{-5}	1.998×10^{-5}	1.980×10^{-5}
0.05	6.262×10^{-4}	5.106×10^{-4}	5.053×10^{-4}	5.002×10^{-4}	4.957×10^{-4}
0.10	2.516×10^{-3}	2.052×10^{-3}	2.030×10^{-3}	2.010×10^{-3}	1.992×10^{-3}
0.15	5.706×10^{-3}	4.651×10^{-3}	4.603×10^{-3}	4.556×10^{-3}	4.516×10^{-3}
0.20	1.026×10^{-2}	8.358×10^{-3}	8.271×10^{-3}	8.187×10^{-3}	8.114×10^{-3}
0.25	1.626×10^{-2}	1.324×10^{-2}	1.311×10^{-2}	1.297×10^{-2}	1.286×10^{-2}
0.30	2.384×10^{-2}	1.941×10^{-2}	1.920×10^{-2}	1.901×10^{-2}	1.884×10^{-2}
0.35	3.315×10^{-2}	2.697×10^{-2}	2.669×10^{-2}	2.642×10^{-2}	2.668×10^{-2}
0.40	4.441×10^{-2}	3.661×10^{-2}	3.573×10^{-2}	3.537×10^{-2}	3.505×10^{-2}
0.45	5.789×10^{-2}	4.705×10^{-2}	4.655×10^{-2}	4.608×10^{-2}	4.566×10^{-2}
0.50	7.395×10^{-2}	6.006×10^{-2}	5.943×10^{-2}	5.882×10^{-2}	5.829×10^{-2}
0.55	9.303×10^{-2}	7.551×10^{-2}	7.472×10^{-2}	7.395×10^{-2}	7.328×10^{-2}
0.60	1.157×10^{-1}	9.389×10^{-2}	9.290×10^{-2}	9.194×10^{-2}	9.111×10^{-2}
0.65	1.428×10^{-1}	1.158×10^{-1}	1.146×10^{-1}	1.134×10^{-1}	1.124×10^{-1}
0.70	1.752×10^{-1}	1.422×10^{-1}	1.407×10^{-1}	1.393×10^{-1}	1.380×10^{-1}
0.75	2.143×10^{-1}	1.743×10^{-1}	1.724×10^{-1}	1.707×10^{-1}	1.692×10^{-1}
0.80	2.620×10^{-1}	2.137×10^{-1}	2.115×10^{-1}	2.094×10^{-1}	2.076×10^{-1}
0.85	3.216×10^{-1}	2.636×10^{-1}	2.609×10^{-1}	2.584×10^{-1}	2.561×10^{-1}
0.90	3.990×10^{-1}	3.295×10^{-1}	3.263×10^{-1}	3.233×10^{-1}	3.206×10^{-1}
0.95	5.124×10^{-1}	4.279×10^{-1}	4.240×10^{-1}	4.202×10^{-1}	4.170×10^{-1}
0.99	7.067×10^{-1}	5.994×10^{-1}	5.994×10^{-1}	5.896×10^{-1}	5.854×10^{-1}
0.01	1.971×10^{-5}	1.960×10^{-5}	1.876×10^{-5}	1.823×10^{-5}	1.752×10^{-5}
0.05	4.933×10^{-4}	4.905×10^{-4}	4.695×10^{-4}	4.564×10^{-4}	4.387×10^{-4}
0.10	1.982×10^{-3}	1.971×10^{-3}	1.886×10^{-3}	1.834×10^{-3}	1.762×10^{-3}
0.15	4.494×10^{-3}	4.468×10^{-3}	4.276×10^{-3}	4.157×10^{-3}	3.995×10^{-3}
0.20	8.075×10^{-3}	8.028×10^{-3}	7.683×10^{-3}	7.468×10^{-3}	7.178×10^{-3}
0.25	1.279×10^{-2}	1.272×10^{-2}	1.217×10^{-2}	1.183×10^{-2}	1.137×10^{-2}
0.30	1.875×10^{-2}	1.864×10^{-2}	1.783×10^{-2}	1.733×10^{-2}	1.666×10^{-2}
0.35	2.605×10^{-2}	2.590×10^{-2}	2.478×10^{-2}	2.409×10^{-2}	2.315×10^{-2}
0.40	3.488×10^{-2}	3.468×10^{-2}	3.318×10^{-2}	3.224×10^{-2}	3.098×10^{-2}
0.45	4.544×10^{-2}	4.517×10^{-2}	4.322×10^{-2}	4.200×10^{-2}	4.035×10^{-2}
0.50	5.800×10^{-2}	5.766×10^{-2}	5.516×10^{-2}	5.360×10^{-2}	5.150×10^{-2}
0.55	7.292×10^{-2}	7.249×10^{-2}	6.934×10^{-2}	6.738×10^{-2}	6.474×10^{-2}
0.60	9.066×10^{-2}	9.014×10^{-2}	8.621×10^{-2}	8.378×10^{-2}	8.049×10^{-2}
0.65	1.119×10^{-1}	1.112×10^{-1}	1.064×10^{-1}	1.034×10^{-1}	9.933×10^{-2}
0.70	1.374×10^{-1}	1.366×10^{-1}	1.306×10^{-1}	1.270×10^{-1}	1.220×10^{-1}
0.75	1.683×10^{-1}	1.674×10^{-1}	1.602×10^{-1}	1.557×10^{-1}	1.497×10^{-1}
0.80	2.066×10^{-1}	2.054×10^{-1}	1.967×10^{-1}	1.913×10^{-1}	1.840×10^{-1}
0.85	2.549×10^{-1}	2.535×10^{-1}	2.431×10^{-1}	2.365×10^{-1}	2.278×10^{-1}
0.90	3.191×10^{-1}	3.174×10^{-1}	3.048×10^{-1}	2.970×10^{-1}	2.863×10^{-1}
0.95	4.152×10^{-1}	4.131×10^{-1}	3.977×10^{-1}	3.880×10^{-1}	3.750×10^{-1}
0.99	5.832×10^{-1}	5.805×10^{-1}	5.607×10^{-1}	5.484×10^{-1}	5.317×10^{-1}

largest ionic size. This behavior may be attributed to the combined effect of the ionic sizes and the hydration energies of these metal ions. It can be said that the pore size of the cavity and stiffness of the matrix may be the controlling parameters, causing the hindrance to the entry of the larger ions and, hence, partial ion exchange. The rubidium and cesium ions are exchanged to a lesser extent, due to their larger ionic radii, while the greater hydration

TABLE 35
Dimensionless Time as a Function of Fractional Attainment of Equilibrium for Some Particle Diffusion-Controlled Metal(II)-H(I) Exchanges Involving Different Mobility Ratios

Fractional attainment of equilibrium, U(T)	Dimensionless time, T				
	$\alpha = \frac{\bar{D}_{Mn}}{\bar{D}_H} = 0.088$	$\alpha = \frac{\bar{D}_{Co}}{\bar{D}_H} = 0.135$	$\alpha = \frac{\bar{D}_{Ni}}{\bar{D}_H} = 0.138$	$\alpha = \frac{\bar{D}_{Fe/Cr}}{\bar{D}_H} = 0.141$	$\alpha = \frac{\bar{D}_{Mg/Cu}}{\bar{D}_H} = 0.144$
0.01	4.360×10^{-6}	4.723×10^{-6}	4.747×10^{-6}	4.770×10^{-6}	4.791×10^{-6}
0.05	1.093×10^{-4}	1.183×10^{-4}	1.189×10^{-4}	1.195×10^{-4}	1.200×10^{-4}
0.10	4.404×10^{-4}	4.768×10^{-4}	4.792×10^{-4}	4.815×10^{-4}	4.836×10^{-4}
0.15	1.004×10^{-3}	1.086×10^{-3}	1.092×10^{-3}	1.097×10^{-3}	1.102×10^{-3}
0.20	1.817×10^{-3}	1.966×10^{-3}	1.975×10^{-3}	1.985×10^{-3}	1.993×10^{-3}
0.25	2.909×10^{-3}	3.143×10^{-3}	3.158×10^{-3}	3.173×10^{-3}	3.187×10^{-3}
0.30	4.319×10^{-3}	4.661×10^{-3}	4.682×10^{-3}	4.704×10^{-3}	4.723×10^{-3}
0.35	6.102×10^{-3}	6.575×10^{-3}	6.604×10^{-3}	6.634×10^{-3}	6.660×10^{-3}
0.40	8.336×10^{-3}	8.964×10^{-3}	9.002×10^{-3}	9.041×10^{-3}	9.075×10^{-3}
0.45	1.113×10^{-2}	1.193×10^{-2}	1.198×10^{-2}	1.203×10^{-2}	1.208×10^{-2}
0.50	1.462×10^{-2}	1.563×10^{-2}	1.569×10^{-2}	1.575×10^{-2}	1.580×10^{-2}
0.55	1.901×10^{-2}	2.024×10^{-2}	2.031×10^{-2}	2.038×10^{-2}	2.045×10^{-2}
0.60	2.449×10^{-2}	2.597×10^{-2}	2.606×10^{-2}	2.615×10^{-2}	2.622×10^{-2}
0.65	3.524×10^{-2}	3.302×10^{-2}	3.312×10^{-2}	3.323×10^{-2}	3.332×10^{-2}
0.70	3.921×10^{-2}	4.177×10^{-2}	4.191×10^{-2}	4.205×10^{-2}	4.218×10^{-2}
0.75	4.981×10^{-2}	5.250×10^{-2}	5.267×10^{-2}	5.284×10^{-2}	5.299×10^{-2}
0.80	6.371×10^{-2}	6.692×10^{-2}	6.712×10^{-2}	6.732×10^{-2}	6.750×10^{-2}
0.85	8.387×10^{-2}	8.761×10^{-2}	8.784×10^{-2}	8.808×10^{-2}	8.829×10^{-2}
0.90	1.169×10^{-1}	1.209×10^{-1}	1.211×10^{-1}	1.214×10^{-1}	1.216×10^{-1}
0.95	1.849×10^{-1}	1.874×10^{-1}	1.875×10^{-1}	1.877×10^{-1}	1.878×10^{-1}
0.99	3.310×10^{-1}	3.338×10^{-1}	3.337×10^{-1}	3.336×10^{-1}	3.336×10^{-1}
0.01	4.803×10^{-6}	4.817×10^{-6}	4.930×10^{-6}	5.009×10^{-6}	5.131×10^{-6}
0.05	1.203×10^{-4}	1.207×10^{-4}	1.235×10^{-4}	1.255×10^{-4}	1.285×10^{-4}
0.10	4.848×10^{-4}	4.862×10^{-4}	4.975×10^{-4}	5.055×10^{-4}	5.177×10^{-4}
0.15	1.104×10^{-3}	1.107×10^{-3}	1.133×10^{-3}	1.151×10^{-3}	1.178×10^{-3}
0.20	1.998×10^{-3}	2.003×10^{-3}	2.049×10^{-3}	2.081×10^{-3}	2.130×10^{-3}
0.25	3.194×10^{-3}	3.203×10^{-3}	3.274×10^{-3}	3.324×10^{-3}	3.400×10^{-3}
0.30	4.734×10^{-3}	4.747×10^{-3}	4.849×10^{-3}	4.925×10^{-3}	5.031×10^{-3}
0.35	6.674×10^{-3}	6.692×10^{-3}	6.832×10^{-3}	6.930×10^{-3}	7.078×10^{-3}
0.40	9.904×10^{-3}	9.117×10^{-3}	9.299×10^{-3}	9.426×10^{-3}	9.618×10^{-3}
0.45	1.210×10^{-2}	1.213×10^{-2}	1.236×10^{-2}	1.252×10^{-2}	1.276×10^{-2}
0.50	1.583×10^{-2}	1.587×10^{-2}	1.615×10^{-2}	1.634×10^{-2}	1.663×10^{-2}
0.55	2.048×10^{-2}	2.052×10^{-2}	2.086×10^{-2}	2.108×10^{-2}	2.143×10^{-2}
0.60	2.626×10^{-2}	2.631×10^{-2}	2.671×10^{-2}	2.698×10^{-2}	2.738×10^{-2}
0.65	3.337×10^{-2}	3.343×10^{-2}	3.391×10^{-2}	3.424×10^{-2}	3.473×10^{-2}
0.70	4.225×10^{-2}	4.233×10^{-2}	4.301×10^{-2}	4.348×10^{-2}	4.420×10^{-2}
0.75	5.308×10^{-2}	5.318×10^{-2}	5.399×10^{-2}	5.457×10^{-2}	5.544×10^{-2}
0.80	6.760×10^{-2}	6.772×10^{-2}	6.870×10^{-2}	6.938×10^{-2}	7.042×10^{-2}
0.85	8.841×10^{-2}	8.855×10^{-2}	$8{,}968 \times 10^{-2}$	9.048×10^{-2}	9.170×10^{-2}
0.90	1.217×10^{-1}	1.219×10^{-1}	1.231×10^{-1}	1.239×10^{-1}	1.253×10^{-1}
0.95	1.878×10^{-1}	1.879×10^{-1}	1.886×10^{-1}	1.892×10^{-1}	1.901×10^{-1}
0.99	3.335×10^{-1}	3.335×10^{-1}	3.329×10^{-1}	3.326×10^{-1}	3.321×10^{-1}

energies cause a lesser exchange of lithium and sodium ions. Since the potassium ion lies in between the two governing conditions in all respects, it is highly preferred by C-NbP and exchanged to a greater extent. This may be the only reason that it has the lowest $|C|$ value. The overall higher values of $|C|$ also indicate a smaller cavity in C-NbP than in those of other zeolites.

In light of the above discussion, it can be inferred that the exchange of alkali metal ions takes place in the unhydrated form. The ions have to shed most of their water of hydration at the surface of the cavity to enter into it, where they again get hydrated. With the progress of exchange, ions accumulate inside the cavity and cause steric hindrance. Also, the steric hindrance is not compensated by the exchanger, due to its inelastic and, hence, rigid matrix.

Steric hindrance is more pronounced in the case of the lithium ion due to its greater hydration energy and the hydrated ionic radius, compared to rubidium and cesium ions, where ionic sizes play a dominating role in a lesser exchange of these ions. The hydration energy and size of the potassium ion are such that it can easily enter and best fit itself into the cavity, and can form the maximum number of ion pairs with the fixed ionogenic groups of the matrix, compared to other group members. The steric effect is also responsible for the overall partial exchange of the cations.

The selectivity sequence for the different alkali metal ions on C-NbP with the progress of exchange is summarized below:

Equivalent ionic fraction ($\bar{X}_M$)	Selectivity sequence at 20°C
0.00—0.02	$Na^+ > Rb^+ > K^+ > Li^+ > Cs^+$
0.02—0.17	$Na^+ > Rb^+ > K^+ > Cs^+ > Li^+$
0.17—0.42	$Na^+ > K^+ > Rb^+ > Cs^+ > Li^+$
0.42—0.57	$K^+ > Na^+ > Rb^+ > Cs^+ > Li^+$
0.57—0.64	$K^+ > Rb^+ > Na^+ > Cs^+ > Li^+$
0.64—0.93	$K^+ > Rb^+ > Cs^+ > Na^+ > Li^+$
0.93—1.00	$K^+ > Cs^+ > Rb^+ > Na^+ > Li^+$

Of the above sequences of exchange, only the fourth one, which is at about half exchange, is parallel to the order of maximum uptake, and again supports the partial exchange of alkali metal ions due to steric hindrance.

The C-NbP has high negative values of ΔS°_{ex} for Na^+ in tracer concentration, thereby indicating the highest stability and, hence, the highest selectivity for this ion in a lower concentration range.

The other parameters, namely, entropy, enthalpy, and free energy, also support the higher selectivity of Na^+ in the tracer concentration range.

On the other hand, the overall thermodynamic parameters and ion-exchange data indicate a stiff three-dimensional structure for C-NbP. It has an overall greater affinity for K^+ because of its suitable size, hydration energy, and other factors which enable it to fit in the cavity of the exchanger easily and, of course, with the least steric hindrance.

2. Selectivity for Alkaline Earths

Varshney and co-workers have studied[270-273] the thermodynamics of alkaline earth-H^+ ion exchanges on antimony(V) silicate (SbSi) and zirconium(IV) phosphosilicate (ZPS) cation exchangers. On SbSi,[270,271] the exchange equilibrium is established in 1 h for the Ba^{2+}-H^+ and Sr^{2+}-H^+ exchanges, while it requires 3 h for Ca^{2+}-H^+ and Mg^{2+}-H^+ exchanges. All the equilibrium studies were performed after shaking the solutions for these time intervals in the respective systems. The equivalent ion fractions of the counter ions in the SbSi and solution phases $\bar{X}_M$, $\bar{X}_H$, X_M, and X_H are calculated by the expressions:

$$\bar{X}_M = \frac{\bar{C}_M}{\bar{C}},\ X_M = \frac{C_M}{C},\ \bar{X}_H = \frac{\bar{C}_H}{\bar{C}},\ X_H = \frac{C_H}{C}$$

where $\bar{C}$ and C are the total electrolyte concentrations in the solid and solution phases, respectively.

To examine the affinity for the interacting ions,the separation factors and selectivity coefficients are calculated[281] by the following equations, assuming the ratio of the activity coefficients in solution to be unity.[282] Under these conditions, K_c ceases to express interactions in solutions.

$$\text{Now, } \alpha_H^M = \frac{\overline{X}_M}{\overline{X}_H}\frac{X_H}{Z_M}$$

$$\text{and } K_c = \frac{\overline{X}_M}{\overline{X}_H^{\,2}} \cdot \frac{X_H^{\,2}}{Z_M}$$

For the Ba^{2+}-H^+ at 25 and 50°C, the exchange isotherms are sigmoid, indicating a selectivity reversal. Barium is preferred by the exchanger up to a certain value of the ionic fraction (0.60 at 25°C and 0.64 at 50°C). Thereafter, it shows a preference for H^+ ions. However, in the Sr^{2+}-H^+ and Ca^{2+}-H^+ exchanges, Sr^{2+} and Ca^{2+} ions are adsorbed by the solid phase at all concentrations, the preference being higher at a higher temperature. In the case of the Mg^{2+}-H^+ exchange, the reverse is true i.e., the material has a preference for H^+ ions at both temperatures over the entire concentration range. These conclusions are substantiated by the separation factors and selectivity coefficients.

The thermodynamic equilibrium constants are obtained[283] from the relationship

$$\ln K = (Z_A - Z_B) + \int_o^1 \ln K_c \, d\overline{X}_M$$

where Z_A and Z_B are the valencies of the competing ions. The integrals are evaluated using the trapezoidal rule.[284]

The standard free energies of exchange (ΔG°) are then calculated[285] using the relation:

$$\Delta G^\circ = -RT \ln K$$

where R is the universal gas constant and T is the temperature in degrees Kelvin.

The standard enthalpy changes (ΔH°) are then calculated from the Van't Hoff isochore

$$\ln\left(\frac{K_{T_2}}{K_{T_1}}\right) = -\frac{\Delta H^\circ}{R}\left(\frac{1}{T_2} - \frac{1}{T_1}\right)$$

and standard entropy changes (ΔS°) by the equation:

$$\Delta G^\circ = \Delta H^\circ - T\,\Delta S^\circ$$

A negative free-energy change is observed from the Ba^{2+}-H^+, Sr^{2+}-H^+, and Ca^{2+}-H^+ interactions, indicating that these metal ions are adsorbed preferably, compared to the H^+ ions on SbSi. However, for the Mg^{2+}-H^+ exchange, the free-energy change is positive, indicating that SbSi has a lower preference for Mg^{2+} than for H^+ ions at both temperatures.

The activity coefficients of the metal and H^+ ions are calculated from the following expressions:

$$\ln f_M = (\overline{X}_M - 1)\ln K_c - \int_o^{\overline{X}_M} \ln K_c \, d\overline{X}_M$$

$$\ln f_H = \overline{X}_M \ln K_c - \int_o^{\overline{X}_M} \ln K_c \, d\overline{X}_M$$

TABLE 36
Thermodynamic Parameters for the Ca^{2+}-H^+, Mg^{2+}-H^+, Ba^{2+}-H^+, and Sr^{2+}-H^+ Exchanges on Antimony(V) Silicate at 25 and 50°C

Thermodynamic parameter	Ca^{2+}-H^+ system		Mg^{2+}-H^+ system		Ba^{2+}-H^+ system		Sr^{2+}-H^+ system	
	25°C	50°C	25°C	50°C	25°C	50°C	25°C	50°C
K	18.33	25.96	0.07	0.12	5.35	6.64	13.68	26.95
ΔG° (kJ/mol)	−7.22	−8.76	6.70	5.63	−4.17	−5.09	−6.49	−8.89
ΔH° (kJ/mol)	11.13		19.45		6.87		21.73	
ΔS° (J mol $^{-1}$ K^{-1})	60	60	40	40	0.04	0.04	0.09	0.09

At both temperatures, the f_M values are less than unity for Ba^{2+}, Sr^{2+}, and Ca^{2+} exchanges with H^+ ions. But for the Mg^{2+}-H^+ exchange, values are greater than unity. A heterogeneity in the distribution of ions on the SbSi surface during the ion-exchange process is similar to that in soils reported by Diest and Talibuddin.[286]

To further examine the deviation of these heterogeneous systems from ideality, the excess thermodynamic functions are calculated[287,288] from the expressions:

$$G^x_m = RT\,(\overline{X}_M \ln f_M + \overline{X}_H \ln f_H)$$

$$H^x_m = -\,RT^2\left[\overline{X}_M\left(\frac{\ln f_M}{\Delta T}\right) + \overline{X}_H\left(\frac{\ln f_H}{\Delta T}\right)\right]$$

$$\text{and } \Delta G^x_m = \Delta H^x_m - T\,\Delta S^x_m$$

where ΔG^x_m, ΔH^x_m, and ΔS^x_m are the excess free energies, enthalpies, and entropies of mixing, respectively. These values indicate that the heterogeneous mixture of ions during the exchange is more stable, compared to the pure forms. Table 36 summarizes the thermodynamic parameters for the Ca^{2+}-H^+, Mg^{2+}-H^+, Ba^{2+}-H^+, and Sr^{2+}-H^+ exchanges on the SbSi exchanger.

The same treatment was done on ZPS[272,273] as described above for the alkaline earths. According to these studies, the exchange isotherms for the Ba^{2+}-H^+ exchange are sigmoid and show a selectivity reversal at both temperatures. It suggests that H^+ ions are strongly preferred by ZPS up to equivalent ionic fractions of 0.60 at 30°C and 0.49 at 50°C. After that, an upward trend is observed for the Ba^{2+} ions. In the case of the Sr^{2+}-H^+ exchange, the isotherms are of the Langmuir type. They show a stronger preference for H^+ ions, compared to Sr^{2+} ions, throughout the range of concentration studied, and the affinity is affected by temperature, being higher at 30°C. It is also supported by the separation factors. For the Ba^{2+}-H^+ exchange, however, some separation factor values are greater than unity, indicating a selectivity reversal. In the case of the Ca^{2+}-H^+ and Mg^{2+}-H^+ exchanges, it appears that the exchanger has a greater preference for H^+ ions, compared to Ca^{2+} and Mg^{2+} ions, except in the X_{Ca} range of 0.65 to 0.85, where the exchanger shows a preference for Ca^{2+} ions at 50°C. Also, there is a marked increase in the adsorption of metal ions in a specified ionic fraction range — from 0.5 to 0.7 for X_{Ca} and from 0.5 to 0.6 for X_{Mg}.

These studies also reveal that ZPS has a lower preference for Ca^{2+} and Mg^{2+} than for H^+. Hence, the replacement of H^+ ions by Ca^{2+} or Mg^{2+} ions is not spontaneous. In both cases, the reaction is accompanied by a negative enthalpy change which presumably is due to the smaller hydrated radii and higher change of Ca^{2+} and Mg^{2+} ions, compared to H^+ ions. It also suggests that the adsorption of Ca^{2+} and Mg^{2+} ions decreases with the rise in temperature.

TABLE 37
Thermodynamic Parameters for the Ca^{2+}-H^+, Mg^{2+}-H^+, Ba^{2+}-H^+, and Sr^{2+}-H^+ Exchanges on Zirconium Phosphosilicate at 30 and 50°C

Thermodynamic parameters	Ca^{2+}-H^+ system		Mg^{2+}-H^+ system		Ba^{2+}-H^+ system		Sr^{2+}-H^+ system	
	30°C	50°C	30°C	50°C	30°C	50°C	30°C	50°C
K	0.87	0.58	0.71	0.50	2.12	2.54	1.08	1.41
ΔG° (kJ/mol)	0.36	1.45	0.87	1.86	−1.89	−2.51	−0.19	−0.92
ΔH° (kJ/mol)	−16.12		−14—17		8—40		10.83	
ΔS° (kJ mol^{-1} degree^{-1})	−0.05	−0.05	−0.05	−0.05	0.03	0.03	0.04	0.04

In the case of Ba^{2+}-H^+ and Sr^{2+}-H^+ exchanges, it is observed that the exchanger has a higher affinity for the Ba^{2+} and Sr^{2+} ions at a higher temperature. Further, the material is more selective for Ba^{2+} ions.

All other observations are similar, as noted above on SbSi[270,271] and soil[286] systems. Table 37 summarizes the thermodynamic parameters for these four systems on ZPS.

3. Adsorption of Pesticides

Other interesting thermodynamic studies relate to the adsorption of pesticides on inorganic ion exchangers. This type of work was recently initiated by the research group of Varshney.[98] To start with, they have studied the adsorption of carbofuran on SbSi.

Adsorption of carbofuran on SbSi at 30 and 50°C is represented by the adsorption isotherms, which are concave to the Y axis up to an equilibrium concentration of $\sim 1.12 \times 10^{-3}$ mmol/ml. Above this concentration, a reversal is obtained. The adsorption deceases with a rise in temperature. The average partial molal free-energy changes are obtained from the thermodynamic relationship:

$$-F = RT \ln \frac{C_e}{C_o}$$

where C_e and C_o are the equilibrium concentrations of carbofuran in the suspension.

The adsorption of carbofuran is in close agreement with the Freundlich equation

$$x/m = KC^{1/n}$$

where x/m = millimoles of carbofuran adsorbed per gram of SbSi, C is the concentration of carbofuran in equilibrium suspension per ml, and K and 1/n are the constants evaluated from the intercept and slope of the Freundlich isotherms.

The thermodynamic equilibrium constant (K_o) for the adsorption reaction was obtained by plotting $\ln (C_s/C_e)$ versus C_s and extrapolating C_s to zero,[289] where C_s = millimoles of carbofuran adsorbed per gram and C_e = millimoles of carbofuran per milliliter in equilibrium suspension.

The standard free energy (ΔG°), enthalpy (ΔH°), and entropy changes (ΔS°) were then calculated as described above for metal ions. Table 38 summarizes the values of the various thermodynamic parameters obtained in this study.

The study reveals the following facts:

1. Adsorption of carbofuran is higher at a lower temperature.
2. Adsorption is easier at a higher equilibrium concentration range of carbofuran.
3. The adsorption is much higher on SbSi, compared to sandy clay loam and silt loam soils.[290]

TABLE 38
Freundlich Isotherm Constants (K and $^1/_n$) and Various Thermodynamic Parameters for the Adsorption of Carbofuran on Antimony(V) Silicate Cation Exchanager at 30 and 50°C

	Constants	Values at 30°C	Values at 50°C
Freundlich isotherm constants	K	3.02×10^{-3}	1.995×10^{-3}
	$^1/_n$	1.1881	1.2000
Thermodynamic constants	K_c	1.5300	1.3648
	Δ° (cal/mol)	−256.3828	−199.9272
	ΔH° (cal/mol)		−1111.6861
	ΔS° (cal/mol/degree)		−2.8228

It is, therefore, clear that the presence of an inorganic material possessing a good ion-exchange capacity may enhance the adsorption characteristics of the soil, thus retarding effectively the movement of a pesticide which may otherwise have a harmful effect on crops. Further studies are, however, required to generalize the findings to all pesticides.

REFERENCES

1. **Clearfield, A., Ed.,** *Inorganic Ion Exchange Materials,* CRC Press, Boca Raton, FL, 1982.
2. **Amphlett, C. B.,** *Inorganic Ion Exchangers,* Elsevier, Amsterdam 1964.
3. **Walton, H. F.,** *Ion Exchange Chromatography,* Dowden, Hutchinson & Ross, Stroudsburg, PA, 1976.
4. **Qureshi, M. and Varshney, K. G.,** Preparation and properties of stannic tungstate, *J. Inorg. Nucl. Chem.,* 30, 3081, 1968.
5. **Rawat, J. P. and Thind, P. S.,** A kinetic study of ion exchange in tantalum arsenate to understand the theoretical aspects of separations, *J. Phys. Chem.,* 80, 2384, 1976.
6. **Varshney, K. G. and Naheed, S.,** Amine Sn(II) hexacyanoferrate(II) as an ion exchanger, *J. Inorg. Nucl. Chem.,* 39, 2075, 1977.
7. **Gill, J. S. and Tandon, S. N.,** Synthesis and ion exchange properties of ceric antimonate: separation of Hg^{2+} from Zn^{2+}, Cd^{2+}, Pb^{2+} and Ti^{4+}; Cu^{2+} from Mn^{2+}; Fe^{3+} from Al^{3+}; and Zr^{4+} from Th^{4+}, *Talanta,* 19, 1335, 1972.
8. **Topp, N. E. and Pepper, K. W.,** Properties of ion-exchange resins in relation to their structure. I. Titration curves, *J. Chem. Soc. A,* 3299, 1949.
9. **Hamaguchi, H., Kuroda, R., Aoki, K., Sugisita, R., and Onuma, N.,** Cation-exchange separation of scandium, *Talanta,* 10, 151, 1963.
10. **Varshney, K. G., Khan, A. A. and Premadas, A.,** pH-titration and distribution studies of alkali metal ions on amorphous Sn(IV) and Cr(III) arsenophosphates: separation of metal ions, *Ann. Chim.,* 579, 1981.
11. **Varshney, K. G. and Khan, A. A.,** Synthesis and ion exchange properties of Sn(IV) and Cr(III) arsenophosphates, *J. Inorg. Nucl. Chem.,* 41, 241, 1979.
12. **Sunandamma, Y.,** Some Studies on Inorganic Ion Exchangers Based on Lead, Antimony and Silicon, Ph.D. thesis, Aligarh Muslim University, India, 1986.
13. **Abe, M.,** Synthetic inorganic ion-exchange materials. XIII. The mutual separation of alkali metals with three different antimonic acids, *Bull. Chem. Soc. Jpn.,* 42, 2683, 1969.
14. **Abe, M., Achmad, E. A., and Hayashi, K.,** Ion exchange separation of lithium from large amounts of sodium, calcium and other elements by a double column of Dowex 50 WX8 and crystalline antimony(V) acid, *Anal. Chem.,* 52, 524, 1980.
15. **Gill, J. S. and Tandon, S. N.,** Ion-exchange properties of titanic antimonate. Separation of aluminum (3+) from chromium (3+) and iron (3+); cobalt (2+) from nickel (2+) and manganese (2+); uranyl ion from thorium (4+); and zinc (2+) from mercury (2+) and cadmium (2+), *J. Radioanal. Chem.,* 20, 5, 1974.
16. **Mathew, J. and Tandon, S. N.,** Synthesis and ion exchange properties of zirconium antimonate; separation of mercury from zinc, cadmium, lead and bismuth; calcium from magnesium; strontium from rubidium, cesium and barium; barium from lanthanum; indium from iron, gallium and lanthanum; and silver from thallium ions, *J. Radioanal. Chem.,* 27, 315, 1975.

17. **Abe, M. and Ito, M.,** Synthetic inorganic ion-exchange materials. IX. The mutual separation of alkali metals with antimonic acid, *Bull. Chem. Soc. Jpn.*, 40, 1013, 1967.
18. **Qureshi, M. and Kumar, V.,** Synthesis and ion-exchange characteristics of titanium antimonates, *J. Chem. Soc. A*, 1488, 1970.
19. **Qureshi, M., Kumar, V. and Zehra, N.,** Synthesis of tin(IV) antimonate of high thermal and chemical stability, *J. Chromatogr.*, 67, 351, 1972.
20. **Phillips, H. O. and Kraus, K. A.,** Adsorption of inorganic materials. IV. Cation exchange properties of zirconium antimonate, *J. Am. Chem. Soc.*, 84, 2267, 1962.
21. **Qureshi, M., Gupta, J. P. and Sharma, V.,** Synthesis of a reproducible and chemically stable tantalum antimonate, *Anal. Chem.*, 45, 190, 1973.
22. **Abe, M. and Hayashi, K.,** Synthetic inorganic ion exchange materials. XXXIII. Selectivities of alkali metal ions on tin(IV) antimonate cation exchanger prepared at different conditions, *Solvent Extr. Ion Exchange*, 1, 97, 1983.
23. **Helfferich, F.,** *Ion Exchange*, McGraw-Hill New York, 1962.
24. **Gill J. S. and Tandon, S. N.,** Study of stoichiometric uptake of cations by some inorganic ion exchangers using radiotracers, *Int. J. Appl. Radiat. Isot.*, 25, 31, 1974.
25. **Gill, J. S. and Tandon, S. N.,** Preparation and ion exchange properties of stannic ferrocyanide, *J. Inorg. Nucl. Chem.*, 34, 3885, 1972.
26. **Gill, J. S. and Tandon, S. N.,** Synthesis and ion exchange properties of zirconium ferrocyanide, *J. Radioanal. Chem.*, 13, 391, 1973.
27. **Gill, J. S. and Tandon, S. N.,** Synthesis and ion exchange properties of ceric antimonate: separation of Hg^{2+} from Zn^{2+}, Cd^{2+}, Pb^{2+} and Tl^{+}, Cu^{2+} from Mn^{2+}; Fe^{3+} from Al^{3+}; and Zr^{4+} from Th^{4+}, *Talanta*, 19, 1355, 1972.
28. **Gill, J. S. and Tandon, S. N.,** Ion exchange properties of titanic antimonate: separation of Al^{3+} from Cr^{3+} and Fe^{3+}; Co^{2+} from Ni^{2+} and Mn^{2+}; UO_2^{2+} from Th^{4+}; and Zn^{2+} from Hg^{2+} and Cd^{2+}, *J. Radioanal. Chem.*, 20, 5, 1974.
29. **Tandon, S. N. and Gill, J. S.,** Synthesis and ion-exchange properties of ceric tungstate, *Talanta*, 20, 585, 1973.
30. **Qureshi, M. and Thakur, J. S.,** The Qureshi-Thakur series for the solubility of synthetic inorganic ion exchangers, *Inorg. Nucl. Chem. Lett.*, 15, 239, 1979.
31. **Varshney, K. G., Agrawal, S., Varshney, K., Sharma, U., and Rani, S.,** Radiation stability of some thermally stable inorganic ion exchangers, *J. Radioanal. Nucl. Chem.*, 82(2), 299, 1984.
32. **Varshney, K. G., Agrawal, S., and Varshney, K.,** Synthesis, ion exchange behavior and analytical applications of a new crystalline and stable zirconium(IV) arsenophosphate cation exchanger: analysis of some silicate rocks, *Sep. Sci. Technol.*, 18, 59, 1983.
33. **Varshney, K. G., Varshney, K., and Agrawal, S.,** Ion exchange and selectivity behavior of thermally treated and γ-irradiated phases of zirconium(IV) arsenophosphate cation exchanger: separation of Al(III) from some metal ions and removal of cations from water, *Sep. Sci. Technol.*, 18, 905, 1983.
34. **Varshney, K. G., Khan, A. A., Maheshwari, A., Anwar, S., and Sharma, U.,** Synthesis of a new thermally stable Sn(IV) arsenosilicate cation exchanger and its application for the column chromatographic separation of metal ions, *Indian J. Technol.*, 22, 99, 1984.
35. **Varshney, K. G., Sharma, U., and Rani, S.,** Synthesis and analytical applications of thorium(IV) phosphosilicate, a new mercury selective cation exchanger: effect of irradiation on its ion exchange behavior, *J. Indian Chem. Soc.*, 61, 220, 1984.
36. **Varshney, K. G., Sharma, U., Rani, S., and Premadas, A.,** Cation exchange study on crystalline and thermally stable phase of antimony(V) silicate: effect of irradiation on its ion exchange behavior and separation of Cd(II) from Zn(II) and Mn(II), and of Mg(II) from Ba(II), Ca(II) and Sr(II), *Sep. Sci. Technol.*, 17, 1527, 1982-83.
37. **Gill, J. S. and Tandon, S. N.,** Investigation of the resistance of some inorganic ion exchangers against gamma radiation, *Radiochem. Radioanal. Lett.*, 14, 379, 1973.
38. **Singh, N. J., Tandon, S. N., and Gill, J. S.,** Resistance of zirconium arsenophosphate ion exchanger against gamma radiation, *Radiochem. Radioanal. Lett.*, 41, 79, 1979.
39. **Mathew, J., Tandon, S. N., and Gill, J. S.,** Effect of gamma radiation on the exchange characteristics of some antimonates, *Radiochem. Radioanal. Lett.*, 30, 381, 1977.
40. **Zsinka, L., Szirtes, L., Mink, J. and Kalama, A.,** Effect of γ-radiation on various synthetic inorganic ion exchangers, *J. Inorg. Nucl. Chem.*, 36, 1147, 1974.
41. **LaGinestra, A., Ferragina, C., Massucci, M. A., Patrono, P., DiRocco, R., and Tomilson, A. A. G.,** Thermal, redox and catalytic characterization of synthetic inorganic ion exchangers, *Gazz. Chim. Ital.*, 113, 357, 1983.

42. **Clearfield, A. and Medina, A. S.,** On the mechanism of ion exchange in crystalline zirconium phosphates. III, *J. Inorg. Nucl. Chem.*, 32, 2775, 1970.
43. **LaGinestra, A., Ferragina, C., and Patrono, F.,** Silver-zirconium phosphate system: characterization of the phases obtained at different temperatures, *Mat. Res. Bull.*, 14, 1099, 1979.
44. **Qureshi, M., Ahmad, A., Shakeel, N. A., and Gupta, A. P.,** Synthesis, dehydration studies and cation exchange behavior of a new phase of niobium(V) phosphate, *Bull. Chem. Soc. Jpn.*, 59, 3247, 1986.
45. **Herman, R. G., Lunsford, J. H., Beyer, H., Jacobs, P. A., and Uytlechoeven, J. B.,** Redox behavior of transition metal ions in zeolites. I. Reversibility of the hydrogen reduction of copper Y zeolites, *J. Phys. Chem.*, 79, 2388, 1975.
46. **Jacobs, P. A. and Uytlerhoeven, J. B.,** Active sites in zeolites. VII. Isopropanol dehydrogenation over alkali cation-exchanged X and Y zeolites, *J. Chem. Soc. Chem. Commun.*, 128, 1977.
47. **Kalman, T. and Clearfield, A.,** *Proc. Int. Symp. Chem. React. Eng. Adv. Chem.*, 3rd Ser., 133, 65, 1974.
48. **Iwarmoto, M., Nomura, Y., and Kagowoe, S.,** Catalytic oxidation of propene over zirconium phosphates, *J. Catal.*, 69, 234, 1981.
49. **Clearfield, A. and Thakur, D. S.,** The acidity of zirconium phosphates in relation to their activity in the dehydration of cyclohexanol, *J. Catal.*, 65, 185, 1980.
50. **Nozaki, F., Itoh, T., and Ueda, S.,** Metal phosphate catalyst: catalytic activity of zirconium phosphate for dehydration of 2-propanol, *Nippon Kagaku Kaishi*, 4, 674, 1973.
51. **Onoue, Y., Mizutani, Y., Kiyama, S. A., Izumi, Y., and Watanabe, Y.,** Why not do it in one step? The case of MIBK, *Chemtechnology*, 36, 1977.
52. **LaGinestra, A., Patrono, P., Massucci, M. A., and Ferragina, C.,** unpublished studies.
53. **Varshney, K. G., Sharma, U., and Rani, S.,** Synthesis and analytical applications of thorium(IV) phosphosilicate, a new mercury selective cation exchanger: effect of irradiation on its ion exchanger behavior, *J. Indian Chem. Soc.*, 61, 220, 1984.
54. **Varshney, K. G., Khan, A. A., and Khan, A. R.,** Antimony(V) arsenophosphate as a thermally stable cation exchanger: selective adsorption of alkaline earth and transition metal ions on its column, *Bull. Chem. Soc. Jpn.*, 61, 3693, 1988.
55. **Varshney, K. G. and Maheshwari, S. M.,** Synthesis, characterization and ion exchange behavior of antimony(V) phosphate: selective adsorption of cadmium and mercury on its column, *Ecotox. Environ. Safety*, 18, 1, 1989.
56. **Rawat, J. P. and Khan, R. A.,** Preparation, properties and analytical applications of titanium tungstoarsenate as thorium selective inorganic ion exchanger, *Indian J. Chem. A*, 19, 925, 1980.
57. **Rawat, J. P., Khan, M. A., and Thind, P. S.,** Preparation and properties of mercury selective ion exchanger cerium selenite, *Bull. Chem. Soc. Jpn.*, 57, 1701, 1984.
58. **Qureshi, M., Kumar, R., and Rathore, H. S.,** Synthesis and properties of stannic arsenate, *J. Chem. Soc. A*, 272, 1970.
59. **Qureshi, M., Rathore, H. S., and Kumar, R.,** Synthesis and ion-exchange properties of stannic arsenates: separation of Fe^{3+} from Ni^{2+}, Co^{2+}, Mn^{2+}, Ca^{2+}, and Al^{3+}: separation of Al^{3+} from Mg^{2+} and In^{3+}, *J. Chem. Soc. A*, p. 1986, 1970.
60. **Qureshi, M., Rathore, H. S., and Kumar, R.,** Influence of the temperature on the ion exchange properties of stannic arsenate: separation of Pb^{2+}, UO_2^{2+}, and Cr^{3+} from numerous metal ions, *J. Chromatogr.*, 54, 269, 1971.
61. **Qureshi, M., Gupta, J. P., and Sharma, V.,** Synthesis of a reproducible and chemically stable tantalum antimonate: quantitative separation of VO^{2+}-Al^{3+}-Ti^{4+}, VO^{2+}-Fe^{3+}-Ti^{4+}, and UO_2^{2+}-Ti^{4+}, *Anal. Chem.*, 45, 1901, 1973.
62. **Qureshi, M. and Rathore, H. S.,** Synthesis and ion exchange properties of titanium molybdates, *J. Chem. Soc. A*, 2515, 1969.
63. **Qureshi, M., Varshney, K. G., and Khan, F.,** Synthesis, ion exchange behavior and analytical applications of stannous ferrocyanide: separation of Mg-Ca, Mn-Ni and Y-Th, *J. Chromatogr.*, 65, 547, 1972.
64. **Qureshi, M., Varshney, K. G., and Kabiruddin, S. K.,** Synthesis and ion exchange properties of thermally stable titanium(IV) vanadate: separation of Sr(II) from Ca(II), Ba(II) and Mg(II), *Can. J. Chem.*, 50, 2071, 1972.
65. **Qureshi, M., Varshney, K. G., Alam, F. M., and Naheed, S.,** Distribution studies of metal ions on stannic arsenate in butanol-HNO_3 systems: quantitative separation of Ni-Fe, Ni-Cu, Cd-Hg and Pb-Hg, *Chromatographia*, 11, 660, 1978.
66. **Qureshi, M., Varshney, K. G., and Fatima, N.,** Synthesis and ion exchange behavior of nickel and cobalt antimonates: quantitative separation of Bi(III) from 48 metal ions, *J. Chromatogr.*, 169, 365, 1979.
67. **Varshney, K. G., Naheed, S., Khan, A. A., Tandon, S. N., and Gupta, C. B.,** Distribution studies of metal ions on arsenophosphates of Sn(IV) and Cr(III) and on amine Sn(II) hexacyanoferrate(II) using radiotracers: separation of Sr^{2+}-Cs^{+}, Hg^{2+}-Ag^{+} and Hg^{2+}-Zn^{2+}, *Chromatographia*, 12, 413, 1979.

68. **Varshney, K. G. and Premadas, A.,** Silica based double salts as cation exchangers. I. Synthesis and analytical applications of Ce(IV) phosphosilicate, *J. Liq. Chromatogr.*, 4, 1245, 1981.
69. **Varshney, K. G. and Premadas, A.,** Silica based double salts as cation exchangers. II. Synthesis and analytical applications of Sn(IV) phosphosilicate, *J. Liq. Chromatogr.*, 4, 915, 1981.
70. **Varshney, K. G. and Premadas, A.,** Synthesis, composition and ion exchange behavior of Zr(IV) and Ti(IV) arsenophosphates: separation of metal ions, *Sep. Sci. Technol.*, 16, 793, 1981.
71. **Varshney, K. G., Khan, A. and Varshney, S. S.,** Synthesis and ion exchange properties of amine tin(IV) hexacyanoferrate(II) from zinc(II), manganese(II), magnesium(II) and aluminium(III), *Indian. J. Chem. A*, 21, 398, 1982.
72. **Varshney, K. G., Khan, A. A., Maheshwari, A., Anwar, S., and Sharma, U.,** Synthesis of a new thermally stable Sn(IV) arsenosilicate cation exchanger and its application for the column chromatographic separation of metal ions, *Indian J. Technol.*, 22, 99, 1984.
73. **Rawat, J. P. and Thind, P. S.,** Studies on inorganic ion exchangers. I. Preparation, properties and applications of ferric phosphate, *Can. J. Chem.*, 54, 1892, 1976.
74. **Rawat, J. P. and Singh, J. P.,** Studies on inorganic ion exchangers. II. Synthesis, ion exchange properties and applications of ferric arsenate, *Can. J. Chem.*, 54, 2534, 1976.
75. **Rawat, J. P. and Singh, J. P.,** Studies on inorganic ion exchangers: separation of Mg-Ba, Zr-Ti, Hg-Cd and V-Ti on the columns of aluminium antimonate, *Ann. Chim.*, 66, 585, 1976.
76. **Rawat, J. P. and Singh, D. K.,** Synthesis, ion exchange properties and analytical applications of iron(III) antimonate, *Anal. Chim. Acta*, 87, 157, 1976.
77. **Qureshi, M., Rawat, J. P., and Gupta, A. P.,** Quantitative separation of Cu^{2+} from Cd^{2+} and Zn^{2+} using Sn(IV) sulphide as an ion exchanger, *Indian J. Technol.*, 15, 80, 1977.
78. **Qureshi, M., Rawat, J. P. and Gupta, A. P.,** Analytical applications of niobium arsenate, *Chromatographia*, 11, 167, 1978.
79. **Thind, P. S., Sandhu, S. S., and Rawat, J. P.,** Stannic vanadoarsenate, a new inorganic ion exchanger: synthesis, ion exchange properties and applications, *Chem. Anal. (Warsaw)*, 24, 65, 1979.
80. **Rawat, J. P. and Khan, M. A.,** Synthesis and ion exchange properties of titanium tungstoarsenate: separation of Pb(II) from Mg(II), Zn(II) and Cu(II), *Ann. Chim.*, 69, 525, 1979.
81. **Rawat, J. P., Singh, D. K., and Muktawat, K. P. S.,** Quantitative separation of some metal ions on iron(III) antimonate and ion exchange equilibria, *Chem. Anal. (Warsaw)*, 24, 801, 1979.
82. **Rawat, J. P. and Aziz, H. M. A. A.,** Separation of anions and cations on thorium tellurite — a new amphoteric ion exchanger, *J. Liq. Chromatogr.*, 7, 1961, 1984.
83. **Varshney, K. G., Agrawal, S., Varshney, K., Premadas, A., Rathi, M. S., and Khanna, P. P.,** Analytical applications of Zr(IV) and Ti(IV) arsenophosphates as ion exchangers, *Talanta*, 30, 955, 1983.
84. **Varshney, K. G., Agrawal, S., and Varshney, K.,** Synthesis, ion exchanger behavior and analytical applications of a new, crystalline and stable zirconium(IV) arsenosilicate cation exchanger: analysis of some silicate rocks, *Sep. Sci. Technol.*, 18, 59, 1983.
85. **Varshney, K. G., Agrawal, S., and Varshney, K.,** Quantitative determination of iron and aluminium in some alloys on zirconium(IV) phospho- and silico-arsenates, *J. Liq. Chromatogr.*, 6, 1535, 1983.
86. **Varshney, K. G., Agrawal, S., and Varshney, K.,** The separation of aluminium and magnesium in some antacid drugs using zirconium(IV) phospho- and silico-arsenates, *Anal. Lett.*, 16, 685, 1983.
87. **Varshney, K. G., Agrawal, S., Varshney, K., and Saxena, V.,** Quantitative separation of iron from some multivitamin-multimineral formulations using zirconium(IV) arsenophosphate columns, *Anal. Lett.*, 17, 2111, 1984.
88. **Rawat, J. P. and Iqbal, M.,** A new chelating material for analytical separations, *Ann. Chim.*, 71, 431, 1981.
89. **Rawat, J. P., Iqbal, M., and Ali, S.,** A new chelating material prepared by the modification of an anion exchanger, *J. Indian Chem. Soc.*, 6, 185, 1985.
90. **Rawat, J. P., Alam, M., Singh, B., and Aziz, H. M. A.,** Zirconium triethylamine — a new class of chelate ion exchangaer, *Bull. Chem. Soc. Jpn.*, 60, 2619, 1987.
91. **Rawat, J. P. and Kamoonpuri, S. I. M.,** Anion exchange and ligand exchange studies on thoriumtetracyclohexylamine, *React. Poly. Ion. Exchange Sorbents*, 8, 153, 1988.
92. **Rawat, J. P. and Alam, M.,** Thorium triethanolamine as a new chelating material and anion exchanger, *J. Indian Chem. Soc.*, 65, 255, 1988.
93. **Rawat, J. P. and Chitra,** Ion exchange rate of chromate and dichromate ion on his (triethylamine) zirconium, *Bull. Chem. Soc. Jpn.*, 61, 2268, 1988.
94. **Rawat, J. P. and Iqbal, M.,** Separation and recovery of some metal ions on PAN sorbed zinc silicate, *J. Liq. Chromatogr.*, 3, 591, 1980.
95. **Rawat, J. P. and Kamoonpuri, S. I. M.,** Recovery of silver from laboratory wastes, *J. Chem. Educ.*, 63, 537, 1986.
96. **Varshney, K. G., Agrawal, S., and Varshney, K.,** A complexo-ion exchange method for the specific detection of iron on zirconium(IV) arsenophosphate beads, *Anal. Lett.*, 16, 1381, 1983.

97. **Varshney, K. G., Rani, S., Anwar, S., and Sharma, U.,** A complexo-ion exchange method for the sensitive and selective detection of iron(III) and molybdenum(VI) on antimony(V) silicate beads, *Anal. Lett.*, 18, 2033, 1985.
98. **Varshney, K. G., Rani, S., and Singh, R. P.,** Adsorption thermodynamics of carbofuran on antimony(V) silicate cation exchanger, *Ecotox. Environ. Safety*, 11, 179, 1986.
99. **Tuckerman, M. M.,** Ion-exchange paper in rapid separation and identification of basic amino acids, arginine, histidine and lysine from casein hydrolyzates, *Anal. Chem.*, 30, 231, 1958.
100. **Lederer, M. and Kertes, S.,** Chromatography on paper impregnated with ion exchange resins. II. Separation of selenite and tellurite, *Anal. Chim. Acta*, 15, 226, 1956.
101. **Hale, D. K.,** Extracting solutes from solutions, *Chem. Ind. (London)*, 1147, 1955.
102. **Flood, H.,** Inorganic chromatographic analysis, *Discuss. Faraday Soc.*, 7, 191, 1949.
103. **Alberti, G.,** Chromatographic separations on papers impregnated with inorganic ion exchangers, *Chromatogr. Rev.*, 8, 246, 1966.
104. **Qureshi, M.,** *Fifth Int. Sym. Chromatogr. Electrophoresis Rev.Presses Acad. Eur. Brussels*, 1969, 197.
105. **Sherma, J. and Fried, B.,** Thin layer chromatography and paper chromatography, *Anal. Rev.*, 56, R48, 1984.
106. **Alberti, G. and Grassini, G.,** Chromatography on papers impregnated with zirconium phosphate, *J. Chromatogr.*, 4, 83, 1960.
107. **Grassini, G. and Padiglione, C.,** Exchange capacity of papers impregnated with zirconium phosphate, *J. Chromatogr.*, 13, 361, 1964.
108. **Alberti, G., Conte, A., and Allulli, S.,** Chromatographic behavior of various metal cations in fused salts on γ-alumina and synthetic inorganic ion exchange materials, *J. Chromatogr.*, 18, 564, 1965.
109. **Sakodinsky, K. and Lederer, M.,** The chromatographic properties of paper impregnated with titanium hydroxides, *J. Chromatogr.*, 20, 358, 1964.
110. **Smit, J. V. R., Jacobs, J. J., and Robb, W.,** Cation exchange properties of the NH_4 heteropoly acid salts, *J. Inorg. Nucl. Chem.*, 12, 95, 104, 1959.
111. **Alberti, G. and Grassini, G.,** Chromatographic separations of alkali metals on paper impregnated with ammonium molybdophosphate, *J. Chromatogr.*, 4, 423, 1960.
112. **Qureshi, M., Varshney, K. G., and Khan, F.,** Chromatographic separation of metal ions on stannous ferrocyanide paper, *Sep. Sci.*, 8, 279, 1973.
113. **Rawat, J. P. and Chitra,** Chromatography of certain metal ions on ligand combined ion exchange papers, *J. Indian Chem.Soc.*, 63, 857, 1986.
114. **Alberti, G.,** Chromatographic separations on paper impregnated with inorganic ion exchangers, *Chromatogr. Rev.*, 8, 259, 1966.
115. **Lederer, M. and Battilotti, M.,** Sensitivity sequence of a series of tripositive Co(III) complexes on inorganic and organic ion exchangers, *J. Chromatogr.*, 95, 81, 1974.
116. **Lederer, M. and Polcaro, C.,** Adsorption of inorganic ions on hydrous oxides, *J. Chromatogr.*, 94, 313, 1974.
117. **Singh, N. J. and Tandon, S. N.,** Rapid separation of metal ions on hydrous zirconium oxide paper, *Chromatographia*, 10, 309, 1977.
118. **Surendranath, K. V. and Tandon, S. N.,** Chromatographic separation of anions on papers impregnated with hydrous titanium oxide, *J. Liq. Chromatogr.*, 11, 1433, 1988.
119. **Singh, N. J., Rajeev, and Tandon, S. N.,** Rapid separation of amino acids on hydrous zirconium oxide paper, *Indian J. Chem.*, 15, 581, 1977.
120. **Singh, N. J. and Rama Rao, N. V.,** Rapid separation of some alkaloids of pharmacological and toxicological interest on hydrous zirconium oxide papers, *Curr.Sci.*, 49, 193, 1980.
121. **Rawat, J. P., Muktawat, K. P. S., and Singh, O.,** Detection and chromatographic separation of phenols on papers impregnated with ferric hydroxide, *J. Indian Chem. Soc.*, 59, 1011, 1982.
122. **Rathore, H. S., Kumari, K. and Garg, M.,** Chromatographic studies of 34 organic acids on papers impregnated with hydroxides of aluminium and cadmium, *J. Liq. Chromatogr.*, 6, 973, 1973.
123. **Nunes Da Costa, M. J. and Jeronimo, M. A. S.,** Behavior of metal ions of IB group on zirconium phosphate papers, *J. Chromatogr.*, 5, 456, 1961.
124. **Alberti, G., Dobici, F., and Grassini, G.** Chromatography on paper impregnated with inorganic ion exchangers. III Chromatography of inorganic ions on zirconium phosphate papers with HCl, H_2SO_4, HNO_3, and CH_3COOH at various concentrations, *J. Chromatogr.*, 8, 103, 1962.
125. **Sastri, M. N. and Rao, A. P.,** Separation of valency states of some elements on paper impregnated with zirconium phosphates, *J. Chromatogr.*, 9, 250, 1962.
126. **Catelli, P.,** Separation of amino acids on papers impregnated with zirconium phosphate, *J. Chromatogr.*, 9, 534, 1962.
127. **Coussio, I. D., Masini Bettolo, G.B., and Moscatelli, V.,** Separation of alkaloids on paper impregnated with zirconium phosphate, *J. Chromatogr.*, 11, 238, 1963.

128. **Grassini, G. and Padigloine, C.,** The exchange capacity of papers impregnated with zirconium phosphate, *J. Chromatogr.*, 13, 56, 1964.
129. **Adloff, J. P.,** Separation of radioelements by chromatography on papers impregnated with ion exchange minerals, *J. Chromatogr.*, 5, 366, 1961.
130. **Alberti, G., Conte, A., Grassini, G., and Lederer, M.,** *J. Electroanal. Chem.*, 4, 301, 1962.
131. **Cabral J. M. P.,** Chromatography on papers impregnated with inorganic ion exchangers, *J. Chromatogr.*, 4, 86, 1960.
132. **Qureshi, M., Akhtar, I., and Mathur, K. N.,** Separation of metal ions on stannic phosphate and stannic tungstate papers: specific separations of gold(III), mercury(II), platinum(IV), magnesium(II), molybdenum(VI) and selenium(IV), *Anal. Chem.*, 39, 1766, 1967.
133. **Qureshi, M. and Qureshi, S. Z.,** Chromatography of some metal ions on paper impregnated with stannic phosphate, *J. Chromatogr.*, 22, 198, 1968.
134. **Qureshi, M. and Qureshi, S.Z.,** Separation of nickel(II) and thalium(I) at different temperatures on stannic phosphate papers, *Sep. Sci.*, 7, 187, 1972.
135. **Alberti, G., Massucci, M. A., and Torracca, S.,** Separation of metal ions on cerium phosphate paper in perchloric acid, *J. Chromatogr.*, 30, 579, 1967.
136. **Qureshi, M. and Sharma, S. D.,** Chromatography of 48 metal ions on stannic arsenate and plain papers in butanol-nitric acid media: prediction of K_{sp} from R_f values, *Anal. Chem.*, 45, 1283, 1973.
137. **Varshney, K. G. and Varshney, S. S.,** Cation chromatography on tin(II) hexacyanoferrate(II) papers in hydrochloric acid and hydrochloric acid-ammonium chloride systems: prediction of K_{sp} values from R_f values, *Chromatographia*, 10, 542, 1977.
138. **Qureshi, M. and Mathur, K. N.,** Quantitative separation of selenium from metal ions on tin(IV) tungstate papers, *Ann. Chim. Acta*, 41, 523, 1968.
139. **Qureshi, M., Varshney, K. G., Gupta, M. P. and Gupta, S. P.,** Cation chromatography on stannic tungstate papers, *Chromatographia*, 10, 29, 1977.
140. **Qureshi, M., Mathur, K. N., and Israili, A. H.,** Separation of metal ions on tin(IV) tungstate and selenite papers, *Talanta*, 16, 503, 1969.
141. **Qureshi, M., Varshney, K. G., Gupta, S. P., and Gupta, M. P.,** Electrochromatographic separation of metal ions on stannic tungstate papers, *Ann. Chem.*, 66, 557, 1976.
142. **De, A. K. and Chowdhury, K.,** Studies on thorium phosphate ion exchangers: paper chromatographic behavior and separation of metal ions on thorium phosphate paper, *Sep. Sci.*, 10, 639, 1975.
143. **Rawat, J. P., Iqbal, M., and Khan, M. A.,** Chromatography on paper impregnated with zinc silicate, new adsorbent for qualitative and quantitative separation of amines, *J. Liq. Chromatogr.*, 6, 959, 1983.
144. **Rawat, J. P., Iqbal, M., and Alam, M.,** Chromatography on paper impregnated with zinc silicate, new adsorbent for the separation of phenols, *J. Liq. Chromatogr.*, 5, 967, 1982.
145. **Singh, D. K. and Darbari, A.,** Ligand exchange separation of amines on copper(II) sorbed on zinc silicate, *Chromatographia*, 23, 93, 1987.
146. **Qureshi, M., Rawat, J. P., and Sharma, V.,** Chromatographic behavior of 47 metal ions on titanium(IV) arsenate papers, *Talanta*, 20, 267, 1973.
147. **Qureshi, M., Nabi, S. A., and Nighat, Z.,** Chromatographic separations of amino acids on titanium arsenate papers, *Sep. Sci.*, 10, 801, 1975.
148. **Qureshi, M., Nabi, S. A., and Nighat, Z.,** Chromatography of alkaloids on titanium arsenate papers and quantitative separation of some alkaloids from nicotine on titanium arsenate columns, *Talanta*, 23, 31, 1976.
149. **Qureshi, M. and Sharma, S. D.,** A systematic study of chromatographic behaviour of 48 metal ions on plain and stannic arsenate impregnated papers in DMSO-HNO_3 systems, *Chromatographia*, 11, 153, 1978.
150. **Qureshi, M. and Sharma, S. D.,** Complexation ion exchange electrochromatography of 52 metal ions on stannic arsenate papers, *Sep. Sci. Technol.*, 13, 723, 1978.
151. **Qureshi, M. and Israili, A. H.,** Electrochromatographic separations of metal ions on tin(IV) phosphate paper, *Anal. Chim. Acta*, 41, 523, 1968.
152. **Rawat, J. P. and Mujtaba, S. Q.,** Separation of metal ions on titanium molybdate papers in sodium form using sodium nitrate solvent of various concentrations, *Sep. Sci.*, 10, 150, 1975.
153. **Tandon, S. N. and Mathew, J.,** Rapid separation of microgram amounts of metal ions on stannic antimonate ion exchange papers, *Mikrochim. Acta*, 1, 5, 1977.
154. **Qureshi, M., Varshney, K. G., and Rajput, R. P. S.,** Chromatography of 49 metal ions on stannic antimonate papers in dimethylsulfoxide-nitric acid systems, *Anal. Chem.*, 47, 1520, 1975.
155. **Qureshi, M., Varshney, K. G., and Rajput, R. P. S.,** Electrochromatographic behavior of metal ions on stannic antimonate paper, *Ann. Chim.*, 66, 356, 1976.
156. **Mathew, J., Rajeev, and Tandon, S. N.,** Rapid separation of alkali metals on zirconium antimonate papers, *Chromatographia*, 10, 45, 1977.
157. **Mathew, J., Rajeev, and Tandon, S. N.,** Rapid chromatographic separation of metal ions on zirconium antimonate impregnated papers, *Chem. Anal. (Warsaw)*, 22, 219, 1977.

158. **Rajput, R. P. S. and Agrawal, S.,** Chromatographic separations on zirconium antimonate paper, *Anal. Lett.*, 11, 1743, 1985.
159. **Seth, N. S., Rajput, R. P. S., Agrawal, N. K., Agrawal, S. K., and Agrawal, S.,** Separations of metal ions on zirconium antimonate paper, *Anal. Lett.*, 18, 1783, 1985.
160. **Qureshi, M., Kishore, J., and Varshney, R. G.,** Electrophoresis studies of metal ions on titanium tungstate paper: specific separation of palladium from various metal ions, *J. Electroanal. Chem. Interfacial Electrochem.*, 76, 383, 1977.
161. **Gopal, N. G. S. and Mishra, N. C.,** Separation of selective radionuclides on pyroantimonic acid papers, *Proc. Chem. Symp.*, 1, 111, 1972.
162. **Rawat, J. P., Mujtaba, S. Q., and Singh, P.,** Separation and identification of phenols on stannic molybdate paper, *Z. Anal. Chem.*, 279, 368, 1976.
163. **Qureshi, M. and Rawat, J. P.,** Specific separation of gallium, gold and magnesium on stannic molybdate papers, *Sep. Sci.*, 7, 297, 1972.
164. **Rawat, J. P. and Singh, P.,** Electrophoretic separations of metal ions on stannic molybdate, *Ann. Chim.*, 64, 1976.
165. **Husain, S. W.,** Separation of metal ions on titanium tungstate paper using solvent mixtures and buffers. *Analusis*, 1, 314, 1972.
166. **Qureshi, M., Varshney, K. G., and Khan, F.,** Quantitative separation of Mg^{2+} and Pd^{2+} from many ions by electrochromatography on titanium tungstate paper, *Sep. Sci.*, 6, 559, 1971.
167. **De, A. K. and Chakraborty, P.,** Synthetic inorganic ion exchangers. XXI. Electrochromatographic separation of metal ions on lanthanum antimonate impregnated papers, *Electrophoresis (Weinhem Fed. Pep. Germ.)*, 2, 330, 1982.
168. **Srivastava, S. P., Jain, A. K., Kumar, S., and Singh, R. P.,** Chromatography of metal ions on pyridinium tungstoarsenate papers, *J. Radioanal. Chem.*, 53, 49, 1979.
169. **Varshney, K. G., Khan, A. A., Jain, J. B., and Sharma, S. D.,** Cation chromatography on tin(IV) hexacyanoferrate(II) papers in carboxylic acids-DMSO-nitric acid systems: separation of metal ions, *J. Indian Chem. Soc.*, 58, 241, 1981.
170. **Varshney, K. G., Khan, A. A., and Jain, J. B.,** Cation chromatography on tin(IV) hexacyanoferrate(II) papers: separation of metal ions, *J. Indian Chem.Soc.*, 58, 1025, 1981.
171. **Shih, N. S., Zhun, Z., and Hnei, V. C.,** Ion exchange properties of ammonium salts of 12-heteropoly acids. I. Chromatographic separations of alkali metals on ammonium molybdophosphate papers, *Acta Chim. Sinica*, 30, 21, 1964.
172. **Zhu, J. Z., Ying, B. H., Ting, Z. B., and Shih, N. S.,** Ion exchange properties of ammonium salts of 12-heteropolyacids. II. Ion exchange chromatography of alkali metal ions on ammonium tungstophosphate papers, *Acta Chim. Sinica*, 31, 218, 1965.
173. **Pe, H. Y.,** Separation of potassium, rubidium and cesium by electrophoresis on paper impregnated with ammonium molybdophosphate, *Acta Chim. Sinica*, 31, 260, 1965.
174. **Marcu, G. and Ghizdavu, L.,** Use of organic decatungstodiniobiosilicate heteropoly compound as an ion-exchanger, *Rev. Chim. (Bucharest)*, 27, 907, 1976.
175. **Maron, G. and Ghizdavu, L.,** Ion exchange separation on some organic heteropolytungstates of uranium and products of its fission, *Stud. Univ. Babes Bolyai Ser. Chem.*, 22, 72, 1977.
176. **Varshney, K. G., Khan, A. A., and Anwar, S.,** Chromatography of some transition and alkaline earth metals on Sn(IV) arsenosilicate and Sn(IV) arsenophosphate papers in EDTA solutions of varying pH, *Proc. Natl. Acad. Sci. A*, 57, 523, 1987.
177. **Varshney, K. G., Anwar, S., and Maheshwari, S. M.,** Chromatography of amino acids on papers impregnated with tin(IV) and thorium(IV) phosphosilicates: separation of basic amino acids from others, *Proc. Natl. Acad. Sci. A*, 59, 364, 1989.
178. **Sherma, J. and Fried, B.,** Paper and thin layer chromatography, *Anal. Rev.*, 56, R48, 1984.
179. **Zabin, B. A. and Rollins, C. B.,** Inorganic ion exchangers for thin layer chromatography, *J. Chromatogr.*, 14, 534, 1964.
180. **Berger, J. A., Meyniel, G., and Petit, J.,** Successful separation of inorganic ions on ion exchange resins and crystalline ion exchangers, *J. Chromatogr.*, 29, 190, 1967.
181. **Berger, J. A., Chabard, J. L., Besse, G., and Voissiere, G.,** Separation of inorganic ions on zirconium oxide, *Bull. Soc. Chim. Fr.*, 1027, 1969.
182. **Kirchner, J. G., Miller, J. M., and Keller, G. J.,** Thin layer chromatography of terpenes, *Anal. Chem.*, 23, 420, 1951.
183. **Sen, A. K., Das, S. B., and Ghosh, U. C.,** Studies on hydrous zirconium oxide. I. Thin layer chromatography of some metal ions on hydrous zirconium oxide. Quantitative separation of bismuth(III) from several other metal ions, *J. Liq. Chromatogr.*, 8, 2999, 1985.
184. **Sen, A. K. and Ghosh, U. C.,** Studies on hydrated stannic oxide. VI. Thin layer chromatography of some anions on hydrated stannic oxide: quantitative separation of chromium(VI) from some ores and alloys, *J. Liq. Chromatogr.*, 3, 71, 1980.

185. **Rathore, H. S., Kumari, K., and Agrawal, M.**, Quantitative separation of citric acid from trichloroacetic acid on plates coated with calcium sulphate and zinc oxide, *J. Liq. Chromatogr.*, 8, 1299, 1985.
186. **Grace, W. & Co.**, British Patent, 1, 181089, 1970.
187. **Cremer, E. and Siedl, E.**, Separation of radioactive anions in the region below 10^{-14} g by thin layer chromatography, *Chromatographia*, 3, 17, 1970.
188. **Keonig, K. H. and Demiel, K.**, Thin Layer chromatography of inorganic ions on zirconium hypophosphate, *J. Chromatogr.*, 39, 101, 1969.
189. **Alberti, G., Giammiri, G., and Grazzinistrazza, G.**, Chromatographic behavior of inorganic ions on the crystalline titanium phosphate and zirconium phosphate thin layers, *J. Chromatogr.*, 28, 188, 1967.
190. **Keonig, K. H. and Graf, H.**, Thin layer chromatography of inorganic cations on crystalline Ce(IV) phosphate sulphate, *J. Chromatogr.*, 67, 200, 1972.
191. **Srivastava, S. P., Dua, V. K., and Gupta, K.**, TLC separation of some inorganic ions on silica gel and ceric molybdate impregnated layers, *J. Indian Chem. Soc.*, 56, 529, 1979.
192. **Yin, B. and Lue, J.**, Thin layer chromatography of noble metal ions on semicrystalline stannic phosphate, *Huaxue Tongboo*, 7, 25, 1985.
193. **Lei, G., Gou, Z., and Yin, B.**, Thin layer electrophoresis on stannic phosphate, *Huaxue Shiji*, 7, 265, 1985.
194. **Qureshi, M., Varshney, K. G., Gupta, S. P., and Gupta, M. P.**, Cation chromatography on stannic tungstate thin layers in DMSO-acid and aqueous DMSO systems. Quantitative separation of gold from numerous metal ions, *Sep. Sci.*, 12, 649, 1977.
195. **Nabi, S. A., Farooqui, W. U., and Rahman, N.**, Semicrystalline ion exchangers for TLC separation of phenolic compounds, *Chromatographia*, 20, 109, 1985.
196. **De, A. K. and Pal, B. K.**, Synthetic inorganic ion exchangers. XV. Thin layer chromatography of metal ions on thorium tungstate. Quantitive separation of Hg(II) from several other metal ions, *J. Liq. Chromatogr.*, 2, 937, 1979.
197. **Husain, S. W. and Fahimeh Eivazi**, Thin layer chromatography of 57 metal ions on an inorganic ion exchanger in mixed solvent systems, *Chromatographia*, 8, 277, 1975.
198. **Husain, S. W. and Kazmi, S. K.**, Thin layer chromatography of metal ions on a new synthetic inorganic ion-exchanger, *Experimentia*, 28, 988, 1972.
199. **Husain, S. W. and Rasheezad, Sh.**, Thin layer chromatographic separation of platinum metals on semicrystalline inorganic ion exchanger, *Mikrochim. Acta*, 11, 1978.
200. **Qureshi, M., Varshney, K. G., and Fatima, N.**, Thin layer chromatography of metal ions on stannic arsenate: quantitative separation of Hg(II) from Cd(II), Zn(II) and Cu(II), *Sep. Sci.*, 12, 321, 1977.
201. **Qureshi, M., Varshney, K. G., and Fatima, N.**, A novel quantitative separation of Cr(III) from numerous metal ions on β-stannic arsenate thin layers in DMSO-HCl systems, *Sep. Sci. Technol.*, 13, 321, 1978.
202. **Qureshi, M., Varshney, K. G., and Sharma, S. D.**, Unusual selectivity of stannic arsenate layer in DMSO-HCl systems: separation of zirconium from numerous metal ions, *Sep. Sci. Technol.*, 13, 917, 1978.
203. **Sharma, S. D., Sharma, T. R., and Sethi, B. M.**, Prediction of K_{sp} (solubility product) from R_f values; TLC of 47 metal ions on stannic arsenate in hydrochloric acid systems, *J. Liq. Chromatogr.*, 6, 1253, 1983.
204. **Seth, N. S. and Rajput, R. P. S.**, TLC of metal ions on titanium(IV) antimonate in aqueous and mixed solvent systems containing dimethyl formamide: quantitative separation of Pd(II) from several metal ions, *Indian J. Chem. A*, 22, 1088, 1983.
205. **Seth, N. S., Rajput, R. P. S., Agrawal, N. K., Agrawal, S. K., and Agrawal, S.**, Quantitative separation of Pd(II) by TLC of metal ions on Ce(IV) antimonate, *Anal. Lett.*, 18, 481, 1985.
206. **Qureshi, M., Varshney, K. G., and Rajput, R. P. S.**, TLC of 49 metal ions on stannic antimonate with dimethyl sulphoxide solvents: separation of uranium from various metal ions, *Sep. Sci.*, 11, 533, 1976.
207. **De, A. K., Rajput, R. P. S., Das, S. K. and Chowdhury, N. D.**, Synthetic inorganic ion exchangers. XIV. TLC of metal ions on thorium phosphate: quantitative separation of gold(III) from several other metal ions, *J. Liq. Chromatogr.*, 2, 117, 1979.
208. **Srivastava, S. P., Bhushan, R., and Chauhan, R. S.**, TLC of some closely related synthetic dyes on impregnated silica gel layers, *J. Liq. Chromatogr.*, 8, 1255, 1985.
209. **Fog, A. G. and Wood, R.**, Thin layer chromatography on zinc ferrocyanide, *J. Chromatogr.*, 20, 613, 1965.
210. **Kawamura, S., Kurotaki, K., Koruku, H. and Izawa, M.**, Rapid separation of Na, Li, Rb, and Cs by thin layer chromatography on zinc ferrocyanide, *J. Chromatogr.*, 28, 557, 1967.
211. **Lesigang, M.**, Chromatographic separations of alkali metals. I. *Mikrochim. Acta*, 34, 1964.
212. **Lesigang, M. and Hecht, F.**, Chromatographic separations of alkali metals, *Mikrochim. Acta*, 508, 1964.
213. **Lesigang, M. and Buchtela, K.**, Thin layer chromatography of inorganic ions. V. Addition to silica gel layers, *Mikrochim. Acta*, 1027, 1969.
214. **Lepri, L. and Desideri, P. G.**, TLC of aromatic amines on ammonium molybdophosphate and tungstophosphate, *J. Chromatogr.*, 207, 29, 1981.

215. **Lepri, L., Desideri, P. G., and Heimler, D.**, TLC of sulphonamides on AMP layers, *J. Chromatogr.*, 176, 181, 1979.
216. **Lepri, L., Desideri, P. G., and Heimler, D.**, Chromatographic behavior of amino acids, inter derivatives and peptides on layers of ATP and silanized silica gel, *J. Chromatogr.*, 268, 493, 1983.
217. **Lepri, L., Desideri, P. G., Heimler, D., and Giannassi, S.**, High performance TLC of diastereomeric di and tripeptides on ready for use plates of silanized silica gel and on ATP, *J. Chromatogr.*, 265, 328, 1983.
218. **Lepri, L., Desideri, P. G., and Heimler, D.**, Chromatographic behavior of small peptides on layers of ATP, *J. Chromatogr.*, 243, 339, 1982.
219. **Lepri, L., Desideri, P. G., and Heimler, D.**, High performance TLC of 2-4-dinitrophenyl amino acids on layers of RP-8, RP-18 and ATP, *J. Chromatogr.*, 235, 411, 1982.
220. **Lepri, D., Desideri, P. G., and Heimler, D.**, High performance thin layer chromatography of phenyl ethyl amines and phenolic acids in silanized silica and on ATP, *J. Chromatogr.*, 347, 303, 1985.
221. **Kaletka, R. and Koneeny, C.**, Report U.J.V. 2643-Ch. 1971, 31.
222. **Srivastava, S. P., Dua, V. K., and Gupta, K.**, TLC of aminoacids on pyridinium tungstoarsenate layers, *Chromatographia*, 12, 605, 1979.
223. **Srivastava, S. P., Dua, V. K., Pal, S., and Gupta, K.**, Chromatographic separation of some inorganic ions on silica gel pyridinium tungstoarsenate impregnated thin layer plates, *Anal. Lett.*, 11, 813, 1978.
224. **Andreev, L. V.**, Method of separation and detection of substances on silica gel thin films impregnated with silver nitrate and tungstophosphoric acid, *J. High Resolution Chromatogr. Commun.*, 6, 575, 1983.
225. **Varshney, K. G., Khan, A. A., and Anwar, S.**, Chromatography of alkaline earths and transitional metals on tin(IV) arsenosilicate and arsenophosphate thin layers in buffered EDTA solutions, *J. Liq. Chromatogr.*, 8, 1347, 1985.
226. **Varshney, K. G., Khan, A. A., and Anwar, S.**, Thin layer chromatography of aminoacids on tin(IV) arsenosilicate cation exchanger in some buffer solutions, *Anal. Lett.*, 19, 543, 1986.
227. **Janauer, G. E.**, Ion exchange in dimethyl sulfoxide media. I. Distribution studies with group IIA, IIIA, and IVA cations, *Microchim. Acta*, 6, 1111, 1968.
228. **Birze, I., Marple, L. W., and Diehl, H.**, Cation exchange separation of metals in dimethyl sulphoxide-aqueous hydrochloric acid media, *Talanta*, 15, 1441, 1968.
229. **Fritz, J. S. and Lehoczky, M.**, Anion-exchange separation of metal ions in dimethyl sulphoxide-methanol-hydrochloric acid, *Talanta*, 15, 287, 1968.
230. **Phipps, A. M.**, *Anal. Chem.*, 40, 1769, 1968.
231. **Qureshi, M., Ed.**, *Handbook of Chromatography (Inorganics)*, Vol. 1, CRC Press, Boca Raton, FL, 1987.
232. **Rawat, J. P. and Singh, D. K.**, Kinetics of Ag^+, Zn^{2+}, Ca^{2+}, Hg^{2+}, La^{3+} and Th^{4+} exchanges in iron(III) antimonate, *J. Inorg. Nucl. Chem.*, 40, 897, 1978.
233. **Singhal, J. P., Rawat, J. P., and Gupta, G. K.**, A kinetic study of zinc exchange on sodium dickite, *Indian J. Chem. A*, 17, 32, 1979.
234. **Rawat, J. P. and Thind, P. S.**, A kinetic study of ion exchange in tantalum arsenate to understand the theoretical aspects of separations, *J. Phys. Chem.*, 80, 1384, 1976.
235. **Nancollas, G. H. and Paterson, R.**, Kinetics of ion exchange on zirconium phosphate and hydrated thoria, *J. Inorg. Nucl. Chem.*, 22, 259, 1961.
236. **Hallaba, E., Mizak, N. Z., and Salania, H. N.**, Exchange characteristics of hydrated zirconia, *Indian J. Chem. A*, 11, 580, 1973.
237. **Alberti, G., Bertrami, R., Cascoila, M., Costantino, U., and Gupta, J. P.**, Crystalline insoluble acid salts of tetravalent metals. XXI. Ion exchange mechanism of alkaline earth metal ions on crystalline Zr HNa $(PO_4)_2{\cdot}5H_2O$, *J. Inorg. Nucl. Chem.*, 38, 843, 1976.
238. **Harman, R. G. and Clearfield, A.**, Crystalline cerium(IV) phosphates. II. The ion exchange characteristics with alkali metal ions, *J. Inorg. Nucl. Chem.*, 38, 853, 1976.
239. **Singh, N. J., Mathew, J., and Tandon, S. N.**, Kinetics of ion exchange: a radio chemical study of Rb^+-H^+ and Ag^+-H^+ exchanges on zirconium arsenophosphate, *J. Phys. chem.*, 84, 21, 1980.
240. **Singh, N. J., Tandon, S. N., and Mathew, J.**, Kinetics of binary ion exchange on titanium vanado-phosphate, *Indian J. Chem. A*, 19, 416, 1980.
241. **Varshney, K. G. and Premadas, A.**, The kinetics of Ag^+, Cu^{2+}, Sr^{2+}, Ba^{2+}, and Y^{3+} exchanges in Sn(IV) arsenophosphate, *Indian, J. Chem. A*, 29, 441, 1981.
242. **Heitner-Wirguin, C. and Allm-Yaron, A.**, Hydrous oxides and their cation exchange properties, *J. Appl. Chem.*, 15, 445, 1965.
243. **Dyer, A. and Gill, J. S.**, Studies on crystalline zirconium phosphate. III. Self diffusion of sodium ion into mono sodium forms of zirconium phosphate, *J. Inorg. Nucl. Chem.*, 39, 665, 1977.
244. **Bunzl, K.**, Competitive ion exchange in mixed cation exchanger systems: kinetics and equilibria, *J. Inorg. Nucl. Chem.*, 39, 1049, 1977.
245. **Rawat, J. P. and Khan, M. A.**, Mechanism of cation exchange on stannic arsenate, *J. Inorg. Nucl. Chem.*, 42, 905, 1980.

246. **Boyd, G. E., Adamson, A. W., and Myers, L. S.,** The exchange adsorption of ions from aqueous solutions by organic zeolites. II. Kinetics, *J. Am. Chem. Soc.*, 69, 2836, 1947.
247. **Helfferch, F.,** *Ion Exchange,* McGraw-Hill, New York, 1962, 266.
248. **Helfferch, F.,** *Ion Exchange,* McGraw-Hill, New York, 1962, 261.
249. **Varshney, K. G., Khan, A. A., Varshney, K., and Agrawal, S.,** A kinetic approach to evaluate the energy and entropy of activation for the exchange of alkaline earth metal ions on tin(IV) tungstate cation exchanger, *Solvent Extr. Ion Exchange,* 2, 923, 1984.
250. **Varshney, K. G. and Rani, S.,** Ion exchange kinetics of some divalent metal ions on crystalline antimonic acid and antimony(V) silicate cation exchangers, *React. Polym.*, 3, 231, 1985.
251. **Varshney, K. G., Agrawal, K., Agrawal, S., Saxena, V., and Khan, A. R.,** Synthetic, kinetic, and analytical studies on titanium(IV) arsenosilicate ion exchanger: separation of lead from its synthetic alloys, *Colloid Surf.*, 29, 175, 1988.
252. **Varshney, K. G. and Sharma, U.,** Kinetics of transition metal ions on zirconium(IV) and thorium(IV) phosphosilicate cation exchangers, *Acta Chim. Hung.*, 116, 103, 1984.
253. **Varshney, K. G. and Sharma, U.,** Kinetics of ion exchange of transition metal ions on tin(IV) arsenosilicate cation exchanger, *React. Kinet. Lett.*, 28, 27, 1985.
254. **Bobman, M. H., Golden, T. C., and Jenkins, R. G.,** Ion exchange selected low rank coals. I. Equilibrium; II. Kinetics, *Solvent Extr. Ion Exchange,* 1, 813, 1983.
255. **Rees, L. V. C. and Rao, A.,** Self diffusion of various cations in natural mordenite, *Trans. Faraday Soc.*, 62, 2103, 1966.
256. **Rao, A. and Rees, L. V. C.,** Kinetics of ion exchange in mordenite, *Trans. Faraday Soc.*, 62, 2506, 1966.
257. **Varshney, K. G., Agrawal, S., and Varshney, K.,** Ion exchange kinetics of some metal pollutants and their separation on zirconium(IV) phospho and silicoarsenate cation exchangers, *Colloid Surf.*, 9, 189, 1984.
258. **Varshney, K. G., Agrawal, S., and Varshney, K.,** Ion exchange kinetics of alkaline earth metals on zirconium(IV) arsenophosphate: a linear variation of energy and entropy of activation with ionic radii and mobility, *Colloid Surf.*, 13, 341, 1985.
259. **Varshney, K. G., Sharma, U., Anwar, S., and Khan, A. A.,** Kinctics of exchange of some divalent metal ions on phosphosilicates of zirconium(IV) and thorium(IV), and arsenosilicate of tin(IV), *Indian J. Chem. A,* 23, 152, 1984.
260. **Varshney, K. G., Agrawal, S., and Varshney, K.,** Ion exchange kinetics of alkaline earths on zirconium(IV) arsenosilicate cation exchanger: activation energy as a linear function of the ionic mobilities and radii, *Acta Chim. Hung.*, 116, 69, 1984.
261. **Varshney, K. G., Khan, A. A., and Rani, S.,** Forward and reverse ion exchange kinetics for Na^+-H^+ and K^+-H^+ exchanges on crystalline antimony(V) silicate, *Colloid Surf.*, 25, 131, 1987.
262. **Cotton, F. A. and Wilkinson, G.,** *Advanced Inorganic Chemistry,* John Wiley & Sons, New York, 1964, 321.
263. **Varshney, K. G., Varshney, K., and Agrawal, S.,** Evaluation of the dimensionless time parameter for some particle diffusion controlled forward and reverse H(I)-metal(II) exchanges, *Colloid Surf.*, 18, 67, 1986.
264. **Larsen, E. M. and Vissers, D. R.,** The exchange of Li^+, Na^+ and K^+ with H^+ on zirconium phosphate, *J. Phys. Chem.*, 64, 1732, 1960.
265. **Baetsle, L.,** Ion exchange properties of zirconyl phosphate. III. Influence of temperature on tracer ion-equilibria, *J. Inorg. Nucl. Chem.*, 25, 271, 1963.
266. **Nancollas, G. H. and Tilak, B. V. K. S. R. A.,** Thermodynamics of cation exchange on semicrystalline zirconium phosphate, *J. Inorg. Nucl. Chem.*, 31, 3643, 1969.
267. **Alberti, G. and Costantino, U.,** Forward and reverse lithium-potassium ion-exchange isotherms of crystalline zirconium phosphate, *J. Inorg. Nucl. Chem.*, 36, 653, 1974.
268. **Rawat, J. P. and Thind, P. S.,** Thermodynamics of cation exchange, *J. Indian Chem. Soc.*, 57, 819, 1980.
269. **Rawat, J. P. and Muktawat, K. P. S.,** Thermodynamics of ion exchange on ferric antimonate, *J. Inorg. Nucl. Chem.*, 43, 2121, 1981.
270. **Varshney, K. G., Singh, R. P., and Rani, S.,** Thermodynamics of Ba^{2+}-H^+ and Sr^{2+}-H^+ exchanges on antimony(V) silicate cation exchanger, *Proc. Indian Natl. Sci. Acad.*, 50, 75, 1984.
271. **Varshney, K. G., Singh, R. P., and Rani, S.,** Thermodynamics of Ca^{2+}-H^+ and Mg^{2+}-H^+ exchanges on antimony(V) silicate cation exchanger, *Acta Chim. Hung.*, 115, 403, 1984.
272. **Varshney, K. G., Singh, R. P., and Sharma, U.,** Thermodynamics of the Ba^{2+}-H^+ and Sr^{2+}-H^+ exchanges on zirconium(IV) phosphosilicate cation exchanger, *Proc. Indian Natl. Sci. Acad.*, 51, 726, 1985.
273. **Varshney, K. G., Singh, R. P., and Sharma, U.,** Thermodynamics of the Ca^{2+}-H^+ and Mg^{2+}-H^+ exchanges on zirconium(IV) phosphosilicate cation exchanger, *Colloid Surf.*, 16, 207, 1985.

274. **Varshney, K. G., Singh, R. P., and Rani, S.,** Adsorption thermodynamics of carbofuran on antimony(V) silicate cation exchanger, *Ecotox. Environ. Safety,* 11, 179, 1986.
275. **Abe, M. and Sudoh, K.,** Synthetic inorganic ion exchange materials. XXIII. Ion exchange equilibria of transition metals and hydrogen ions on crystalline antimonic(V) acid, *J. Inorg. Nucl. Chem.*, 42, 1051, 1980.
276. **Qureshi, M. and Ahmad, A.,** Some studies on thermodynamics of ion exchange of alkali metal ions on crystalline niobium(V) phosphate, *Solvent Extr. Ion Exchange,* in press.
277. **Kullberg, L. and Clearfield, A.,** Mechanism of ion exchange in zirconium phosphates 31. Thermodynamics of alkali metal ion exchange on amorphous ZrP, *J. Phys. Chem.*, 85, 1578, 1981.
278. **Kullberg, L. and Clearfield, A.,** Mechanism of ion exchange in zirconium phosphates 32. Thermodynamics of alkali metal ion exchange on crystalline A ZrP, *J. Phys. Chem.*, 85, 1585, 1981.
279. **Barrer, R. M. and Meler, W. M.,** Exchange equilibria in a synthetic crystalline exchanger, *Trans. Faraday Soc.*, 54, 1074, 1958.
280. **Barrer, R. M., Papadopoulos, R. and Rees, L. V. C.,** Exchange of sodium in clinoptilolite by organic cations, *J. Inorg. Nucl. Chem.*, 29, 2047, 1967.
281. **Robinson, R. A. and Stokes, R. H.,** *Electrolyte Solutions,* Butterworths, London, 1959, 481.
282. **Singhal, J. P., Singh, R. P., Singh, C. P., and Gupta, G. K.,** Thermodynamics of the exchange of nicotine on alluminium, monmorillonite, *J. Soil Sci.*, 27, 42, 1976.
283. **Gains, G. L. and Thomas, H. C.,** Adsorption studies on clay minerals. II. A formation of the thermodynamics of exchange adsorption, *J. Chem. Phys.*, 21, 714, 1953.
284. **Singhal, J. P. and Singh, R. P.,** Aluminium-nicotine exchange equilibria. III. Kaolinite, *Acta Chim. Acad.*, 93, 307, 1977.
285. **Glasstone, S.,** *Textbook of Physical Chemistry,* McMillan, London, 1960, 827.
286. **Diest, J. and Talibuddin, O.,** Ion exchange in soils for the ion-pairs K-Ca, Ca-Rb and K-Na, *J. Soil Sci.*, 18, 125, 1967.
287. **Gast, R. G. and Klobe, W. D.,** Sodium-lithium exchange equilibria on vermiculite at 25° and 50°C, *Clays Clay Min.*, 19, 311.
288. **Vanselow, A. P.,** Equilibrium of the base exchange reactions of bentonites, permutites, soil colloids and zeolites, *Soil Sci.*, 33, 95, 1932.
289. **Bigger, J. W. and Cheung, M. W.,** Adsorption of picloram (4-amino-3, 5,6-trichloropicolinic acid) on panoche, ephrate and palous soils: a thermodynamic approach to the adsorption mechanism, *Soil Sci. Soc. Am. Proc.*, 37, 863, 1973.
290. **Singh, R. P., Varshney, K. G., and Rani, S.,** Adsorption thermodynamics of carbofuran on sandy clay loam and silt loam soils, *Ecotox. Environ. Safety,* 10, 309, 1985.

INDEX

A

D

E

F

G

H

I

K

R

S

T

U